Protective and Detrimental Role of Heme Oxygenase-1

Protective and Detrimental Role of Heme Oxygenase-1

Special Issue Editor

Valeria Sorrenti

MDPI • Basel • Beijing • Wuhan • Barcelona • Belgrade

MDPI

Special Issue Editor
Valeria Sorrenti
University of Catania
Italy

Editorial Office
MDPI
St. Alban-Anlage 66
4052 Basel, Switzerland

This is a reprint of articles from the Special Issue published online in the open access journal *International Journal of Molecular Sciences* (ISSN 1422-0067) in 2019 (available at: https://www.mdpi.com/journal/ijms/special_issues/Heme_Oxygenase).

For citation purposes, cite each article independently as indicated on the article page online and as indicated below:

LastName, A.A.; LastName, B.B.; LastName, C.C. Article Title. *Journal Name* **Year**, *Article Number, Page Range.*

ISBN 978-3-03921-806-6 (Pbk)
ISBN 978-3-03921-807-3 (PDF)

Contents

About the Special Issue Editor

Valeria Sorrenti is Associate Professor of Biochemistry at University of Catania, where she teaches Industrial Biochemistry, Biochemistry, Nutritional Biochemistry, and Biology. Pr. V. Sorrenti is a Pharmacy graduate. During her career, Pr. Sorrenti obtained her doctorate (Ph.D.) in Biology and Medical Biochemistry. The research fields of Pr. V. Sorrenti include oxygen and nitric oxide toxicity and cellular defenses in various pathophysiological conditions; the pathway ADMA/DDAH/NOS in various pathophysiological conditions; evaluation of the activity of natural drugs and new synthetic compounds; the evaluation of synthetic compounds as heme oxygenase inhibitors. Pr. Sorrenti is a referee for *Nitric Oxide: Biology and Chemistry, Life Sciences, Journal of Enzyme Inhibition & Medicinal Chemistry, Nutrition and Cancer, J. Pathology, Evidence-Based Complementary and Alternative Medicine, BMC Cancer, Molecules, BioMed Research International, PLOS One, Medicinal Chemistry, Stem Cells International, Cancers, Toxins, Oncology Reports, Molecular Medicine Reports* as well as Peer Reviewer for Philip Morris Scientific grants. Pr. V.Sorrenti has published 106 scientific publications, co-authored 2 books ("Aspetti molecolari dell'apoptosi e ruolo fisiopatologico" and "Flessibilmente: un modello sistemico di approccio al tema della flessibilità": Flessibilità e biologia-Organismi viventi come esseri flessibili") and co-edited "Recent Research Developments in Chemistry and Biology of Nitric Oxide—2008".

International Journal of
Molecular Sciences

MDPI

Editorial

Editorial of Special Issue "Protective and Detrimental Role of Heme Oxygenase-1"

Valeria Sorrenti

Department of Drug Science, Biochemistry Section, University of Catania, 95125 Catania, Italy; sorrenti@unict.it; Tel.:+39-0957-3741-15

Received: 16 September 2019; Accepted: 21 September 2019; Published: 24 September 2019

The Special Issue, "Protective and Detrimental Role of Heme Oxygenase-1", of the *International Journal of Molecular Sciences*, includes original research papers and reviews, some of which were aimed to understanding the dual role (protective and detrimental) of HO-1 and the signaling pathway involved. Heme oxygenase (HO)-1 is known to metabolize heme into biliverdin/bilirubin, carbon monoxide, and ferrous iron, and it has been suggested to demonstrate cytoprotective effects against various stress-related conditions. HO-1 is commonly regarded as a survival molecule, exerting an important role in cancer progression and its inhibition is considered beneficial in a number of cancers. However, increasing studies have shown a dark side of HO-1, in which HO-1 acts as a critical mediator in ferroptosis induction and plays a causative factor for the progression of several diseases [1]. Lackani et al. demonstrated for the first time that HO-1 has the ability to restore cellular redox, rescue SIRT1, and prevent Ang II-induced impaired effects on adipocytes and the systemic metabolic profile [2]. The study of Fujiwara et al. demonstrated that the physiological effects of the HO-1/CO system were employed for preserving donor lungs with unique characteristics via the high-pressure gas (HPG) preservation method. This approach has significant potential to be used as a new preservation method for lungs [3]. The pharmacological activation of HO-1 activity mimics the effect of caloric restriction (CR), while the HO-1 inhibitor Tin-mesoporphyrin IX (SnMP) increased oxidative stress and cardiac hypertrophy. These data suggest the critical role of HO-1 in protecting the diabetic heart [4]. Bilirubin (BR), the end product of the heme degradation pathway is an important endogenous antioxidant, and it plays a crucial role in protection against oxidative stress. OH-1 activity can modulate BR levels. Decreased inflammatory status has been reported in subjects with mild unconjugated hyperbilirubinemia. Valaskova et al. reported that hyperbilirubinemia in Gunn rats is associated with an attenuated systemic inflammatory response and decreased liver damage upon exposure to Lipopolysaccharide (LPS) [5]. Antigen-presenting cells (APCs) including dendritic cells (DCs) play a critical role in the development of autoimmune diseases by presenting self-antigen to T-cells. It has been reported that the protective effect and the reduction of lesions in the pancreas were due to the inhibition of oxidative stress mediated by HO-1 activity. Data obtained by Pogu et al. demonstrated the potential of induction of HO-1 expression in DCs as a preventive treatment, and potential as a curative approach for Type I diabetes [6]. Given the association between inflammation and prostate cancer (PCa), and the anti-inflammatory role of heme oxygenase 1 (HO-1), the study of Leonardi et al. identified an interaction between HO-1 and glucocorticoid receptor (GR). The modulation between HO-1 and GR pathways may represent a therapeutic strategy in PCa therapy [7]. Gall et al. review the heme–heme oxygenase–endoplasmic reticulum (ER) stress relationship; the major mechanisms of their interactions by which ER stress contributes to the cell and organ damage in diabetes, atherosclerosis, and brain hemorrhage. Since HO-1 presents a unique Janus-faced character in brain pathologies, this issue has received special attention [8]. The review by Kishimoto et al. summarizes the roles of HO-1 in atherosclerosis and focuses on the clinical studies that examined the relationships between HO-1 levels and atherosclerotic diseases [9].

Other original research papers of the Special issue were aimed at the identification of natural molecules or new synthetic compounds able to modulate HO-1 activity/expression. These articles will help make HO-1 a potential therapeutic target for the amelioration of various diseases. It has been reported that hepatoprotective effect of *Myristica fragrans* kernels in the livers of rats exposed to Acetaminophen (APAP)-induced hepatotoxicity could be linked to their ability to promote the NF-E2-related factor 2 (Nrf2)/ antioxidant responsive element (ARE) pathway. Hepatoprotection effects were mediated via suppressing oxidative stress, inflammation, and apoptosis [10]. A lot of evidence showed that HO-1 induces ferroptosis through an increase of ROS production mediated by iron accumulation and accompanied by augmentation lipid peroxidation and glutathione depletion. Results obtained in the study of Acquaviva et al. demonstrated that, highest concentration of *Betula etnensis Raf. (Birch Etna)* extract, was able to induce ferroptotic cancer cell death. HO-1 mediated ferroptosis may represent a chemotherapeutic strategy against tumor [11]. Metformin (MET), a drug widely used for type 2 diabetes, has recently gained interest for treating several cancers. Disrupting antioxidant HO-1 activity, especially under low glucose concentrations, could be an attractive approach to potentiate metformin antineoplastic effects, and could provide a biochemical basis for developing HO-1-targeting drugs against solid tumors [12]. Data obtained by Sorrenti et al., demonstrated that inducible nitric oxide synthase/gamma-Glutamyl-cysteine ligase (iNOS/GGCL) and dimethylarginine dimethylaminohydrolase (DDAH) dysregulation may play a key role in high glucose mediated oxidative stress, whereas HO-1 inducers such as Caffeic acid phenethyl ester (CAPE) or its more potent derivatives may be useful in diabetes and other stress-induced pathological conditions [13]. The study of Moreno et al. reveals an interaction between HO-1 and nitric oxide synthase-1 (NOS1)/ nitric oxide synthase-2 (NOS2) during peripheral inflammation and shows that Cobalt protoporphyrin (CoPP) and CO-releasing molecules-2 (CORM-2) improved HO-1 expression and modulated the inflammatory and/or plasticity changes caused by peripheral inflammation in the *locus coeruleus* [14].

Overall, the 14 contributions published in this Special Issue highlight the dual role (protective and detrimental) of HO-1 and the signaling pathways involved. HO-1 may represent a potential therapeutic target for the amelioration of various diseases. Natural molecules or new synthetic compounds able to modulate HO-1 activity/expression may represent a therapeutic strategy against various diseases.

Funding: This research received no external funding.

Conflicts of Interest: The authors declare no conflicts of interest.

References

1. Chiang, S.C.; Chen, S.E.; Chang, L.C. A Dual Role of Heme Oxygenase-1 in Cancer Cells. *Int. J. Mol. Sci.* **2019**, *20*, 39. [CrossRef] [PubMed]
2. Lakhani, H.V.; Zehra, M.; Pillai, S.S.; Puri, N.; Shapiro, J.I.; Abraham, N.G.; Sodhi, K. Beneficial Role of HO-1-SIRT1 Axis in Attenuating Angiotensin II-Induced Adipocyte Dysfunction. *Int. J. Mol. Sci.* **2019**, *20*, 3205. [CrossRef] [PubMed]
3. Fujiwara, A.; Hatayama, H.; Matsuura, N.; Yokota, N.; Fukushige, K.; Yakura, T.; Tarumi, S.; Go, T.; Hirai, S.; Naito, M.; et al. High-Pressure Carbon Monoxide and Oxygen Mixture is Effective for Lung Preservation. *Int. J. Mol. Sci.* **2019**, *20*, 2719. [CrossRef] [PubMed]
4. Waldman, M.; Nudelman, V.; Shainberg, A.; Zemel, R.; Kornwoski, R.; Dan Aravot, D.; Peterson, S.J.; Arad, M.; Hochhauser, E. The Role of Heme Oxygenase 1 in the Protective Effect of Caloric Restriction against Diabetic Cardiomyopathy. *Int. J. Mol. Sci.* **2019**, *20*, 2427. [CrossRef] [PubMed]
5. Valaskova, P.; Dvorak, A.; Lenicek, M.; Zizalova, K.; Kutinova-Canova, N.; Zelenka, J.; Cahova, M.; Vitek, L.; Muchova, L. Hyperbilirubinemia in Gunn Rats Is Associated with Decreased Inflammatory Response in LPS-Mediated Systemic Inflammation. *Int. J. Mol. Sci.* **2019**, *20*, 2306. [CrossRef] [PubMed]
6. Pogu, J.; Sotiria Tzima, S.; Georges Kollias, K.; Ignacio Anegon, I.; Philippe Blancou, P.; Simon, T. Genetic Restoration of Heme Oxygenase-1 Expression Protects from Type 1 Diabetes in NOD Mice. *Int. J. Mol. Sci.* **2019**, *20*, 1676. [CrossRef] [PubMed]

7.   Leonardi, D.B.; Anselmino, N.; Brandani, J.N.; Jaworski, F.M.; Páez, A.V.; Mazaira, G.; Meiss, R.P.; Nuñez, M.; Nemirovsky, S.I.; Giudice, J.; et al. Heme Oxygenase 1 Impairs Glucocorticoid Receptor Activity in Prostate Cancer. *Int. J. Mol. Sci.* **2019**, *20*, 1006. [CrossRef] [PubMed]

8.   Gáll, T.; Balla, G.; József Balla, J. Heme, Heme Oxygenase, and Endoplasmic Reticulum Stress—A New Insight into the Pathophysiology of Vascular Diseases. *Int. J. Mol. Sci.* **2019**, *20*, 3675. [CrossRef] [PubMed]

9.   Kishimoto, Y.; Kazuo Kondo, K.; Momiyama, Y. The Protective Role of Heme Oxygenase-1 in Atherosclerotic Diseases. *Int. J. Mol. Sci.* **2019**, *20*, 3628. [CrossRef] [PubMed]

10. Dkhil, M.A.; Abdel Moneim, A.E.; Hafez, T.A.; Mubaraki, M.A.; Mohamed, W.F.; Thagfan, F.A.; Al-Quraishy, S. *Myristica fragrans* Kernels Prevent Paracetamol-Induced Hepatotoxicity by Inducing Anti-Apoptotic Genes and Nrf2/HO-1 Pathway. *Int. J. Mol. Sci.* **2019**, *20*, 993. [CrossRef] [PubMed]

11. Malfa, G.A.; Tomasello, B.; Acquaviva, R.; Genovese, C.; Mantia, A.L.; Cammarata, F.P.; Ragusa, M.; Renis, M.; Giacomo, C.D. *Betula etnensis Raf. (Betulaceae)* Extract Induced HO-1 Expression and Ferroptosis Cell Death in Human Colon Cancer Cells. *Int. J. Mol. Sci.* **2019**, *20*, 2723.

12. Raffaele, M.; Pittalà, V.; Zingales, V.; Barbagallo, I.; Salerno, L.; Volti, G.L.; Romeo, G.; Carota, G.; Sorrenti, V.; Luca Vanella, L. Heme Oxygenase-1 Inhibition Sensitizes Human Prostate Cancer Cells towards Glucose Deprivation and Metformin-Mediated Cell Death. *Int. J. Mol. Sci.* **2019**, *20*, 2593. [CrossRef] [PubMed]

13. Sorrenti, V.; Raffaele, M.; Vanella, L.; Acquaviva, R.; Salerno, L.; Pittalà, V.; Intagliata, S.; Giacomo, C.D. Protective Effects of Caffeic Acid Phenethyl Ester (CAPE) and Novel Cape Analogue as Inducers of Heme Oxygenase-1 in Streptozotocin-Induced Type 1 Diabetic Rats. *Int. J. Mol. Sci.* **2019**, *20*, 2441. [CrossRef] [PubMed]

14. Moreno, P.; Cazuza, R.A.; Mendes-Gomes, J.; Díaz, A.F.; Polo, S.; Leánez, S.; Leite-Panissi, C.R.A.; Pol, O. The Effects of Cobalt Protoporphyrin IX and Tricarbonyldichlororuthenium (II) Dimer Treatments and Its Interaction with Nitric Oxide in the *Locus Coeruleus* of Mice with Peripheral Inflammation. *Int. J. Mol. Sci.* **2019**, *20*, 2211. [CrossRef] [PubMed]

International Journal of
Molecular Sciences

MDPI

Review

A Dual Role of Heme Oxygenase-1 in Cancer Cells

Shih-Kai Chiang [1], Shuen-Ei Chen [1,2,3,4] and Ling-Chu Chang [5,*]

[1] Department of Animal Science, National Chung Hsing University, Taichung 40227, Taiwan;
    shihkaichiang@gmail.com (S.K.C.); chenshuenei@hotmail.com (S.E.C.)
[2] Innovation and Development Center of Sustainable Agriculture (IDCSA), National Chung Hsing University,
    Taichung 40277, Taiwan
[3] The iEGG and Animal Biotechnology Center, National Chung Hsing University, Taichung 40277, Taiwan
[4] Research Center for Sustainable Energy and Nanotechnology, National Chung Hsing University,
    Taichung 40277, Taiwan
[5] Chinese Medicinal Research and Development Center, China Medical University Hospital,
    Taichung 40447, Taiwan
[*] Correspondence: t27602@mail.cmuh.org.tw; Tel.: +886-4-22052121 (ext. 7913); Fax: +886-4-22333496

Received: 9 November 2018; Accepted: 19 December 2018; Published: 21 December 2018

Abstract: Heme oxygenase (HO)-1 is known to metabolize heme into biliverdin/bilirubin, carbon monoxide, and ferrous iron, and it has been suggested to demonstrate cytoprotective effects against various stress-related conditions. HO-1 is commonly regarded as a survival molecule, exerting an important role in cancer progression and its inhibition is considered beneficial in a number of cancers. However, increasing studies have shown a dark side of HO-1, in which HO-1 acts as a critical mediator in ferroptosis induction and plays a causative factor for the progression of several diseases. Ferroptosis is a newly identified iron- and lipid peroxidation-dependent cell death. The critical role of HO-1 in heme metabolism makes it an important candidate to mediate protective or detrimental effects via ferroptosis induction. This review summarizes the current understanding on the regulatory mechanisms of HO-1 in ferroptosis. The amount of cellular iron and reactive oxygen species (ROS) is the determinative momentum for the role of HO-1, in which excessive cellular iron and ROS tend to enforce HO-1 from a protective role to a perpetrator. Despite the dark side that is related to cell death, there is a prospective application of HO-1 to mediate ferroptosis for cancer therapy as a chemotherapeutic strategy against tumors.

Keywords: ferroptosis; heme oxygenase-1; iron; reactive oxygen species; glutathione; chemotherapy

1. Introduction

Oxidative stress is caused by an imbalance between cellular oxidants and antioxidants. Reactive oxygen species (ROS) are the major cellular oxidants, which are normally generated as by-products in oxygen metabolism. However, under some circumstances, extracellular insults (e.g., ionizing radiation and UV light), xenobiotics, and pathogens also greatly provoke ROS production, leading to an imbalance of the intracellular reduction-oxidation (redox) status [1]. Excessive ROS can induce oxidative damage of DNA, and, to a higher degree, gene mutation and carcinogenesis [2–4]. Moreover, lipid peroxidation by excessive ROS may damage cellular structures and eventually induce cell death [1]. In fact, the augmentation of ROS is a useful approach for clinical cancer treatment. Various chemotherapeutic agents, such as cisplatin, doxorubicin, and 5-fluorouracil, have been shown to exert their antitumor activity via ROS-dependent activation of apoptosis [4,5]. Therefore, oxidation therapy becomes a possible strategy by provoking ROS production and diminishing antioxidant enzymes in cancer cells. Ferroptosis is a newly identified non-programmed cell death, characterized by excessive accumulation of free cellular iron and severe lipid peroxidation [6]. This ROS- and iron-overload cell death became a new therapeutic strategy in several diseases, especially in cancer treatment. Indeed, ferroptosis-inducing agents (erastin, RSL3, and sorafenib) have demonstrated therapeutic effects against cancers [6,7].

Heme oxygenase-1 (HO-1) is a phase II enzyme that responds to electrophilic stimuli, such as oxidative stress, cellular injury, and diseases. HO-1 is elevated in various human malignancies, implicating its contribution to settle the tumor microenvironment for cancer cell growth, angiogenesis, and metastasis, as well as resistance to chemotherapy and radiation therapy. By contrast, augmented expression of HO-1 in tumor cells can enhance cell death in many cancers. [8–11]. Its multiple pleiotropies in tumorigenesis, including tumor initiation, angiogenesis, and metastasis, have been well reviewed [8–11]. Although the bright and dark sides are both discussed in different studies, HO-1 has been widely recognized to play a cytoprotective role in tumor cells to conquer the assault of augmented oxidative stress by chemotherapeutic agents, thus preventing the cancer cells from apoptosis and autophagy, and even promoting cell proliferation and metastasis [8,9,11]. The protective or detrimental effects of HO-1 were also reported in different diseases, including kidney injury and neurodegeneration [12–14]. Emerging evidence has revealed another dark side of HO-1, showing that HO-1 induces ferroptosis through iron accumulation [15–17] or other unknown mechanisms. Based on the current findings, this review provides a brief background on the biological functions of HO-1, as well as its metabolites, namely biliverdin/bilirubin, carbon monoxide, and ferrous iron, to delineate how HO-1 mediates ferroptosis.

2. Ferroptosis and Cancer

Ferroptosis is a recently identified type of cell death that is morphologically, genetically, and mechanistically distinct from regulated cell death, including apoptosis, necroptosis, and autophagy [6]. Ferroptotic cells are morphologically characterized by small mitochondria, collapsed mitochondrial cristae, and increased mitochondrial membrane density [6]. Mechanistically, ferroptosis is induced by iron accumulation and lipid peroxidation, accompanied by glutathione depletion. Excessive lipid peroxidation impairs the cellular membrane fluidity, permeability, and cellular integrity, eventually leading to cell death [6,18–20]. Ferroptosis can be induced by the overloading of iron (ferric ammonium citrate), glutathione/glutamine antiporter system Xc^- inhibition (e.g., erastin, sorafenib, sulfasalazine, and lanperisone), and glutathione peroxidase 4 (GPx4) inactivation (RSL3, DPI7). Pharmacological manipulations, such as iron chelators (deferoxamine, ciclipirox, and deferiprone), glutathione replenishment (N-acetyl-L-cysteine, β-mercaptoethanol, cysteine/cystine, intracellular glutathione), and inhibitors of ROS production and lipid peroxidation (liporxstatin-1, ferrostatin-1, zileuton) that modulate the ferroptotic process have been shown to function against diseases, including cancer, neurotoxicity, and ischemia/reperfusion-induced injury [18–20].

The interrelationship between ferroptosis and cancer progression has been validated using ferroptotic agents. Some small molecules (e.g., erastin and RSL3) and clinical cancer drugs (e.g., sorafenib, sulfasalazine, and artesunate) induced cell death via the inhibition of system Xc^- and GPx4 in various types of cancer cells [6,7,19]. A delay of ferroptosis protects cancer cells from metabolic oxidative stress, and thus increases their survival and distal metastasis [21]. Besides, the induction of ferroptosis was shown to overcome artesunate-induced resistance in head and neck cancer cells [22], and the induction of ferroptosis that contributes to anticancer activity has been identified in different cancer types [23]. Diffuse large B-cell lymphomas and renal cell carcinomas strongly rely on GPx4 availability to maintain redox status, and thus may suggest a high sensitivity to ferroptosis [7]. Certain cancer cells, such as pancreatic cancer cell lines (MIA PaCa-2, PANC-1, and BxPC-3) and a subset of triple-negative cancer cells also greatly depend on system Xc^- to mediate cysteine uptake for growth, as well as survival under oxidative stress conditions [24,25], suggesting that system Xc^- might serve as a good chemotherapeutic target. These results delineate the potential of ferroptotic process in clinical applications.

3. HO-1-mediated Ferroptosis in Cancer Cell Survival

As a dual regulator in iron and ROS homeostasis [8,26,27], HO-1 was suggested to serve a dominant role in ferroptosis [15–17,22,28–31]. Alzheimer's patients exhibited enhanced lipid peroxidation, which may be associated with HO-1 elevation and iron accumulation [32]. In HT-1080 fibrosarcoma cells, erastin induces a time- and dose-dependent increase of HO-1 expression [15]. Evidences from HO-1 knockdown mice and by the use of HO-1 inhibitor zinc protoporphyrin IX showed that HO-1 promotes erastin-induced ferroptosis and it is associated with iron bioavailability, but not with biliverdin and bilirubin [15]. However, HO-1 also functions as a negative regulator in erastin- and sorafenib-induced hepatocellular carcinoma since knockdown of HO-1 expression enhanced cell growth inhibition by erastin and sorafenib [31]. A similar result of HO-1 to ameliorate ferroptosis induction was also observed in renal proximal tubule cells [28]. Immortalized renal proximal tubule cells that were obtained from HO-1$^{-/-}$ mice exhibited more pronounced cell death induced by erastin and RSL3 than those from wild type mice [28]. In contrast to the negative role in ferroptosis, several recent studies have demonstrated that enhanced HO-1 expression can augment or mediate anti-cancer-agent (Bay117085 and withaferin A) induced ferroptosis by promoting iron accumulation and ROS production [16,17]. Genetic knockdown and the pharmacological inhibition of HO-1 also validated that HO-1 activation triggers ferroptosis through iron overloading and subsequently excessive ROS generation and lipid peroxidation [16,17]. The silencing of HO-1 by siRNA also reversed the resistance to artesunate-induced ferroptosis in cisplatin-resistant head and neck cancer cells [22]. Based on the contradictory results, it appears that HO-1 activation as a cytoprotective defense or governing ferroptotic progression depends on the degree of ROS production and following oxidative damage in response to stimulatory cues.

4. HO-1 Activation and Heme Metabolites

Heme oxygenases, including HO-1 [also called heat shock protein 32 (Hsp32)], are rate-limiting enzymes in the breakdown of heme (iron protoporphyrin IX). Degradation of heme produces biliverdin, carbon monoxide (CO), and iron (ferrous iron, Fe^{2+}) (Figure 1). Biliverdin is subsequently converted to bilirubin by biliverdin reductase. Oxygen, nicotinamide adenine dinucleotide phosphate (NADPH), and cytochrome p450 reductase are required in this catalytic reaction [26,27]. Cellular iron accumulation upregulates the expression of ferritin, which sequesters the pro-oxidant effect of iron [33]. Three types of heme oxygenases are found in mammalian cells, the inducible form HO-1, constitutive form HO-2, and HO-3, which is mostly inactive. HO-1 can be induced by a wide spectrum of cues, including oxidants, inflammatory mediators, chemicals, physical stimuli, and its own substrate, heme [26,27]. After synthesis, the HO-1 protein is normally anchored in the endoplasmic reticulum [34]. The subcellular location of HO-1 is dynamic. Some pathogenic stimuli may induce the translocation of HO-1 into the plasma membranes, nucleus, and/or mitochondria, which might allow the enhancement of HO-1 activity for heme degradation within the target compartment [35,36].

The most important activation of HO-1 is mediated by nuclear factor erythroid 2-related factor 2 (Nrf2). Under resting conditions, Nrf2 activity is inhibited by physical interaction with Kelch-like ECH-associated protein 1 (Keap1), leading to the recruitment of Cullin-3-dependent E3 ubiquitin ligase for proteasomal degradation, thereby maintaining Nrf2 at a low level [37]. Under oxidative stress, Keap1 undergoes a conformational change and releases Nrf2. Free Nrf2 then translocates into the nuclei where it interacts with small Maf protein and further binds onto the antioxidant-response element (ARE) or electrophile-response element (EpRE), to transactivate various genes encoding antioxidant enzymes, including HO-1 [38]. Increased cellular heme level hampers the induction of HO-1 through Bach1, a Nrf2 antagonist, due to the competition for the promoter binding site [39]. Depletion of cellular glutathione has been shown to increase HO-1 gene transcription in the mouse motor neuron-like hybrid cells, NSC34 cells [40]. HO-1 abundance is also regulated by an endoplasmic reticulum-associated degradation pathway [41]. In HIV-infected astrocytes, HO-1 was degraded in an immunoproteasome-dependent pathway in response to IFNγ and TNFα/LPS stimulation [42].

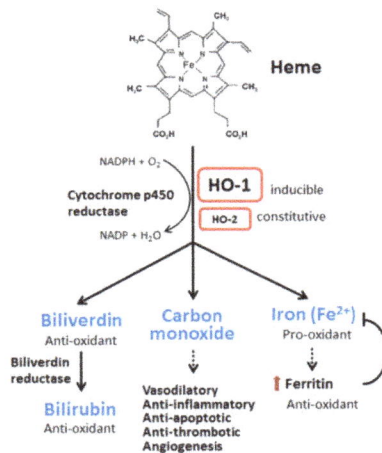

Figure 1. Heme metabolism. Heme is degraded by heme oxygenase (HO), leading to the generation of biliverdin, carbon monoxide, and ferrous iron. Biliverdin is subsequently converted to bilirubin by biliverdin reductase. Under most conditions, biliverdin and bilirubin act as anti-oxidants by scavenging or neutralizing reactive oxygen species (ROS). Carbon monoxide, a gaseous product, mainly functions in signaling transduction, including the vasodilation of blood vessels, production of anti-inflammatory cytokines, upregulation of anti-apoptotic effectors, and thrombosis. Ferrous iron is the major pro-oxidant in all metabolites of heme. However, heme oxygenase-1 (HO-1) activation also increases ferritin expression, which can bind to ferrous iron and detoxify its pro-oxidant effect. The black arrows indicate that biliverdin metabolize into bilirubin. The dotted arrow indicates that carbon monoxide serves a regulator in vasodilatory, anti-inflammatory, anti-apoptotic, anti-thromobtic, and angiogenesis activities. The dotted arrow below iron indicates the iron increase will increase ferritin, which neutralizes the pro-oxidant effect of iron.

The pleiotropic effects of HO-1 and metabolites from heme on tumor growth, neurodegenerative diseases, ischemia/reperfusion injury, and renal injury have been thoroughly reviewed [8,9,11,13,14]. Most of the evidence has suggested that HO-1 functions in cytoprotective defense mechanisms against oxidative attacks through its metabolites biliverdin/bilirubin and CO. However, those metabolites also have demonstrated the detrimental effects, especially in neuronal damage and degeneration [14]. Both biliverdin and bilirubin can inhibit the peroxidation of lipid and protein through scavenging ROS [43]. Biliverdin also shows an ability to modulate the activation of endothelial nitric oxide synthase, leading to a decrease in nitric oxide production [44]. Another protective effect of biliverdin and bilirubin is to interfere with the apoptotic process [45]. Moreover, biliverdin provides a neutralizing activity of ROS, contributing to a proapoptotic effect and the suppression of tumor growth in head and neck squamous cell carcinoma [46]. The cytoprotective or detrimental effects of heme metabolites are determined by or are attributed to their intracellular levels. A high concentration of biliverdin has been shown to cause apoptosis in cancer cells [47]. Overproduction of bilirubin by hemolytic hyperbilirubinemia is associated with bilirubin neurotoxicity in newborns [48].

Another heme metabolite, CO, a gaseous product, is an important signaling molecule, possessing the vasodilatory, anti-inflammatory, anti-proliferative, anti-apoptotic, thrombosis, and angiogenesis activities in various cell types [8,9]. The mechanisms of intracellular events impacted by CO are complicated. CO also exerts both beneficial and deleterious effects, depending on its targeted molecules. CO can activate soluble guanylyl cyclase, followed by cGMP generation, linking cellular proliferation, thrombosis, and vasodilation. CO can also modulate single kinases, including p38 MAP kinase, ERK, and JNK. The activation of p38 can lead to the downregulation of pro-inflammatory cytokines and the upregulation of anti-inflammatory cytokine production, contributing to the anti-inflammatory protection of tissue [8,9]. CO can cooperate with NF-κB to modulate the expression levels of several anti-apoptotic proteins [8,9].

The last metabolite of HO-1, ferrous iron, is toxic due to the ability to interact with cell oxidants to generate ROS [49]. The details are discussed in the next section. In addition to the significant impact on signaling pathways by heme metabolites, HO-1 can mediate various signaling pathways per se, rather than depending on the enzyme activity. A mutated form of HO-1 protein that is defective in catalytic activity could protect cells against oxidative injury [50]. The benefits of enhanced antioxidant activity by HO-1 were associated with increased catalase activity and glutathione levels [50].

5. HO-1 and Iron

Iron (Fe), an essential metal for biological activities, participates in electron transport of the respiration chain, heme synthesis, erythropoiesis, and enzyme systems. However, iron is a potential toxicant to cells due to its pro-oxidant activity, which can lead to oxidant DNA damage, causing neurodegenerative diseases and promoting oncogenesis [20,49,51]. Ferroptosis, a form of iron-mediated oxidative cell death, has been shown to play a critical role in the pathogenesis involving iron-overload, such as cancer and neurodegenerative diseases [18–20,51], thus implying a harmful role of HO-1.

Iron is involved in the transfer of electrons via oxidation-reduction reactions to transition between the ferric (Fe^{3+}) and ferrous (Fe^{2+}) states [50]. The same mechanism is employed during intracellular transport and toxicity production of iron. Transferrin is responsible for iron transport in the bloodstream. Iron binds to transferrin in an oxidized ferric state (Fe^{3+}). Iron can enter cells by two modes, transferrin receptor-mediated endocytosis and independent transport of non-protein-bound iron (NPBI). In the NPBI system, ferrous irons diffuse into cells through binding with the low-molecular-weight complexes, such as adenosine triphosphate, citrate, ascorbate, peptides, or phosphatases [52,53]. After acidification within the endosome, iron disassociates from transferrin and is reduced by ferric reductase into ferrous iron, which is then transported into the cytosol by divalent metal transporter 1 (DMT1). In the cytosol, ferrous iron can be used, stored into ferritin, or effluxed from cells by the iron exporter ferroportin [53]. Intracellular iron mainly binds to specific proteins, such as ferritin, hemoproteins, and various iron-containing proteins for further utilization. During iron deficiency, iron-binding ferritin can undergo recycling by autophagic turnover in the lysosome [54]. Nuclear Co-Activator 4 (NCOA4) is an autophagic cargo receptor, which can bind to ferritin and carry it into the autophagosome [55].

Few irons in the cytosol are deposited in the labile iron pool where the redox-active iron (Fe^{2+}) is oxidized by hydrogen peroxide to Fe^{3+} (ferric) and therefore generates ROS, including soluble radical (HO·), lipid alkoxy (RO·), and hydroxide ion (OH⁻) via the Fenton reaction. Free iron in the redox-active form is easily accessed as a pro-oxidant [49]. Thus, the NPBI system is crucial for iron overload to induce the lipid peroxidation [52,53]. Overloading of iron can promote the Fenton reaction and ROS generation [49]. Excessive ROS production consequently results in the peroxidation of adjacent lipid and oxidative damage of DNA and proteins, eventually inducing ferroptosis [6,7,18–20].

Cellular iron homeostasis and distribution are regulated by specific iron-regulating proteins. At a low iron level, these regulatory proteins can bind to the iron-response element of target genes to inhibit the expression of iron-binding proteins, such as ferritin and ferroportin, but increase the expression of transferrin receptor [56]. Enhanced HO-1 activity was shown to increase the cellular iron level and also promote ferritin production to sequester iron and following pro-oxidant effects in the meanwhile [33,57]. However, the iron-binding ability of ferritin is disturbed by oxidative stress, causing an uncontrolled release of iron, finally resulting in excessive accumulation of iron in the cytosol and enhancement of lipid peroxidation. The importance of iron in ferroptosis was thoroughly confirmed by increased iron uptake and the presence of iron chelators, such as deferoxamine and ciclopirox, and deferiprone [6,18–20]. Similarly, supplementation with an exogenous source of iron, such as ferric ammonium citrate, ferric citrate, and iron chloride hexahydrate enhanced ferroptosis [6]. Activation of iron metabolism-related proteins also contributes to ferroptosis. Upregulation of transferrin receptor 1 resulted in a higher sensitivity to ferroptosis induction [58,59]. The iron-response element binding protein 2 (IREB2) is essential for erastin-induced ferroptosis [6]. Moreover, a decrease in ferritin for iron storage may cause a free iron overload [58], thereby enhancing ferroptosis induction by stimulatory agents [6]. Increased degradation of NCOA4, a receptor cargo protein of ferritin, resulted in ferritinophagy and ferritin degradation, leading to an increase in free iron and ROS generation [60].

Ferroptosis induction by HO-1 is mediated by iron augmentation and lipid peroxidation [15–17]. The induction is tightly associated with oxidative stress, due to its predominant sensitivity to oxidative inducers, such as redox-active iron, heme, hemoglobin, and heme-containing proteins [61,62]. The upregulation of HO-1 can enhance heme degradation and ferritin synthesis and change the intracellular iron distribution [57].

6. HO-1 Modulation in Ferroptosis

Due to highly reactive iron and oxidative stress, the role of HO-1 in ferroptosis was recently re-proposed, in which both the protective and detrimental roles have been demonstrated [15–17,22,28–31]. Thus, HO-1 may provide redox-active iron production for the re-modification of biomolecules and structures, including lipid and protein peroxidation. The pro-oxidative activity of HO-1 contributes to ferroptosis induction [15–17,29], which relies on iron accumulation, not biliverdin/bilirubin [15]. Under pro-oxidant circumstances, more iron is released from iron-storing proteins, further increasing the production of ROS to accelerate oxidative stress. It can be explained that a moderate level of HO-1 activation exerts a cytoprotective effect, while the over-activation of HO-1 becomes cytotoxic due to excessive increase of labile Fe^{2+} behind the buffering capacity of ferritin [4,17,63]. A cascade of increased iron release and ROS production resulting in extensive oxidative damage to cells has been reported [64], exemplifying the potential role of HO-1 in iron accumulation and ROS generation, followed by lipid peroxidation and ferroptosis induction. Intriguingly, HO-1 is also transcriptionally upregulated by lipid peroxidation products (4-hydroxynonenal) and phospholipase metabolites (diacylglycerol and arachidonic acid), for example, in response to UV light induction [65].

Several small molecules have been identified to trigger ferroptosis through the regulation of HO-1 expression and activity (Table 1, Figure 2). These small molecules possess similar properties. Most of them can trigger iron release and generate massive production of ROS. Some molecules have been identified to induce HO-1-associated ferroptosis, including heme [66], erastin/sorafenib/RSL3 [15,22,31], magnesium isoglycyrrhizinate [29], BAY117085 [16], and withaferin A [17].

6.1. Heme

Heme is a large complex, comprising iron and protoporphyrin IX. Heme is regarded as a pro-oxidant molecule that participates in the formation of oxidative radicals and leads to oxidative injury [67]. However, because of free iron with different oxidation states, heme-containing enzymes can exert its catalysis in both reductive and oxidative reaction [67]. Therefore, several heme-containing proteins are involved in the electron transport chain in mitochondria and redox reactions, such as catalase, peroxidase, and cytochrome p450. Heme can also transfer and store oxygen while it binds to globin [67]. Hemolysis causes heme release into the circulation, leading to the generation of ROS and oxidative stress [67]. In plasma, free heme can be scavenged by hemopexin or degraded by HO-1 to generate biliverdin, CO, and ferrous iron [68]. Nevertheless, in severe hemolysis, the extensive release of heme from hemoglobin or decreased hemopexin level leads to an increase in the iron level in the circulation. Heme induces apoptosis and necroptosis in red blood cells and cerebral microvascular cells by increasing the cellular calcium level and depleting the glutathione reservoir [69]. These observations suggest that heme serves as a pro-oxidant molecule, not only functioning as an activator of HO-1, but also acting on the ferroptotic process.

Table 1. Role of increased HO-1 in the ferroptosis and anti-ferroptosis response to different inducers.

Drugs	Cell Types	HO-1 Activity	Significance of HO-1 Increase in Ferroptosis	References
Heme	Platelets	Upregulated	Pro-ferroptotic effect ↑Iron ↑ROS ↓GSH ↑Lipid peroxidation	[66]
Hydrogen peroxide	A549 lung adenocarcinoma	Upregulated	Anti-ferroptotic effect ↑Transferrin ↑Transferrin receptor ↑Ferritin ↓Iron	[57]
Erastin RSL3	HT-1080 fibrosarcoma cells	Upregulated	Pro-ferroptotic effect ↑Iron ↑CO	[15]
Erastin Sorafenib	Hepatocellular carcinoma, HepG2 and Hepa1-6 cells	Upregulated	Anti-ferroptotic effect ↑ROS ↑Iron ↑Ferritin	[31]
Magnesium isoglycyrrhizinate	CCL4-induced liver fibrosis rat model Hepatic stellate cell line HSC-T6	Upregulated	Pro-ferroptotic effect ↑Iron ↑Lipid peroxidation	[29]
Artesunate	Cisplatin-resistant head and neck cancer cell	Upregulated	Anti-ferroptotic effect HO-1 co-works with NQO-1 to serve as antioxidant	[22]
BAY117085	Triple-negative breast cancer, MDA-MB-231 cells; *Glioblastoma, DBTRG-05MG*	Upregulated	Pro-ferroptotic effect Inhibit system Xc⁻ ↓GSH ↑ROS ↑Iron	[16]
Withaferin A	Neuroblastoma, IMR-32 cells.	Upregulated	Pro-ferroptotic effect ↑ROS ↑Iron	[17]

The up arrows indicate the increased expression or enhanced activity. The down arrows indicate decreased expression or decreased levels.

Hemin, an iron-containing porphyrin with chlorine, promotes erastin-induced ferroptotic cell death in a HO-1-dependent manner in HT-1080 fibrosarcoma cells [15]. Similar to hemin-induced ferroptosis, iron-induced ferroptosis through increased production of ROS is also observed in neuroblastoma IMR-32 cells [17]. Interestingly, HO-1 induction might protect cells from oxidative assaults. In cultured alveolar epithelial cells, HO-1 overexpression increased the expression of ferritin and transferrin receptor, resulting in the alteration of the intracellular iron distribution and providing a protective effect against iron cytotoxicity from heme degradation [70]. Additionally, the protective effect by heme was demonstrated in oxidant hydrogen peroxide-treated human lung adenocarcinoma A549 cells. HO-1 upregulation by hydrogen peroxide enhanced heme synthesis and the expression of ferritin and transferrin receptor, leading to the capture of intracellular redox-active iron and the elimination of oxidative stress [59].

6.2. Erastin, Sorafenib and RSL3

Erasin, sorafenib, and RSL3 are known as ferroptosis-inducing agents to inhibit system Xc⁻ and GPx4 [18–20]. Both erastin and RSL3 were shown to upregulate HO-1 in renal proximal tubule cells [28] and HT-1080 fibrosarcoma cells [15]. In renal proximal tubule cells, HO-1$^{-/-}$ cells showed more sensitivity to erastin- and RLS3-induced cell death than HO-1$^{+/+}$ cells [28]. Thus, HO-1 played an anti-ferroptotic role due to its anti-oxidant effect against ROS. By contrast, in HT-1080 fibrosarcoma cells, HO-1$^{+/+}$ cells exhibited a more profound effect on ferroptotic induction than HO-1$^{-/-}$ cells.

The metabolites of heme by HO-1, including iron and CO, contribute to the pro-ferroptotic effect of HO-1 [15]. In hepatocellular carcinoma, HepG2 and Hepa1-6 cells, genetic knockdown of Nrf2, HO-1, quinone oxidoreductase 1 (NQO-1), or ferritin also promoted erastin- and sorafenib-induced ferroptosis [31]. In hepatocellular carcinoma cells, Nrf2 expression contributed to ferroptosis resistance, in association with its downstream effector proteins, such as HO-1, NQO-1, and ferritin. HO-1 exerts a detoxification and antioxidant effect in mediating the anti-ferroptosis of Nrf2 [31]. Additionally, Nrf2, HO-1, and ferritin are transcriptionally regulated by p62, a ubiquitin binding protein that is involved in cell signaling pathways, such as the oxidative response and autophagy [71,72]. It is particularly noteworthy that the p62-Keap1-Nrf2 pathway plays an important role in ferroptosis through the upregulation of multiple genes, including HO-1, NQO-1, and ferritin [31]. The protective role of Nrf2-derived genes in erastin-induced ferroptosis was also proposed in glioblastoma [30].

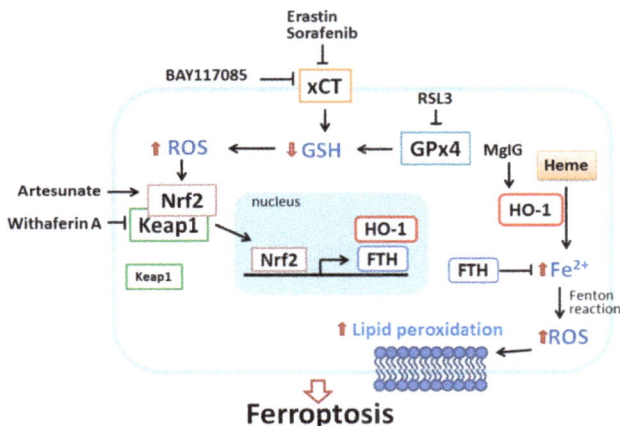

Figure 2. Scheme of HO-1-regulated ferroptosis. HO-1 plays a dual role in ferroptosis, pro-ferroptotic and anti-ferroptotic effects. Erastin and sorafenib (xCT inhibitor) and RSL3 (glutathione peroxidase 4 (GPx4) inhibitor) can deplete glutathione, leading to ROS generation. In response to oxidative stress, nuclear factor erythroid 2-related factor 2 (Nrf2) disassociates from Kelch-like ECH-associated protein 1 (Keap1), and then migrates into nuclei, where it binds the antioxidant-response element (ARE) site of target genes such as HO-1 and ferritin. HO-1 catalyzes heme degradation to generate ferrous iron (Fe^{2+}). Ferrous iron is highly reactive as a pro-oxidant and, thus, produces ROS. Excessive ROS damage intracellular structures and DNA, causing the peroxidation of lipid and protein and eventually cell death. Nrf2 induces ferritin expression to chelate ferrous iron, avoiding ROS overload. Recently, some small molecules were identified to possess a pro-ferroptosis effect through HO-1. Heme can directly activate HO-1 expression. Similar to erastin and sorafenib, BAY117089 can deplete GSH and increase ROS production, resulting in Nrf2−HO-1 activation and ferroptosis. Withaferin A directly targets Keap1 and releases Nrf2, followed by HO-1 activation, iron accumulation, and cell death. Magnesium isoglycyrrhizinate (MgIG) increases HO-1 expression and free cellular iron level. By contrast, the activation of HO-1 might provide a cytoprotective effect. For example, in erastin-, sorafenib-, and RSL-stimulated cells, ferritin expression is increased through the Nrf2−HO-1 pathway and neutralize iron toxicity. Nrf2-targeted antioxidant gene expression also benefits the acquisition of drug resistance. Artesunate also induces the Nrf2−HO-1 signal to assist cells to acquire drug resistance.

6.3. Magnesium Isoglycyrrhizinate

Magnesium isoglycyrrhizinate (MgIG) is a hepatoprotective drug with a potential to alleviate the inflammation and accelerate the recovery of injured liver [73]. Sui et al. found that MgIG markedly attenuated liver injury and reduced fibrotic scar formation in a rat model of CCL4-induced liver fibrosis [29]. MgIG significantly inhibited the growth of hepatic stellate cells, the main effector cell in liver fibrosis [74], by promoting ferroptotic induction, as evidenced by the elevated iron accumulation and lipid peroxidation and the suppression of the cellular glutathione levels [66]. HO-1 deficiency of hepatic stellate cells HSC-T6 made them less sensitive to MgIG-induced ferroptosis. Interestingly, transferrin, transferrin receptor, and ferritin were all upregulated by MgIG in HSC-T6 cells [66]. However, MgIG ameliorated liver injury by ethanol through the inhibition of ROS and neutrophil infiltration, as well as by reduced expression of proinflammatory cytokines and chemokines [75].

6.4. Artesunate

Artesunate is a semisynthetic derivative of artemisinin, isolated from *Artemisia annua* L., which has been used in the treatment of *P. falciparum* malaria [76]. The main mechanism of the antimalarial action of artesunate involves NADPH activation, ROS generation, and DNA damage [77]. In MCF7 breast cancer cells, artesunate impacts the endolysosomal and autophagosomal compartments, leading to the blockade of autophagosome turnover and perinuclear clustering of autophagosomes, endosomes, and lysosomes. Free iron is thereby accumulated and serves as the major cause of ROS production, which turns out to be a critical prerequisite for artesunate-mediate cell death in MCF-7 breast cancer cells [76]. Artesunate can induce ferroptosis in head and neck cancer cells, but cisplatin-resistant cells are less sensitive to artesunate. In cisplatin-resistant head and neck cancer cells, the activation of Nrf2 and downstream targets HO-1 and NQO-1 by artesunate contributed to the resistance against ferroptosis. Inactivation of the Nrf2 pathway using a siRNA genetic approach reversed the ferroptotic induction by artesunate, and this ferroptosis resistance was further blocked by deferoxamine, an iron chelator, and by antioxidant Trolox [22].

6.5. BAY117085

BAY117085 was identified as an NF-κB inhibitor by blocking the phosphorylation and nuclear translocation of IκBα [78]. However, BAY117085 can induce ferroptosis in an NF-κB-independent manner [16]. In triple-negative breast cancer cells, MDA-MB-231 cells, and glioblastoma multiforme DBTRG-05MG cells, BAY117085 upregulated HO-1 expression through the Nrf2−SLC7A11 pathway, which, in turn, depleted the cellular glutathione reservoir and provoked ROS generation, resulting in iron accumulation and ultimately ferroptosis. Enforced expression of HO-1 substantially promoted ROS production and iron release, leading to endoplasmic reticulum stress, as evidenced by increased Chop and spliced XBP1 transcripts. Interestingly, BAY117085 also caused the compartmentalization of HO-1 within the nucleus and mitochondria, and subsequently caused mitochondrial dysfunction, leading to lysosome targeting for mitophagy [16]. Mitochondria-targeted HO-1 was further shown to induce higher ROS production, leading to mitochondrial dysfunction, such as fission and later development into cytotoxicity, as observed in macrophages, kidney fibroblasts, and chronic alcohol hepatotoxicity [79]. Mitochondrial targeting of HO-1 also enhanced autophagy by increasing the translocalization of LC3 and Drp1 into the mitochondria [79]. In doxorubicin-induced cardiomyopathy, however, mitochondrial targeting of HO-1 demonstrated a cytoprotective role to improve mitochondrial quality [80]. In contrast to that in wild-type mice, enforced expression of HO-1 remarkably ameliorated doxorubicin-mediated dilation of cardiac sarcoplasmic reticulum, mitochondrial fragmentation, and the number of damaged mitochondria in autophagic vacuoles. The amelioration was attributed to the increase in mitochondrial biogenesis, as evidenced by the upregulation of Nrf1, PGC1α, and TFAM, as well as by the attenuated changes in the expression of the mitochondrial fission mediator Fis1 and fusion mediators, Mfn1 and Mfn2 [80].

6.6. Withaferin A

Withaferin A is a steroidal lactone extracted from the roots and leaves of *Withania somnifera* Dunal, commonly known as Ashwagandha, Indian ginseng, or Indian winter cherry [81]. Due to its anti-proliferative and pro-apoptotic activities, withaferin A demonstrates its therapeutic potential for chemoprevention in various cancer types. Mechanistically, withaferin A can disturb the cell cycle, inhibit the activation of proliferation-related kinases (EGFR, Akt, and NF-κB), alter the ratio of pro-apoptotic/anti-apoptotic proteins, and provoke ROS generation. With the predominant oxidative cytotoxic effect, withaferin A was shown to cause mitochondrial dysfunction, apoptosis, and paraptosis [82]. Hassannia et al. demonstrated that withaferin A induced ferroptosis in neuroblastoma IMR-32 cells via two different pathways—repressing the protein level and activity of GPx4 and upregulating HO-1 expression [17]. Withaferin A decreases Keap1, leading to the activation of Nrf2 and upregulation of HO-1, followed by an increase in intracellular labile ferrous iron and consequently ferroptosis. Administration of withaferin A significantly suppressed the tumor growth of neuroblastoma through increased HO-1 expression and decreased GPx4 expression [17]. A detailed mechanism regarding upregulated HO-1 via directly targeting Keap1 by withaferin A has also been identified in endothelial cells [83].

7. Manipulation of HO-1 in Ferroptosis

Due to a high proliferation and fast metabolic rate, cancer cells tend to exhibit higher ROS production [2–4], and thus it may promote tumorigenesis and render cancer cells more vulnerable to oxidative stress-induced cell death. Manipulation of the intracellular ROS level may be a useful approach for cancer treatment [3,4]. Various ROS-modulating agents are currently developed for more precise and specific effectiveness to kill cancer cells [2]. On the other hand, cancer cells also develop redox adaption through the upregulation of anti-apoptotic and antioxidant molecules, allowing for cancer cells to survive and increase resistance against anticancer drugs. However, cancer cells remain to maintain higher ROS production than normal cells [2], and thus a dramatic increase of intracellular ROS still functions to kill cancer cells by shutting down antioxidant systems. Normal cells with a lower basal ROS level are less dependent on antioxidants, making normal cells less sensitive to oxidative insults. It is well-accepted that a moderate availability of ROS serves an oncogenic function to stimulate mutagenesis facilitating cancer cells to respond to microenvironment changes. In the meantime, the use of antioxidants should prevent tumorigenesis. By contrast, a high ROS level is toxic to cancer cells, which may induce oxidative stress to suppress cancer cell survival [2,4]. However, how to define a toxicity threshold of ROS availability to kill cancer cells or a level to help cancer cells to acquire mutations and resistance is difficult, particularly, among the various types of cancers.

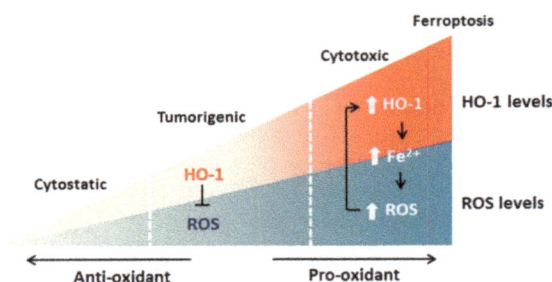

Figure 3. Model of HO-1-mediated ferroptosis. HO-1 exerts a cytoprotective effect by scavenging ROS during moderate activation. By contrast, excessive activation of HO-1 increases labile Fe^{2+}, leading to ROS overload and death of cancer cells.

Ferroptosis employs iron-dependent ROS production to kill cells [6] and HO-1 functions in ferroptosis by operating at cellular iron level and ROS generation [15–17]. Accordingly, when HO-1 is activated moderately, Nrf2-derived HO-1 exerts a cytoprotective effect by neutralizing ROS. Since cancer cells express a higher level of HO-1 [84–90], a high degree of HO-1 activation can increase labile Fe^{2+}, leading to ROS overload, and thereby oxidative-cell death (Figure 3) [2,63]. The induction of ferroptosis via HO-1 activation could create a new chemotherapeutic strategy for cancer treatment.

Undoubtedly, how the precise activation of HO-1 to selectively kill cancer cells and overcome drugs' resistances needs to be defined.

8. HO-1 Modulators in Cancer Treatment

Despite the dual role in ferroptosis modulation, HO-1 is commonly regarded as a survival molecule, exerting an important role in cancer progression that is involved in the cell proliferation, metastasis, angiogenesis, as well as resistance against chemotherapy, radiation, and photodynamic therapy [9,11]. High levels of HO-1 expression have been shown in different cancer types [84–90], and the resistance to anticancer therapy was associated with a responsive upregulation of HO-1 [9,11]. Despite the facts, genetic and pharmacological approaches demonstrated the critical role of HO-1 activation in ferroptotic process in cancer cell death [16,17]. In clinical cases, patients with higher HO-1 expression showed the lower survival rate and poor outcomes [91]. Thereby, the modulation of HO-1 and HO-1 related pathway may serve as a potential therapeutic strategy. Several HO-1 inhibitors are developed and validated further. HO-1 inhibitors are categorized into two types, metalloporphyrins and imidazole-based compounds [92–94]. Metalloporphyrins are heme analogues composing of a protoporphyrin IX and a metal, such as zinc protoporphyrin (ZnPPIX) and tin protoporphyrin (SnPPIX). With the similar structure, the chemicals occupy the heme binding pocket of HO and competitively inhibit the enzyme activity [93]. Metalloporphyrin are widely used in HO-1-related studies and they have been confirmed with the efficiency against cancers. ZnPPIX was showed to inhibit the proliferation of leukemic cells [95] and suppress the resistance against panobinostat [96]. Treatment with pegylated-ZnPPIX significantly attenuated tumor growth and increased the sensitivity to chemotherapy in human colon cancer SW480 xenograft mouse model [97]. Besides, ZnPPIX significantly increased the chemotherapeutic response of cisplatin in hepatoma cells [98] and laryngeal squamous cancer cells [99], as well as the efficiency of photodynamic therapy in cultured melanoma tumor cells [100]. In vivo studies, ZnPPIX reduced tumor growth of LL/2 lung cancer in C57BL mice, by suppressing vascular endothelial growth factor concentration in tumors [101]. In addition, ZnPPIX inhibited peritoneal metastasis of gastric cancer via anti-angiogenesis and improved the survival of tumor-bearing mice, which were related to the suppression of ROS production and ERK activation [102]. However, the non-selective activity of metalloporphyrin limits its application, due to the similarity of structure to heme. Metalloporphyrins can affect other heme-containing enzymes, such as nitric oxide synthases (NOS), soluble guanylyl cyclase (sGCs), and cytochromes P450 (CYP450) [103,104]. Azole-based compounds are structurally distinctive from metalloporphyrins and highly selective for the HO-1 inhibition [91,94], exhibiting low or no inhibitory activity on NOS, sGCs, or CYP450 [104,105]. They bind to the distal side of heme through an azole anchor which coordinates with the heme iron [106]. Imidazole-derived HO-1 inhibitors showed a better profile in HO-1 inhibition in prostate and breast cancer cell lines [107]. A series of hybrid compounds of imatinib and imidazole-based HO-1 inhibitors were able to inhibit both tyrosine kinase and HO-1/HO-2 activities and reduced the viability in imatinib-resistant chronic myeloid leukemia cells [108].

Specific inhibition of HO-1 can be achieved through genetic tools. Small interfering RNA (siRNA) and short hairpin RNA (shRNA) are two common approaches on gene silencing via the cleavage of target gene or inhibition of protein synthesis. An intervention of HO-1 expression by shRNA was shown to induce apoptosis in lung cancer cells [109], colon carcinoma cells [45], leukemic cells [95,110], esophageal squamous carcinoma cells [111], and breast cancer cells [112]. In addition, siRNA/shRNA of HO-1 also enhanced the sensitivity to chemotherapy in pancreatic cancers [113,114], lung cancer [109], and breast cancer [112]. In an orthotopic model of hepatocellular carcinoma, intraperitoneal injection of HO-1 siRNA attenuated the growth of tumor [115]. Moreover, in the 5-fluorouracil resistant human colon cancer cells xenografted subcutaneously, HO-1 knockdown by shRNA significantly reduced tumor size and increased the sensitivity to 5-fluorouracil treatment [116]. Furthermore, the CRISPR/Cas9 knockout system also gave positive results, showing HO-1 depletion to mediate a decrease of 293T cell viability, growth, and an increase of sensitivity to H_2O_2 treatment [117].

9. Conclusion and Prospective

With the genetically heterogeneous and labyrinthine gene expression, anti-drug regulation of survival and metastatic metabolism of tumors against cancer treatment, most aggressive cancer treatments have less favorable outcomes, reflecting the lack of sufficient promising therapies that are capable of curing the most aggressive cancers. Chemotherapy is being developed against cancer with new chemotherapeutic drugs and strategies being tested in preclinical and clinical trials. In this review, we discussed how HO-1 regulates ferroptosis, the therapeutic strategy by manipulating HO-1 to mediate ferroptosis, and prospective chemotherapeutic drugs against cancer via HO-1-mediated ferroptosis. Several prospective chemodrugs (e.g., BAY117085, withaferin A, erastin, RSL3, and sorafenib) involved in the HO-1-mediated ferroptosis for chemotherapeutic strategy are discussed, providing some representative examples of the application in killing different types of cancer cells. Up to date, few studies have focused on the role of HO-1 in ferroptosis. Results from these studies implicate that the dual role of HO-1 in ferroptosis regulation might depend on different pathological conditions. The precise mechanism of this phenomenon needs to be further investigated. Although iron-dependent and ROS-promoted ferroptosis was redefined at 2012 [6], a large part of the mechanisms underlying the regulation by HO-1 remains elusive, particularly in respect to applications of chemotherapy in cancers.

Author Contributions: S.K.C. and L.C.C. wrote the manuscript. S.E.C. corrected the manuscript.

Funding: This work was supported by the Ministry of Science and Technology, Taiwan (MOST 104-2314-B-039-034 and MOST 105-2628-B-039-004-MY3 to L.-C. Chang) and China Medical University Hospital, Taiwan (DMR104-097 and DMR 105-108 to L.-C. Chang), and by the iEGG and Animal Biotechnology Center and the Innovation and Development Center of Sustainable Agriculture from The Featured Areas Research Center Program within the framework of the Higher Education Sprout Project by the Ministry of Education (MOE) in Taiwan (S.E. Chen).

Conflicts of Interest: The authors declare no conflict of interest.

Abbreviations

CO	Carbon monoxide
GPx4	Glutathione peroxidase 4
GSH	Glutathione
HO-1	Heme oxygenase-1
Keap1	Kelch-like ECH-associated protein 1
MgIG	*Magnesium isoglycyrrhizinate*
NCOA4	Nuclear Co-Activator 4
NPBI,	Non-protein-bound iron
NQO-1	Quinone oxidoreductase 1
Nrf2	Nuclear factor erythroid 2-related factor 2
ROS	Reactive oxygen species
Redox	Reduction-oxidation

References

1. Valko, M.; Leibfritz, D.; Moncol, J.; Cronin, M.T.D.; Mazur, M.; Telser, J. Free radicals and antioxidants in normal physiological functions and human disease. *Int. J. Biochem. Cell Biol.* **2007**, *39*, 44–84. [CrossRef] [PubMed]
2. Trachootham, D.; Alexandre, J.; Huang, P. Targeting cancer cells by ROS-mediated mechanisms: A radical therapeutic approach? *Nat. Rev. Drug Discov.* **2009**, *8*, 579–591. [CrossRef]
3. Liou, G.Y.; Storz, P. Reactive oxygen species in cancer. *Free Radic. Res.* **2014**, *44*, 1–5. [CrossRef]
4. Gorrini, C.; Harris, I.S.; Mak, T.W. Modulation of oxidative stress as an anticancer strategy. *Nat. Rev. Drug Discov.* **2013**, *12*, 931–947. [CrossRef] [PubMed]
5. Wang, J.; Yi, J. Cancer cell killing via ROS: To increase or decrease, that is the question. *Cancer Biol. Ther.* **2008**, *12*, 1875–1884. [CrossRef]

6.   Dixon, S.J.; Lemberg, K.M.; Lamprecht, M.R.; Skouta, R.; Zaitsev, E.M.; Gleason, C.E.; Patel, D.N.; Bauer, A.J.; Cantley, A.M.; Yang, W.S.; et al. Ferroptosis: An iron-dependent form of nonapoptotic cell death. *Cell* **2012**, *149*, 1060–1072. [CrossRef]

7.   Yang, W.S.; SriRamaratnam, R.; Welsch, M.E.; Shimada, K.; Skouta, R.; Viswanathan, V.S.; Cheah, J.H.; Clemons, P.A.; Shamji, A.F.; Clish, C.B.; et al. Regulation of ferroptotic cancer cell death by GPX4. *Cell* **2014**, *156*, 317–331. [CrossRef]

8.   Ryter, S.W.; Alam, J.; Choi, A.M. Heme oxygenase-1/carbon monoxide: From basic science to therapeutic applications. *Physiol. Rev.* **2006**, *86*, 583–650. [CrossRef]

9.   Loboda, A.; Jozkowicz, A.; Dulak, J. HO-1/CO system in tumor growth, angiogenesis and metabolism—Targeting HO-1 as an anti-tumor therapy. *Vascul. Pharmacol.* **2015**, *74*, 11–22. [CrossRef]

10. Loboda, A.; Damulewicz, M.; Pyza, E.; Jozkowicz, A.; Dulak, J. Role of Nrf2/HO-1 system in development, oxidative stress response and diseases: An evolutionarily conserved mechanism. *Cell. Mol. Life Sci.* **2016**, *73*, 3221–3247. [CrossRef]

11. Nitti, M.; Piras, S.; Marinari, U.M.; Moretta, L.; Pronzato, M.A.; Furfaro, A.L. HO-1 induction in cancer progression: A matter of cell adaptation. *Antioxidants* **2017**, *6*, 29. [CrossRef] [PubMed]

12. Gozzelino, R.; Jeney, V.; Soares, M.P. Mechanisms of cell protection by heme oxygenase-1. *Annu. Rev. Pharmacol. Toxicol.* **2010**, *50*, 323–354. [CrossRef]

13. Bolisetty, S.; Zarjou, A.; Agarwal, A. Heme oxygenase 1 as a therapeutic target in acute kidney injury. *Am. J. Kidney Dis.* **2017**, *69*, 531–545. [CrossRef] [PubMed]

14. Nitti, M.; Piras, S.; Brondolo, L.; Marinari, U.M.; Pronzato, M.A.; Furfaro, A.L. Heme oxygenase 1 in the nervous system: Does it favor neuronal cell survival or induce neurodegeneration? *Int. J. Mol. Sci.* **2018**, *19*, 2260. [CrossRef] [PubMed]

15. Kwon, M.Y.; Park, E.; Lee, S.J.; Chung, S.W. Heme oxygenase-1 accelerates Erastin induced Ferroptotic cell death. *Oncotarget* **2015**, *6*, 24393–24403. [CrossRef]

16. Chang, L.C.; Chiang, S.K.; Chen, S.E.; Yu, Y.L.; Chou, R.H.; Chang, W.C. Heme oxygenase-1 mediates BAY 11-7085 induced ferroptosis. *Cancer Lett.* **2018**, *416*, 124–137. [CrossRef]

17. Hassannia, B.; Wiernicki, B.; Ingold, I.; Qu, F.; Van Herck, S.; Tyurina, Y.Y.; Bayır, H.; Abhari, B.A.; Angeli, J.P.F.; Choi, S.M.; et al. Nano-targeted induction of dual ferroptotic mechanisms eradicates high-risk neuroblastoma. *J. Clin. Investig.* **2018**, *128*, 3341–3355. [CrossRef]

18. Cao, J.Y.; Dixon, S.J. Mechanisms of ferroptosis. *Cell. Mol. Life Sci.* **2016**, *73*, 2195–2209. [CrossRef]

19. Xie, Y.; Hou, W.; Song, X.; Yu, Y.; Huang, J.; Sun, X.; Kang, R.; Tang, D. Ferroptosis: Process and function. *Cell Death Differ.* **2016**, *23*, 369–379. [CrossRef]

20. Stockwell, B.R.; Friedmann Angeli, J.P.; Bayir, H.; Bush, A.I.; Conrad, M.; Dixon, S.J.; Fulda, S.; Gascon, S.; Hatzios, S.K.; Kagan, V.E.; et al. Ferroptosis: A regulated cell death nexus linking metabolism, redox biology, and disease. *Cell* **2017**, *171*, 273–285. [CrossRef]

21. Piskounova, E.; Agathocleous, M.; Murphy, M.M.; Hu, Z.; Huddlestun, S.E.; Zhao, Z.; Leitch, A.M.; Johnson, T.M.; DeBerardinis, R.J.; Morrison, S.J. Oxidative stress inhibits distant metastasis by human melanoma cells. *Nature* **2015**, *52*, 186–191. [CrossRef]

22. Roh, J.L.; Kim, E.H.; Jang, H.; Shin, D. Nrf2 inhibition reverses the resistance of cisplatin-resistant head and neck cancer cells to artesunate-induced ferroptosis. *Redox Biol.* **2017**, *11*, 254–262. [CrossRef] [PubMed]

23. Lu, B.; Chen, X.B.; Ying, M.D.; He, Q.J.; Cao, J.; Yang, B. The role of ferroptosis in cancer development and treatment. *Front. Pharmacol.* **2018**, *8*, 992. [CrossRef]

24. Lo, M.; Ling, V.; Wang, Y.Z.; Gout, P.W. The Xc⁻ cystine/glutamate antiporter: A mediator of pancreatic cancer growth with a role in drug resistance. *Br. J. Cancer* **2008**, *99*, 464–472. [CrossRef] [PubMed]

25. Timmerman, L.A.; Holton, T.; Yuneva, M.; Louie, R.J.; Padro, M.; Daemen, A.; Hu, M.; Chang, D.A.; Ethier, S.P.; van't Veer, L.J.; et al. Glutamine sensitivity analysis identifies the xCT antiporter as a common triple-negative breast tumor therapeutic target. *Cancer Cell* **2013**, *24*, 450–465. [CrossRef] [PubMed]

26. Vitek, L.; Schwertner, H.A. The heme catabolic pathway and its protective effects on oxidative stress-mediated diseases. *Adv. Clin. Chem.* **2007**, *43*, 1–57. [PubMed]

27. Abraham, N.G.; Kappas, A. Pharmacological and clinical aspects of heme oxygenase. *Pharmacol. Rev.* **2008**, *60*, 79–127. [CrossRef]

28. Adedoyin, O.; Boddu, R.; Traylor, A.; Lever, J.M.; Bolisetty, S.; George, J.F.; Agarwal, A. Heme oxygenase-1 mitigates ferroptosis in renal proximal tubule cells. *Am. J. Physiol. Renal Physiol.* **2018**, *314*, F702–F714. [CrossRef]

29. Sui, M.; Jiang, X.; Chen, J.; Yang, H.; Zhu, Y. Magnesium isoglycyrrhizinate ameliorates liver fibrosis and hepatic stellate cell activation by regulating ferroptosis signaling pathway. *Biomed. Pharmacother.* **2018**, *106*, 125–133. [CrossRef]
30. Fan, Z.; Wirth, A.K.; Chen, D.; Wruch, C.J.; Rauh, M.; Buchfelder, M.; Savaskan, N. Nrf2-Keap1 pathway promotes cell proliferation and diminishes ferroptosis. *Oncogenesis* **2017**, *6*, e371. [CrossRef]
31. Sun, X.; Ou, Z.; Chen, R.; Niu, X.; Chen, D.; Kang, R.; Tang, D. Activation of the p62-Keap1-NRF2 pathway protects against ferroptosis in hepatocellular carcinoma cells. *Hepatology* **2016**, *63*, 173–184. [CrossRef] [PubMed]
32. Jomova, K.; Vondrakova, D.; Lawson, M.; Valko, M. Metals, oxidative stress and neurodegenerative disorders. *Mol. Cell. Biochem.* **2010**, *345*, 91–104. [CrossRef] [PubMed]
33. Gonzales, S.; Erario, M.A.; Tomaro, M.L. Heme oxygenase-1 induction and dependent increase in ferritin. A protective antioxidant stratagem in hemin-treated rat brain. *Dev. Neurosci.* **2002**, *24*, 161–168. [CrossRef] [PubMed]
34. Gottlieb, Y.; Truman, M.; Cohen, L.A.; Leichtmann-Bardoogo, Y.; Meyron-Holtz, E.G. Endoplasmic reticulum anchored heme-oxygenase 1 faces the cytosol. *Haematologica* **2012**, *97*, 1489–1493. [CrossRef] [PubMed]
35. Lin, Q.; Weis, S.; Yang, G.; Weng, Y.H.; Helston, R.; Rish, K.; Smith, A.; Bordner, J.; Polte, T.; Gaunitz, F.; et al. Heme oxygenase-1 protein localizes to the nucleus and activates transcription factors important in oxidative stress. *J. Biol. Chem.* **2007**, *282*, 20621–20633. [CrossRef] [PubMed]
36. Converso, D.P.; Taille, C.; Carreras, M.C.; Jaitovich, A.; Poderoso, J.J.; Boczkowski, J. HO-1 is located in liver mitochondria and modulates mitochondrial heme content and metabolism. *FASEB J.* **2006**, *20*, 1236–1238. [CrossRef] [PubMed]
37. Zhang, D.D.; Hannink, M. Distinct cysteine residues in Keap1 are required for Keap1-dependent ubiquitination of Nrf2 and for stabilization of Nrf2 by chemopreventive agents and oxidative stress. *Mol. Cell. Biol.* **2003**, *23*, 8137–8151. [CrossRef]
38. Itoh, K.; Chiba, T.; Takahashi, S.; Ishii, T.; Igarashi, K.; Katoh, Y.; Oyake, T.; Hayashi, N.; Satoh, K.; Hatayama, I.; et al. AnNrf2/ small Maf heterodimer mediates the induction of phase II detoxifying enzyme genes through antioxidant response elements. *Biochem. Biophys. Res. Commun.* **1997**, *236*, 313–322. [CrossRef]
39. Reichard, J.F.; Motz, G.T.; Puga, A. Heme oxygenase-1 induction by NRF2 requires inactivation of the transcriptional repressor BACH1. *Nuclei Acids Res.* **2007**, *35*, 7074–7086. [CrossRef]
40. Chi, L.; Ke, Y.; Luo, C.; Gozal, D.; Liu, R. Depletion of reduced glutathione enhances motor neuron degeneration in vitro and in vivo. *Neuroscience* **2007**, *144*, 991–1003. [CrossRef]
41. Lin, P.H.; Chiang, M.T.; Chau, L.Y. Ubiquitin–proteasome system mediates heme oxygenase-1 degradation through endoplasmic reticulum-associated degradation pathway. *Biochim. Biophys. Acta* **2008**, *1783*, 1826–1834. [CrossRef] [PubMed]
42. Kovacsics, C.E.; Gill, A.J.; Ambegaokar, S.S.; Gelman, B.B.; Kolson, D.L. Degradation of heme oxygenase-1 by the immunoproteasome in astrocytes: A potential interferon-γ-dependent mechanism contributing to HIV neuropathogenesis. *Glia* **2017**, *65*, 1264–1277. [CrossRef] [PubMed]
43. Sugimoto, R.; Tanaka, Y.; Noda, K.; Kawamura, T.; Toyoda, Y.; Billiar, T.R.; McCurry, K.R.; Nakao, A. Preservation solution supplemented with biliverdin prevents lung cold ischaemia/reperfusion injury. *Eur. J. Cardiothorac. Surg.* **2012**, *42*, 1035–1041. [CrossRef] [PubMed]
44. Pae, H.O.; Son, Y.; Kim, N.H.; Jeong, H.J.; Chang, K.C.; Chung, H.T. Role of heme oxygenase in preserving vascular bioactive NO. *Nitric Oxide* **2010**, *23*, 251–257. [CrossRef]
45. Busserolles, J.; Megias, J.; Terencio, M.C.; Alcaraz, M.J. Heme oxygenase-1 inhibits apoptosis in Caco-2 cells via activation of Akt pathway. *Int. J. Biochem. Cell Biol.* **2006**, *38*, 1510–1517. [CrossRef] [PubMed]
46. Zheng, J.; Nagda, D.A.; Lajud, S.A.; Kumar, S.; Mouchli, A.; Bezpalko, O.; O'Malley, B.W., Jr.; Li, D. Biliverdin's regulation of reactive oxygen species signalling leads to potent inhibition of proliferative and angiogenic pathways in head and neck cancer. *Br. J. Cancer* **2014**, *110*, 2116–2122. [CrossRef]
47. Ollinger, R.; Kogler, P.; Troppmair, J.; Hermann, M.; Wurm, M.; Drasche, A.; Konigsrainer, I.; Amberger, A.; Weiss, H.; Ofner, D.; et al. Bilirubin inhibits tumor cell growth via activation of ERK. *Cell Cycle* **2007**, *6*, 3078–3085. [CrossRef] [PubMed]
48. Kaplan, M.; Bromiker, R.; Hammerman, C. Hyperbilirubinemia, hemolysis, and increased bilirubin neurotoxicity. *Semin. Perinatol.* **2014**, *38*, 429–437. [CrossRef]

49. Papanikolaou, G.; Pantopoulos, K. Iron metabolism and toxicity. *Toxicol. Appl. Pharmacol.* **2005**, *202*, 199–211. [CrossRef]

50. Hori, R.; Kashiba, M.; Toma, T.; Yachie, A.; Goda, N.; Makino, N.; Soejima, A.; Nagasawa, T.; Nakabayashi, K.; Suematsu, M. Gene transfection of H25A mutant heme oxygenase-1 protects cells against hydroperoxide-induced cytotoxicity. *J. Biol. Chem.* **2002**, *277*, 10712–10718. [CrossRef]

51. Stephenson, E.; Nathoo, N.; Mahjoub, Y.; Dunn, J.F.; Yong, V.W. Iron in multiple sclerosis: Roles in neurodegeneration and repair. *Nat. Rev. Neurosci.* **2014**, *10*, 459–468. [CrossRef]

52. Paffetti, P.; Perrone, S.; Longini, M.; Ferrari, A.; Tanganelli, D.; Marzocchi, B.; Buonocore, G. Non-protein-bound iron detection in small samples of biological fluids and tissues. *Biol. Trace Elem. Res.* **2006**, *112*, 221–232. [CrossRef]

53. MacKenzie, E.L.; Iwasaki, K.; Tsuji, Y. Intracellular iron transport and storage: From molecular mechanisms to health implications. *Antioxid. Redox Signal.* **2008**, *10*, 997–1030. [CrossRef] [PubMed]

54. Zhang, Y.; Mikhael, M.; Xu, D.; Li, Y.; Soe-Lin, S.; Ning, B.; Li, W.; Nie, G.; Zhao, Y.; Ponka, P. Lysosomal proteolysis is the primary degradation pathway for cytosolic ferritin and cytosolic ferritin degradation is necessary for iron exit. *Antioxid. Redox Signal.* **2010**, *13*, 999–1009. [CrossRef] [PubMed]

55. Dowdle, W.E.; Nyfeler, B.; Nagel, J.; Elling, R.A.; Liu, S.; Triantafellow, E.; Menon, S.; Wang, Z.; Honda, A.; Pardee, G. Selective VPS34 inhibitor blocks autophagy and uncovers a role for NCOA4 in ferritin degradation and iron homeostasis in vivo. *Nat. Cell Biol.* **2014**, *16*, 1069–1079. [CrossRef]

56. Cairo, G.; Racalcati, S. Iron-regulatory proteins: Molecular biology and pathophysiological implications. *Expert Rev. Mol. Med.* **2007**, *9*, 1–13. [CrossRef]

57. Lanceta, L.; Li, C.; Choi, A.M.; Eaton, J.W. Haem oxygenase-1 overexpression alters intracellular iron distribution. *Biochem. J.* **2013**, *449*, 189–194. [CrossRef]

58. Yang, W.S.; Stockwell, B.R. Synthetic lethal screening identifies compounds activating iron-dependent, nonapoptotic cell death in oncogenic-RAS-harboring cancer cells. *Chem. Biol.* **2008**, *15*, 234–245. [CrossRef]

59. Mai, T.T.; Hamai, A.; Hienzsch, A.; Caneque, T.; Muller, S.; Wicinski, J.; Cabaud, O.; Leroy, C.; David, A.; Acevedo, V.; et al. Salinomycin kills cancer stem cells by sequestering iron in lysosomes. *Nat. Chem.* **2017**, *9*, 1025–1033. [CrossRef]

60. Gao, M.; Monian, P.; Pan, Q.; Zhang, W.; Xiang, J.; Jiang, X. Ferroptosis is an autophagic cell death process. *Cell Res.* **2016**, *26*, 1021–1032. [CrossRef]

61. Lamb, N.J.; Quinlan, G.J.; Mumby, S.; Evans, T.W.; Gutteridge, J.M.C. Haem oxygenase shows pro-oxidant activity in microsomal and cellular systems: Implications for the release of low-molecular-mass iron. *Biochem. J.* **1999**, *344*, 153–158. [CrossRef] [PubMed]

62. Clark, J.E.; Foresti, R.; Green, C.J.; Motterlini, R. Dynamics of haem oxygenase-1 expression and bilirubin production in cellular protection against oxidative stress. *Biochem. J.* **2000**, *348*, 615–619. [CrossRef] [PubMed]

63. Suttner, D.M.; Dennery, P.A. Reversal of HO-1 related cytoprotection with increased expression is due to reactive iron. *FASEB J.* **1999**, *13*, 1800–1809. [CrossRef] [PubMed]

64. Piloni, N.E.; Fermandez, V.; Videla, L.A.; Puntarulo, S. Acute iron overload and oxidative stress in brain. *Toxicology* **2013**, *314*, 174–182. [CrossRef] [PubMed]

65. Basu-Modak, S.; Luscher, P.; Tyrrell, R.M. Lipid metabolite involvement in the activation of the human heme oxygenase-1 gene. *Free Radic. Biol. Med.* **1996**, *20*, 887–897. [CrossRef]

66. NaveenKumar, S.; SharathBabu, B.N.; Hemskekhar, M.; Kemparaju, K.; Girish, K.S.; Mugesh, G. The role of reactive oxygen species and ferroptosis in heme-mediated activation of human platelets. *ACS Chem. Biol.* **2018**, *13*, 1996–2002. [CrossRef] [PubMed]

67. Poulos, T.L. Heme enzyme structure and function. *Chem. Rev.* **2014**, *114*, 3919–3962. [CrossRef]

68. Schaer, D.J.; Buehler, P.W.; Alayash, A.I.; Belcher, J.D.; Vercellotti, G.M. Hemolysis and free hemoglobin revisited: Exploring hemoglobin and hemin scavengers as a novel class of therapeutic proteins. *Blood* **2013**, *121*, 1276–1284. [CrossRef]

69. Gladwin, M.T. Cardiovascular complications and risk of death in sickle-cell disease. *Lancet* **2016**, *387*, 2565–2574. [CrossRef]

70. Fang, R.; Aust, A.E. Induction of ferritin synthesis in human lung epithelial cells treated with crocidolite asbestos. *Arch. Biochem. Biophys.* **1997**, *340*, 369–375. [CrossRef]

71. Bjorkoy, G.; Lamark, T.; Johansen, T. p62/SQSTM1: A missing link between protein aggregates and the autophagy machinery. *Autophagy* **2006**, *2*, 138–139. [CrossRef] [PubMed]

72. Komatsu, M.; Kurokawa, H.; Waguri, S.; Taguchi, K.; Kobayashi, A.; Ichimura, Y.; Sou, Y.S.; Ueno, I.; Sakamoto, A.; Tong, K.I. The selective autophagy substrate p62 activates the stress responsive transcription factor Nrf2 through inactivation of Keap1. *Nat. Cell Biol.* **2010**, *12*, 213–223. [CrossRef] [PubMed]

73. Vincenzi, B.; Armento, G.; Spalato Ceruso, M.; Catania, G.; Leakos, M.; Santini, D.; Minotti, G.; Tonini, G. Drug-induced hepatotoxicity in cancer patients—Implication for treatment. *Expert Opin. Drug Saf.* **2016**, *15*, 1219–1238. [CrossRef] [PubMed]

74. Weiskirchen, R. Cellular and molecular functions of hepatic stellate cells in inflammatory responses and liver immunology. *Hepatobiliary Surg. Nutr.* **2014**, *3*, 344–363.

75. Wang, Y.; Zhang, Z.; Wang, X.; Qi, D.; Qu, A.; Wang, G. Amelioration of ethanol-induced hepatitis by magnesium isoglycyrrhizinate through inhibition of neutrophil cell infiltration and oxidative damage. *Mediators Inflamm.* **2017**, *2017*, 3526903. [CrossRef]

76. Hamacher-Brady, A.; Stein, H.A.; Turschner, S.; Toegel, I.; Mora, R.; Jennewein, N.; Efferth, T.; Eils, R.; Brady, N.R. Artesunate activates mitochondrial apoptosis in breast cancer cells via iron-catalyzed lysosomal reactive oxygen species production. *J. Biol. Chem.* **2011**, *286*, 6587–6601. [CrossRef] [PubMed]

77. Gopalakrishnan, A.M.; Kumar, N. Antimalarial action of artesunate involves DNA damage mediated by reactive oxygen species. *Antimicrob. Agents Chemother.* **2015**, *59*, 317–325. [CrossRef] [PubMed]

78. Pierce, J.W.; Schoenleber, R.; Jesmok, G.; Best, J.; Moore, S.A.; Collins, T.; Gerritsen, M.E. Novel inhibitors of cytokine-induced I kappa B alpha phosphorylation and endothelial cell adhesion molecule expression show anti-inflammatory effects in vivo. *J. Biol. Chem.* **1997**, *272*, 21096–21103. [CrossRef]

79. Bansal, S.; Biswas, G.; Avadhani, N.G. Mitochondria-targeted heme oxygenase-1 induces oxidative stress and mitochondrial dysfunction in macrophages, kidney fibroblasts and in chronic alcohol hepatotoxicity. *Redox Biol.* **2013**, *2*, 273–283. [CrossRef]

80. Hull, T.D.; Boddu, R.; Gou, L.; Tisher, C.C.; Traylor, A.M.; Patel, B.; Joseph, R.; Prabhu, S.D.; Suliman, H.B.; Piantadosi, C.A.; et al. Heme oxygenase-1 regulates mitochondrial quality control in the heart. *JCI Insight* **2016**, *1*, e85817. [CrossRef]

81. Vyas, A.R.; Singh, S.V. Molecular targets and mechanisms of cancer prevention and treatment by withaferin-A, a naturally occurring steroidal lactone. *AAPS J.* **2014**, *16*, 1–10. [CrossRef]

82. Lee, I.C.; Choi, B.Y. Withaferin-A—A Natural anticancer agent with Pleitropic mechanisms of action. *Int. J. Mol. Sci.* **2016**, *17*, 290. [CrossRef]

83. Heyninck, K.; Sabbe, L.; Chirumamilla, C.S.; Szarc Vel Szic, K.; Vander Veken, P.; Lemmens, K.J.A.; Lahtela-Kakkonen, M.; Naulaerts, S.; Op de Beeck, K.; Laukens, K.; et al. Withaferin A induces heme oxygenase (HO-1) expression in endothelial cells via activation of the Keap1/Nrf2 pathway. *Biochem. Pharmacol.* **2016**, *109*, 48–61. [CrossRef] [PubMed]

84. Shi, L.; Fang, J. Implication of heme oxygenase-1 in the sensitivity of nasopharyngeal carcinomas to radiotherapy. *J. Exp. Clin. Cancer Res.* **2008**, *27*, 13. [CrossRef] [PubMed]

85. Nuhn, P.; Kunzli, B.M.; Hennig, R.; Mitkus, T.; Ramanauskas, T.; Nobiling, R.; Meuer, S.C.; Friess, H.; Berberat, P. Heme oxygenase-1 and its metabolites affect pancreatic tumor growth in vivo. *Mol. Cancer* **2009**, *8*, 37. [CrossRef]

86. Banerjee, P.; Basu, A.; Wegiel, B.; Otterbein, L.E.; Mizumura, K.; Gasser, M.; Waaga-Gasser, A.M.; Choi, A.M.; Pal, S. Heme oxygenase-1 promotes survival of renal cancer cells through modulation of apoptosis and autophagy-regulating molecules. *J. Biol. Chem.* **2012**, *287*, 32113–32123. [CrossRef] [PubMed]

87. Hjortso, M.D.; Andersen, M.H. The expression, function and targeting of HAEM oxygenase-1 in cancer. *Curr. Cancer Drug Targets* **2014**, *14*, 337–347. [CrossRef]

88. Tibullo, D.; Barbagallo, I.; Giallongo, C.; Vanella, L.; Conticello, C.; Romano, A.; Saccone, S.; Godos, J.; Di Raimondo, F.; Li Volti, G. Heme oxygenase-1 nuclear translocation regulates bortezomib-induced cytotoxicity and mediates genomic instability in myeloma cells. *Oncotarget* **2016**, *7*, 28868–28880. [CrossRef]

89. Salerno, L.; Romeo, G.; Modica, M.N.; Amata, E.; Sorrenti, V.; Barbagallo, I.; Pittala, V. Heme oxygenase-1: A new druggable target in the management of chronic and acute myeloid leukemia. *Eur. J. Med. Chem.* **2017**, *142*, 163–178. [CrossRef]

90. Bekeschus, S.; Freund, E.; Wende, K.; Gandhirajan, R.K.; Schmidt, A. Hmox1 upregulation is a mutual marker in human tumor cells exposed to physical plasma-derived oxidants. *Antioxidants* **2018**, *7*, 151. [CrossRef]

91. Na, H.K.; Surh, Y.J. Oncogenic potential of Nrf2 and its principal target protein heme oxygenase-1. *Free Radic. Biol. Med.* **2014**, *67*, 353–365. [CrossRef] [PubMed]

92. Vlahakis, J.Z.; Kinobe, R.T.; Bowers, R.J.; Brien, J.F.; Nakatsu, K.; Szarek, W.A. Imidazole-dioxolane compounds as isozyme-selective heme oxygenase inhibitors. *J. Med. Chem.* **2006**, *49*, 4437–4441. [CrossRef] [PubMed]

93. Schulz, S.; Wong, R.J.; Vreman, H.J.; Stevenson, D.K. Metalloporphyrins—An update. *Front. Pharmacol.* **2012**, *3*, 68. [CrossRef] [PubMed]

94. Greish, K.F.; Salerno, L.; Al Zahrani, R.; Amata, E.; Modica, M.N.; Romeo, G.; Marrazzo, A.; Prezzavento, O.; Sorrenti, V.; Rescifina, A.; et al. Novel structural insight into inhibitors of heme oxygenase-1 (HO-1) by new imidazole-based compounds: Biochemical and in vitro anticancer activity evaluation. *Molecules* **2018**, *23*, 1209. [CrossRef] [PubMed]

95. Mayerhofer, M.; Gleixner, K.V.; Mayerhofer, J.; Hoermann, G.; Jaeger, E.; Aichberger, K.J.; Ott, R.G.; Greish, K.; Nakamura, H.; Derdak, S.; et al. Targeting of heat shock protein 32 (Hsp32)/heme oxygenase-1 (HO-1) in leukemic cells in chronic myeloid leukemia: A novel approach to overcome resistance against imatinib. *Blood* **2008**, *111*, 2200–2210. [CrossRef] [PubMed]

96. Cheng, B.; Tang, S.; Zhe, N.; Ma, D.; Yu, K.; Wei, D.; Zhou, Z.; Lu, T.; Wang, J.; Fang, Q. Low expression of GFI-1 Gene is associated with Panobinostat-resistance in acute myeloid leukemia through influencing the level of HO-1. *Biomed. Pharmacother.* **2018**, *100*, 509–520. [CrossRef] [PubMed]

97. Fang, J.; Sawa, T.; Akaike, T.; Greish, K.; Maeda, H. Enhancement of chemotherapeutic response of tumor cells by a heme oxygenase inhibitor, pegylated zinc protoporphyrin. *Int. J. Cancer.* **2004**, *109*, 1–8. [CrossRef]

98. Liu, Y.S.; Li, H.S.; Qi, D.F.; Zhang, J.; Jiang, X.C.; Shi, K.; Zhang, X.J.; Zhang, X.H. Zinc protoporphyrin IX enhances chemotherapeutic response of hepatoma cells to cisplatin. *World J. Gastroenterol.* **2014**, *20*, 8572–8582. [CrossRef]

99. Lv, X.; Song, D.M.; Niu, Y.H.; Wang, B.S. Inhibition of heme oxygenase-1 enhances the chemosensitivity of laryngeal squamous cell cancer Hep-2 cells to cisplatin. *Apoptosis* **2016**, *21*, 489–501. [CrossRef]

100. Frank, J.; Lornejad-Schafer, M.R.; Schoffl, H.; Flaccus, A.; Lambert, C.; Biesalski, H.K. Inhibition of heme oxygenase-1 increases responsiveness of melanoma cells to ALA-based photodynamic therapy. *Int. J. Oncol.* **2007**, *31*, 1539–1545. [CrossRef]

101. Hirai, K.; Sasahira, T.; Ohmori, H.; Fujii, K.; Kuniyasu, H. Inhibition of heme oxygenase-1 by zinc protoporphyrin IX reduces tumor growth of LL/2 lung cancer in C57BL mice. *Int. J. Cancer* **2007**, *120*, 500–505. [CrossRef] [PubMed]

102. Shang, F.T.; Hui, L.L.; An, X.S.; Zhang, X.C.; Guo, S.G.; Kui, Z. ZnPPIX inhibits peritoneal metastasis of gastric cancer via its antiangiogenic activity. *Biomed. Pharmacother.* **2015**, *71*, 240–246. [CrossRef]

103. Appleton, S.D.; Chretien, M.L.; Mclaughlin, B.E.; Vreman, H.J.; Stevenson, D.K.; Brien, J.F.; Nakatsu, K.; Maurice, D.H.; Marks, G.S. Selective inhibition of heme oxygenase, without inhibition of nitric oxide synthase or soluble guanylyl cyclase, by metalloporphyrins at low concentrations. *Drug Metab. Dispos.* **1999**, *27*, 1214–1219. [PubMed]

104. Kinobe, R.T.; Vlahakis, J.Z.; Vreman, H.J.; Stevenson, D.K.; Brien, J.F.; Szarek, W.A.; Nakatsu, K. Selectivity of imidazole-dioxolane compounds for in vitro inhibition of microsomal haem oxygenase isoforms. *Br. J. Pharmacol.* **2006**, *147*, 307–315. [CrossRef] [PubMed]

105. Podkalicka, P.; Mucha, O.; Jozkowicz, A.; Dulak, J.; Loboda, A. Heme oxygenase inhibition in cancer, possible tools and targets. *Contemp. Oncol.* **2018**, *22*, 23–32. [CrossRef]

106. Rahman, M.N.; Vlahakis, J.Z.; Roman, G.; Vukomanovic, D.; Szarek, W.A.; Nakatsu, K.; Jia, Z. Structural characterization of human heme oxygenase-1 in complex with azole-based inhibitors. *J. Inorg. Biochem.* **2010**, *104*, 324–330. [CrossRef]

107. Salerno, L.; Pittala, V.; Romeo, G.; Modica, M.N.; Marrazzo, A.; Siracusa, M.A.; Sorrenti, V.; Di Giacomo, C.; Vanella, L.; Parayath, N.N.; et al. Novel imidazole derivatives as heme oxygenase-1 (HO-1) and heme oxygenase-2 (HO-2) inhibitors and their cytotoxic activity in uman-derived cancer cell lines. *Eur. J. Med. Chem.* **2015**, *96*, 162–172. [CrossRef]

108. Sorrenti, V.; Pittala, V.; Romeo, G.; Amata, E.; Dichiara, M.; Marrazzo, A.; Turnaturi, R.; Prezzavento, O.; Barbagallo, I.; Vanella, L.; et al. Targeting heme oxygenase-1 with hybrid compounds to overcome imatinib resistance in chronic myeloid leukemia cell lines. *Eur. J. Med. Chem.* **2018**, *158*, 937–950. [CrossRef]

109. Kim, H.R.; Kim, S.; Kim, E.J.; Park, J.H.; Yang, S.H.; Jeong, E.T.; Park, C.; Youn, M.J.; So, H.S.; Park, R. Suppression of Nrf2-driven heme oxygenase-1 enhances the chemosensitivity of lung cancer A549 cells toward cisplatin. *Lung Cancer* **2008**, *60*, 47–56. [CrossRef]

110. Wei, S.; Wang, Y.; Chai, Q.; Fang, Q.; Zhang, Y.; Wang, J. Potential crosstalk of Ca2+-ROS-dependent mechanism involved in apoptosis of Kasumi-1 cells mediated by heme oxygenase-1 small interfering RNA. *Int. J. Oncol.* **2014**, *45*, 2373–2384. [CrossRef]

111. Ren, Q.G.; Yang, S.L.; Hu, J.L.; Li, P.D.; Chen, Y.S.; Wang, Q.S. Evaluation of HO-1 expression, cellular ROS production, cellular proliferation and cellular apoptosis in human esophageal squamous cell carcinoma tumors and cell lines. *Oncol. Rep.* **2016**, *35*, 2270–2276. [CrossRef] [PubMed]

112. Zhu, X.F.; Li, W.; Ma, J.Y.; Shao, N.; Zhang, Y.J.; Liu, R.M.; Wu, W.B.; Lin, Y.; Wang, S.M. Knockdown of heme oxygenase-1 promotes apoptosis and autophagy and enhances the cytotoxicity of doxorubicin in breast cancer cells. *Oncol. Lett.* **2015**, *10*, 2974–2980. [CrossRef] [PubMed]

113. Berberat, P.O.; Dambrauskas, Z.; Gulbinas, A.; Giese, T.; Giese, N.; Kunzli, B.; Autschbach, F.; Meuer, S.; Buchler, M.W.; Friess, H. Inhibition of Heme Oxygenase-1 Increases Responsiveness of Pancreatic Cancer Cells to Anticancer Treatment. *Clin. Cancer Res.* **2005**, *11*, 3790–3798. [CrossRef] [PubMed]

114. Han, L.; Jiang, J.; Ma, Q.; Wu, Z.; Wang, Z. The inhibition of heme oxygenase-1 enhances the chemosensitivity and suppresses the proliferation of pancreatic cancer cells through the SHH signaling pathway. *Int. J. Oncol.* **2018**, *52*, 2101–2109. [CrossRef] [PubMed]

115. Sass, G.; Leukel, P.; Schmitz, V.; Raskopf, E.; Ocker, M.; Neureiter, D.; Meissnitzer, M.; Tasika, E.; Tannapfel, A.; Tiegs, G. Inhibition of heme oxygenase 1 expression by small interfering RNA decreases orthotopic tumor growth in livers of mice. *Int. J. Cancer* **2008**, *123*, 1269–1277. [CrossRef]

116. Kang, K.A.; Piao, M.J.; Kim, K.C.; Kang, H.H.; Chang, W.Y.; Park, I.C.; Keum, Y.S.; Surh, Y.J.; Hyun, J.W. Epigenetic modification of Nrf2 in 5-fluorouracil-resistant colon cancer cells: Involvement of TET-dependent DNA demethylation. *Cell Death Dis.* **2014**, *5*, e1183. [CrossRef]

117. Mucha, O.; Podkalicka, P.; Czarnek, M.; Biela, A.; Mieczkowski, M.; Kachamakova-Trojanowska, N.; Stepniewski, J.; Jozkowicz, A.; Dulak, J.; Loboda, A. Pharmacological versus genetic inhibition of heme oxygenase-1—The comparison of metalloporphyrins, shRNA and CRISPR/Cas9 system. *Acta Biochim. Pol.* **2018**, *65*, 277–286. [CrossRef]

International Journal of
Molecular Sciences

MDPI

Article

Myristica fragrans Kernels Prevent Paracetamol-Induced Hepatotoxicity by Inducing Anti-Apoptotic Genes and Nrf2/HO-1 Pathway

Mohamed A. Dkhil [1,2,*], Ahmed E. Abdel Moneim [2,*], Taghreed A. Hafez [3],
Murad A. Mubaraki [3], Walid F. Mohamed [4], Felwa A. Thagfan [5] and Saleh Al-Quraishy [1]

[1] Department of Zoology, College of Science, King Saud University, Riyadh 11451, Saudi Arabia; guraishi@yahoo.com
[2] Department of Zoology and Entomology, Faculty of Science, Helwan University, Cairo 11795, Egypt
[3] Clinical Laboratory Sciences Department, College of Applied Medical Sciences, King Saud University, Riyadh 11433, Saudi Arabia; thafiz@KSU.EDU.SA (T.A.H.); mmubaraki@KSU.EDU.SA (M.A.M.)
[4] Department of Biological and Geological Sciences, Faculty of Education, Ain Shams University, Cairo 11341, Egypt; walidfathy72@yahoo.com
[5] Department of Biology, College of Science, Princess Nourah bint Abdulrahman University, Riyadh 11671, Saudi Arabia; fafa-85@hotmail.com
[*] Correspondence: mohameddkhil@yahoo.com (M.A.D.); aest1977@hotmail.com (A.E.A.M.)

Received: 21 December 2018; Accepted: 20 February 2019; Published: 25 February 2019

Abstract: Paracetamol is responsible for acute liver failure in humans and experimental animals when taken at high doses and transformed into a reactive metabolite by the liver cytochrome P450. On the other hand, nutmeg is rich with many phytochemical ingredients that are known for their ability to inhibit cytochrome P450. Hence, the present experiment was aimed at studying the hepatoprotective effect of *Myristica fragrans* (nutmeg), kernel extract (MFKE) in respect to paracetamol (acetaminophen; *N*-acetyl-p-amino-phenol (APAP))-induced hepatotoxicity in rats, focusing on its antioxidant, anti-inflammatory, and anti-apoptotic activities. Liver toxicity was induced in rats by a single oral administration of APAP (2 g/kg). To evaluate the hepatoprotective effect of MFKE against this APAP-induced hepatotoxicity, rats were pre-treated with either oral administration of MFKE at 300 mg/kg daily for seven days or silymarin at 50 mg/kg as a standard hepatoprotective agent. APAP intoxication caused a drastic elevation in liver function markers (transaminases, alkaline phosphatase, and total bilirubin), oxidative stress indicators (lipid peroxidation and nitric oxide), inflammatory biomarkers (tumour necrosis factor-α, interleukin-1β, inducible nitric oxide synthase, and nuclear factor kB) and the pro-apoptotic BCL2 Associated X (Bax) and caspases-3 genes. Furthermore, analyses of rat liver tissue revealed that APAP significantly depleted glutathione and inhibited the activities of antioxidant enzymes in addition to downregulating two key anti-apoptotic genes: Cellular FLICE (FADD-like IL-1β-converting enzyme)-inhibitory protein (c-FLIP) and B-cell lymphoma 2 (Bcl-2). Pre-treatment with MFKE, however, attenuated APAP-induced liver toxicity by reversing all of these toxicity biomarkers. This hepatoprotective effect of MFKE was further confirmed by improvement in histopathological findings. Interestingly, the hepatoprotective effect of MFKE was comparable to that offered by the reference hepatoprotector, silymarin. In conclusion, our results revealed that MFKE had antioxidant, anti-inflammatory, and anti-apoptotic properties, and it is suggested that this hepatoprotective effect could be linked to its ability to promote the nuclear factor erythroid 2–related factor 2 (Nrf2)/antioxidant responsive element (ARE) pathway.

Keywords: paracetamol; *Myristica fragrans* kernels; heme oxygenase 1; liver

Int. J. Mol. Sci. **2019**, *20*, 993

1. Introduction

The liver is one of the most important organs in the body, performing up to 500 functions. It metabolizes most ingested substances and detoxifies toxic substances [1]. Hepatic injury, however, can be caused by the hepatotoxic effects of xenobiotics. Paracetamol (APAP; acetaminophen; N-acetyl-p-aminophenol) is a widely used drug worldwide due to its antipyretic property and is accessible in various formulations [2]. At therapeutic doses, APAP is considered a safe drug. It is also well-recognized, however, that APAP is responsible for acute liver failure [3]. In the United States, for example, about 30,000 patients annually are treated as a result of APAP hepatotoxicity [4].

In normal therapeutic doses, APAP is biotransformed in the liver to form glucuronide and water-soluble sulphate metabolites, which are excreted in the urine. In an overdose, however, APAP is oxidized by cytochrome p450 (CYP450) to form a highly reactive metabolite, *N*-acetyl-*p*-benzoquinone imine (NAPQI) that binds mainly to the sulfhydryl (-SH) group of GSH to form a complex of APAP-GSH, which excreted by the kidney as cysteine and mercapturic acid conjugates (APAP-cys). Furthermore, at toxic doses, APAP quickly depletes GSH, resulting in the accumulation of NAPQI, which then forms NAPQI-protein adducts as a result of covalent bond formation between NAPQI and –SH group of hepatocyte proteins [5]. The accumulation of these NAPQI-protein adducts results in turn in the formation of reactive oxygen species (ROS) that attack the cellular micro molecules causing lipid peroxidation and hepatic necrosis [6]. This damage to the hepatocytes ultimately leads to their leakage and a rise in hepatocyte enzymes in circulation in the blood.

Several experimental studies reported on the protective role of medicinal plants against the hepatotoxicity induced by APAP [7,8], but there is still a need to evaluate other medicinal plants to see if they might offer effective alternatives. Nutmeg is a dry kernel of *Myristica fragrans* belonging to the family Myristicaceae, which is widely distributed in tropical countries. The seed is used as a spice or a flavour to food [9]. The non-volatile part of *M. fragrans* is rich in dimeric phenyl propanoids, lignans, and neolignans [10]. The anti-inflammatory [11], antihyperglycemic [12], and hepatoprotective role of *M. fragrans* [9] against isoproterenol (ISO)-hepatotoxicity have been studied previously. Furthermore, active constituents of nutmeg such as neolignan and myristicin have been reported to inhibit cytochrome CYP3A4 and CYP2C9 [13].

In this study, therefore, we explored the protective role of *M. fragrans* kernel extract against hepatotoxicity induced by APAP. Furthermore, the effect of *M. fragrans* kernel extract was compared with that of silymarin, a standard hepatoprotective agent.

2. Results

The polyphenol and flavonoid fingerprint of the *Myristica fragrans* kernel extract detected at 280 nm is illustrated in Figure 1. The HPLC profile of MFKE shows the presence of 25 peaks with retention times ranging from 2.768 min to 40.842 min. Based on the UV-Visible spectral data and their retention times, the *Myristica fragrans* kernel extract has a UV band at 280 nm characteristic for polyphenol and flavonoid compounds, possibly caftaric acid and its derivatives, ellagic acid, rutin and catechin, and gallic acid and its derivatives, quercetin, and kaempferol.

During the study, there was no prevalence of mortality and APAP administration did not cause any abnormal clinical signs, with no change in food and water consumption in the groups that received APAP. That said, the activity levels of serum liver function markers, namely, ALT, AST, ALP, and Total Bil, were significantly elevated in APAP-treated rats compared to the control rats (Figure 2). On the other hand, in the rats that had been pre-administered MFKE, the elevation in liver function markers was significantly less pronounced, indicating the hepatoprotective effect of *Myristica fragrans* kernels. Indeed, the hepatoprotective effect of MFKE was comparable to silymarin (SLY), which is widely used on account of its hepatoprotective properties.

In order to evaluate the antioxidant effect of MFKE, lipid peroxidation, nitric oxide, and enzymatic and non-enzymatic molecules (GSH, SOD, CAT, GSH-Px, and GSH-R) were examined. SLY was used as a comparator. APAP treatment disturbed the redox status of hepatocytes confirmed by the elevation of LPO and NO (Figure 3), the depletion of GSH and the inhibition of the activities of antioxidant enzymes (Figure 4). On the contrary, MFKE pre-treatment significantly reversed this disturbance in the redox status. As expected, SLY pre-treatment also protected hepatocytes from oxidative stress induction by reversing the disturbance in the redox status.

The study also examined *Nfe2l2* and the expression of its downstream target genes *Hmox1*, *Nqo1*, *Gclc*, and *Utg1a1*. APAP treatment in rats induced a significant decrease in *Nfe2l2* and its putative target genes compared to the control group, but *Hmox1* expression was significantly upregulated. In contrast, MFKE pre-treatment negated the APAP-induced impairment in the cellular detoxification system by enhancing *Nfe2l2*, *Hmox1*, *Nqo1*, *Gclc*, and *Utg1a1* mRNA expressions, and the treatment reduced the severity of the reduction compared to the APAP group (Figure 5). As expected, SLY pre-treatment was effective in protecting hepatic tissue from APAP-mediate toxicity.

In the APAP group, inflammation was initiated and propagated through increases in the protein levels of TNF-α and IL-1β and in NF-κB and iNOS expressions (Figure 6). Pre-treatment with MFKE or SLY, however, prevented the overproduction of proinflammatory cytokines (TNF-α and IL-1β), and the protein expression of NF-κB and *Nos2* mRNA, suggesting that MFKE, like SLY, has an anti-inflammatory effect.

The histopathological examination of the liver sections of the control and the rats treated with alone revealed a normal architecture of hepatic lobules with hepatic parenchyma radiating from the central veins, with narrow sinusoids and prominent nuclei. By contrast, microscopic examination of the APAP rats showed disruption of the normal architecture of hepatic lobules associated with granular degeneration of hepatocytes and infiltration of inflammatory cells accompanied with various degrees of centrilobular hepatocyte necrosis and central veins that showed severe congestion and dilation (Figure 7A). The rats pre-treated with MFKE (300 mg/kg), however, had much healthier hepatic tissue than the rats that had received only APAP, with scanty apoptotic hepatocytes and slight hepatocellular vacuolation. Rats pre-treated with SLY (50 mg/kg) also showed less severe hepatic injury than APAP with slight activation of Kupffer cells. Furthermore, APAP intoxication resulted in abundant activation of HSCs (hepatic stellate cells), as evidenced by an increase in alpha-smooth muscle actin (α-SMA) immunoreactive positive cells (Figure 7B). Pre-treatment with MFKE or SLY, however, successfully restrained the activation of HSCs by preventing α-SMA expression (Supplementary data: Table S2).

In order to uncover the mechanism through which MFKE mitigates APAP-induced liver injury, we used qRT-PCR to examine the expression levels of anti-apoptotic markers, including *Bcl2* and *Cflar*, and pro-apoptotic markers *Bax* and *Casp3* in liver tissue. Our results showed a significant downregulation in *Bcl2* and *Cflar* expressions after APAP injection, but a significant upregulation in *Bax* and *Casp3* expressions (Figure 8). In contrast, the MFKE pre-treatment obviously reduced the number of hepatocytes dying as a result of APAP-injection. The hepatoprotective effect of MFKE was comparable to SLY. Overall, our results indicate that MFKE pre-treatment ameliorates damage to hepatocytes by inducing anti-apoptotic genes and restraining pro-apoptotic genes.

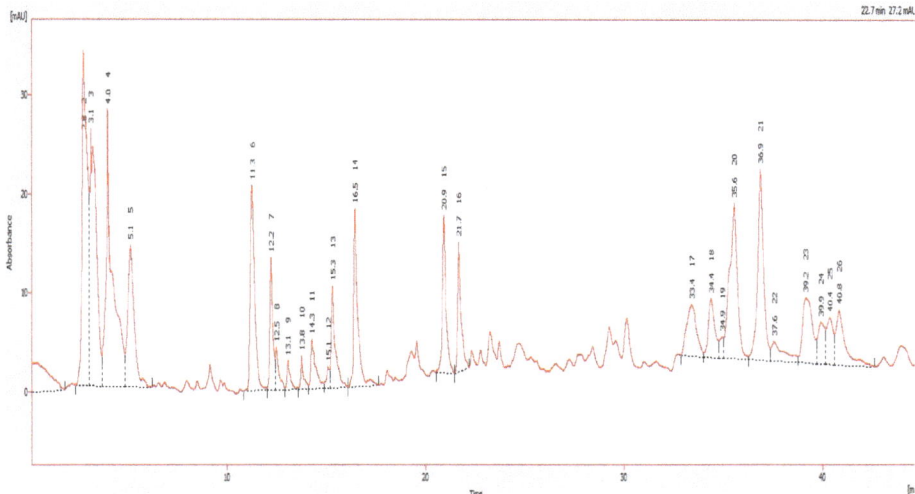

Figure 1. HPLC chromatogram of *Myristica fragrans* kernel extract at 280 nm. A mobile phase consisting of mixture of solvent A (0.2% acetic acid) and B (acetonitrile) and employing a gradient elution (from 10:90 to 100:0, v/v) at a flow rate of 1 mL/min.

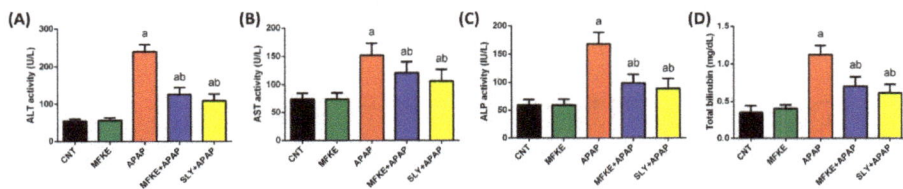

Figure 2. The effect of *Myristica fragrans* kernel extract (MFKE) on serum liver function markers in rats treated with paracetamol (APAP)-induced liver toxicity. Data are expressed as mean $\pm$ SD ($n = 7$); [a] $p < 0.05$ vs. control rats; [b] $p < 0.05$ vs. APAP-treated rats using Tukey's post hoc test. (**A**) Alanine aminotransferase, (**B**) Aspartate aminotransferase, (**C**) Alkaline phosphatase and (**D**) Total bilirubin.

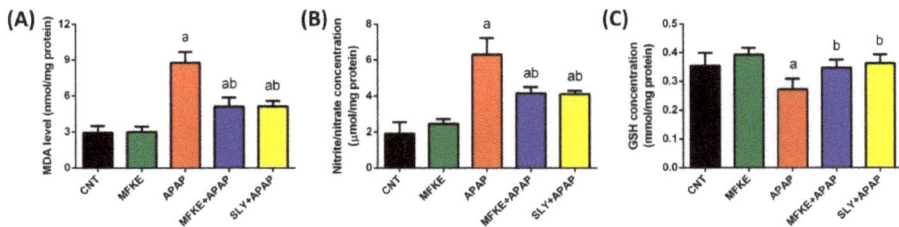

Figure 3. Effects of *Myristica fragrans* kernel extract (MFKE) on oxidative stress markers in rats treated with paracetamol (APAP)-induced liver toxicity. Data are expressed as mean $\pm$ SD ($n = 7$); [a] $p < 0.05$ vs. control rats; [b] $p < 0.05$ vs. APAP-treated rats using Tukey's post hoc test. (**A**) lipid peroxidation, (**B**) nitric oxide, and (**C**) glutathione.

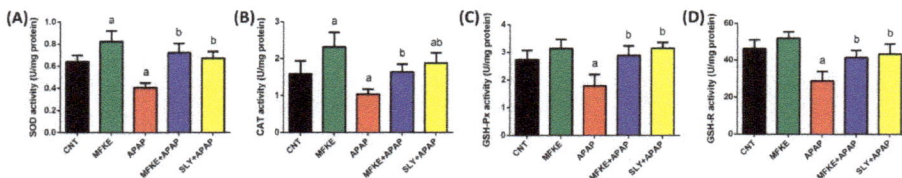

Figure 4. Effects of MFKE on the activity of antioxidant enzymes in rats treated with APAP-induced liver toxicity. Data are expressed as mean $\pm$ SD ($n = 7$); [a] $p < 0.05$ vs. control rats; [b] $p < 0.05$ vs. APAP-treated rats using Tukey's post hoc test. (**A**) Superoxide dismutase, (**B**) Catalase, (**C**) Glutathione peroxidase, and (**D**) Glutathione reductase.

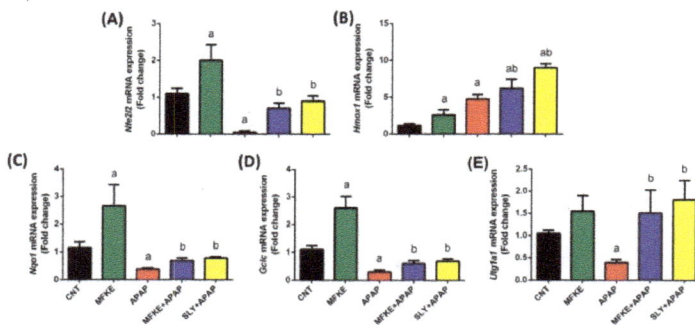

Figure 5. Effects of MFKE on Nrf2/ARE antioxidant signalling pathway gene expression in rats treated with APAP-induced liver toxicity. Results are presented as means $\pm$ SD of triplicate assays and normalized to *Gapdh* and expressed as fold change (log2 scale), relative to mRNA levels in controls; [a] $p < 0.05$ vs. control rats; [b] $p < 0.05$ vs. APAP-treated rats using Tukey's post hoc test. (**A**) Nuclear factor erythroid 2–related factor 2, (**B**) Heme oxygenase 1, (**C**) NAD(P)H quinone oxidoreductase 1, (**D**) glutamate-cysteine ligase, catalytic, and (**E**) UDP glucuronosyltransferase family 1 member A1.

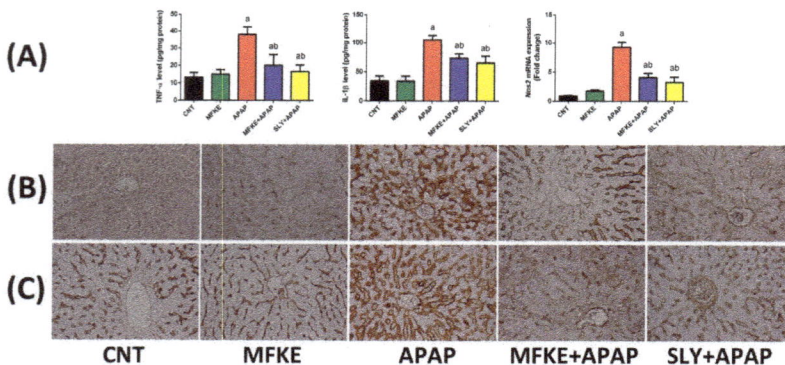

Figure 6. Effects of MFKE on proinflammatory biomarkers in rats treated with APAP-induced liver toxicity. (**A**) MRNA expressions of tumour necrosis factor-α, interlukin-1β and inducible nitric oxide synthase, (**B**) Nuclear factor kB protein expression in liver sections of different treated groups, and (**C**) Inducible nitric oxide synthase protein expression in liver sections of different treated groups. Original magnification ×40 for iNOS and NF-kB expression. Results for gene expression are presented as means $\pm$ SD of triplicate assays and normalized to *Gapdh* and expressed as fold change (log2 scale), relative to mRNA levels in controls; [a] $p < 0.05$ vs. control rats; [b] $p < 0.05$ vs. APAP-treated rats using Tukey's post hoc test.

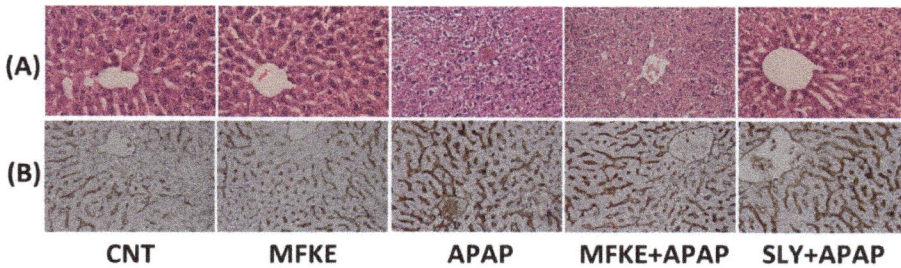

Figure 7. Effects of MFKE on liver pathology and α-SMA in rats treated with APAP-induced liver toxicity. (**A**) H&E stained photomicrographs of liver, and (**B**) Photomicrographs of α-SMA immunoreactivity in liver. Original magnification ×40.

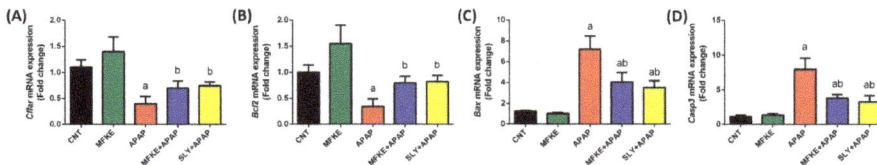

Figure 8. Effects of MFKE on apoptosis related gene expressions in rats treated with APAP-induced liver toxicity. Results are presented as means ± SD of triplicate assays and normalized to *Gapdh* and expressed as fold change (log2 scale), relative to mRNA levels in controls; [a] $p < 0.05$ vs. control rats; [b] $p < 0.05$ vs. APAP-treated rats using Tukey's post hoc test. (**A**) Cellular FLICE (FADD-like IL-1β-converting enzyme)-inhibitory protein, (**B**) B-cell lymphoma 2, (**C**) BCL2 Associated X, and (**D**) Caspases 3.

3. Discussion

APAP is an over-the-counter (OTC) drug commonly prescribed for its antipyretic and analgesic effects. In overdose, however, APAP can cause acute liver failure [6]. In the present study, injection with APAP caused liver injury, evidenced by an increase in liver function markers and pathological changes to liver tissue. In fact, ALT, AST, ALP, and total Bil are routinely monitored to assess liver function and any increase of these enzymes in serum is considered as indicative of liver injury and dysfunction. The increase in these markers indicates necrosis of hepatocytes or an increase in the permeability of their membranes. However, Stec et al. [14] reported that transaminases reduced significantly with the antioxidant properties of bilirubin. In addition, Hinds et al. [15] considered bilirubin as an antioxidant as well as an anti-inflammatory biomolecule and mice expressing the human Gilbert's HuUGT*28 polymorphism are safe from high-fat diet-induced hepatic steatosis and insulin resistance due to an elevation of bilirubin. Our biochemical data were also confirmed by histopathological observations which showed obvious hepatocyte degeneration in APAP-treated rats, as well as severe haemorrhage and dilation in sinusoids accompanied by Kupffer and inflammatory cells. MFKE pre-treatment, however, successfully prevented this liver damage. The hepatoprotective effect of MFKE indicates that MFKE is capable of maintaining the membrane integrity of hepatocytes and preventing the leakage of liver enzymes, thereby safeguarding liver function even when challenged by APAP overdose. The hepatoprotective efficacy of MFKE was previously shown in lipopolysaccharide/D-galactosamine-induced liver injury and inflammation [16] and in radiation-induced liver toxicity and oxidative stress [17]. Interestingly, the hepatoprotection offered by MFKE is comparable to the standard hepatoprotector SLY, with both of them showing clear preventative effects histologically.

Furthermore, oxidative stress induced by APAP injection was evidenced by significant elevation in LPO and NO, concurrently with a significant depletion in GSH and inhibition of the antioxidant defence system. APAP is oxidized mainly by the P450 enzyme, and it is this enzyme that is responsible for oxidative stress induction in the liver [5]. Moreover, the previous study reported that many mitochondrial proteins are targeted by APAP metabolites, including GSH-Px and ATP-synthase alpha-subunit, which are adducted by NAPQI. As a result, 60% of GSH-Px was modified and ATP-synthase function was impaired causing cessation in ATP production. Additionally, NAPQI interferes with the mitochondrial electron transport chain causing the release of electrons from the chain which triggers ROS generation and lipid peroxidation [18]. Furthermore, the escaped electrons from the chain react with oxygen to form superoxide anion, which could either dismutate to form H_2O_2 or react with NO to form peroxynitrite, a strong oxidant. The formed H_2O_2 can be detoxified by GSH, while peroxynitrite is harsher and reacts with a variety of biomolecules and alerted biomolecules' structure [19]. Thus, the overproduction of H_2O_2 and peroxynitrite, in addition to the other ROS, can overwhelm the antioxidant defence system.

The overproduction of NO in the present study is attributed to the upregulation of iNOS in liver tissue. iNOS is another source of ROS in inflammatory cells and iNOS blockers have been found to inhibit APAP-induced hepatotoxicity [20]. Interestingly, promoting the replenishment of endogenous antioxidant molecules with natural products is effective at protecting the liver from APAP overdose-induced hepatotoxicity [21].

In this study, the induction of hepatotoxicity by APAP caused significant increases in TNF-α and IL-1β level and iNOS expression. There is growing evidence to show that inflammation is also a possible mechanism responsible for the pathogenesis resulting from APAP-induced hepatotoxicity. Furthermore, both inflammation and oxidative stress are tightly linked to each other in that, at the site of inflammation, inflamed cells like macrophages are releasing ROS, causing exaggerated oxidative damage. Furthermore, overproduction of ROS initiates a cellular signalling cascade that promotes the gene expression of proinflammatory cytokines [22]. Moreover, Posadas et al. [23] demonstrated that APAP promoted NF-κB p65 phosphorylation at Ser536 and enhanced NF-κB p65 translocation to the nucleus, a contention that supports our findings. Indeed, NF-κB p65 nuclear translocation has been found to induce a variety of proinflammatory cytokines and mediators. The results obtained in our study in respect to the effect of APAP on the liver therefore are generally in line with those of previous studies, and support the existing understanding of the mechanism of action of APAP on liver tissue and cells [20,24].

Meanwhile, the current study demonstrated that pre-treatment with MKFE restrained the oxidative stress and inflammation otherwise induced by APAP, as evidenced by significant decreases in LPO, NO and proinflammatory cytokine levels, as well in NF-κB and iNOS expressions, and significant increases in enzymatic and nonenzymatic molecules. This effect suggests that MFKE has antioxidant properties that might be attributed to the presence of different phenolic compounds, flavonoids, alkaloids, and anthraquinones [25]. It was reported that argenteane, a dilignan which was isolated from MFKE, is as powerful as vitamin E. Indeed, argenteane shows a typical free radical scavenging activity by releasing one or two H atom(s) to the medium [26]. Furthermore, Kapoor et al. [27] demonstrated that essential oil and oleoresins from nutmeg are antioxidants similar to the synthetic butylated hydroxyanisole (BHA) and butylated hydroxytoluene (BHT). Most recently, Erukainure et al. [28] found that more than 14% of nutmeg is oleic acid, which is known for its antioxidant activity, and more than of 46% of the extract is 17-octadecynoic acid, which inhibits cytochrome P450. Additionally, MFKE contains myristicin, which was found to possess antioxidant and hepatoprotective activities [16] and to suppress lipid peroxidation in the liver by trapping free radicals or ROS directly [29]. Furthermore, at high doses (96 mg/kg), pure myristicin is a potent enhancer of hepatic cytochrome P enzymes, causing a 2–20 fold increase in the activity and expression of these enzymes [30].

The myristicin in MFKE does not only act as an antioxidant agent, but also as an anti-inflammatory agent. The anti-inflammatory effect of this constituent was observed in an in vitro study using RAW 264.7 mouse macrophages. In this study, myristicin suppressed the production of different proinflammatory cytokines [31]. Further consideration indicated that this molecule played a regulatory role in chronic autoimmune diseases and that it also attenuated lung inflammatory disease by inhibiting NO production [32]. Moreover, MFKE is a rich source of neolignans that have showed potent NF-κB inhibitory activity [33].

Paracetamol-induced liver fibrosis received growing attention since the reports of occurrences of liver fibrosis and cirrhosis in patients using APAP to reduce pain [34], and the activation of HSCs plays an important role in liver fibrosis. Indeed, HSCs are considered to be the main extracellular matrix–producing cells in the liver [35]. α-SMA expression is a typical marker for activated HSCs [36], and Shen et al. [37] demonstrated that α-SMA expression caused HSCs activation induced acute liver failure. Our study is in line with this work in that we reported that APAP intoxication caused α-SMA expression in the liver.

It is therefore significant that are our results also show that MFKE successfully restrained the activation of HSCs by preventing α-SMA expression. The possible mechanism for this effect might be linked to the presence of meso-dihydroguaiaretic acid, which directly inhibits transactivation of HSCs through downregulating transforming growth factor beta1 and inhibiting activator protein 1 activity [38].

In addition, in the current study, APAP treatment caused a significant upregulation in the HO-1 gene and downregulation in Nrf2 expression. Nrf2 is a prime controller of intracellular redox homeostasis by controlling the antioxidant response element (ARE), which orchestrates adaptability to cellular redox disruption. Chan et al. found that, in Nrf2-knocout mice, APAP intoxication enhanced liver injury and mortality compared with wild-type mice [39]. In this context, the ability of MFKE to promote Nrf2 expression in the present study implied the cytoprotective effect of nutmeg. One way in which it may achieve this is through nectandrin B. Song et al. [40] and Kim et al. [41] demonstrated that nectandrin B, which was isolated from nutmeg, protected hepatocytes against oxidative perturbation through the activation of Nrf2 and the consequent induction of the detoxifying antioxidant enzymes such as NQO1, GCLC, and UTGs. NQO1 is a highly inducible enzyme responsible, among others, for a single-step two-electron reduction of quinones and quinone imines, thus inhibiting the formation of reactive and toxic semiquinone intermediates [42]. Meanwhile, GCLC is the first rate-limiting enzyme complex for GSH synthesis [43]. UGTs play predominant roles in the detoxification of many exogenous and endogenous agents by forming more polar and water-soluble glucuronides [44] and deficiencies in expression of UGTs are responsible for APAP hepatotoxicity observed in rats [45]. HO-1, meanwhile, is the primary rate-limiting enzyme in heme catabolism and is well known to have a cytoprotection effect against liver damage by restraining oxidative stress and inflammation. HO-1 is also involved in maintaining the oxidants/antioxidants balance [46] by increased formation of the antioxidant, bilirubin [15]. In some conditions, however, induction of HO-1 may cause damage and enhance cytotoxicity by increasing the products of heme degradation, such as iron ions, biliverdin, and carbon monoxide [47]. Furthermore, biliverdin is subsequently converted to bilirubin this has been recognized to be an antioxidant and can inhibit lipid peroxidation. The higher plasma bilirubin levels within the normal range were linked with a significant and marked reduction in cardiovascular diseases risk and decreased the incidence of ischemic heart disease [48]. In the present study, APAP treatment induced upregulation in HO-1 expression. Our results are in harmony with those of Bauer et al. [49] and Gao et al. [50]. MFKE pre-treatment, however, modulated HO-1 expression and induced significant down-regulation in the gene compared to the APAP group. When comparing the HO-1 expression with the control group, however, HO-1 showed significant increase in the MFKE and SLY treatment groups. As we expected that silymarin and nutmeg extract as targeted Nrf2 would lead to UGT1A1 upregulation, and this enzyme is the sole enzyme responsible for the bilirubin metabolism, that may explain the decrease in bilirubin in APAP-treated groups. However, silymarin and many flavonoids are known to inhibit UTGs in vitro assays. Those in vitro results turned out to be ineffective in vivo studies due to the poor cell permeability and poor metabolic stability of these natural compounds, together leading to poor bioavailability, are probably the major causes of their ineffectiveness in vivo [44] that seen herein.

The results obtained in the present study revealed that APAP intoxication caused severe apoptosis and necrosis in liver tissue. The death of hepatocytes was evidenced by a significant elevation of pro-apoptotic genes (Bax and caspases-3) and a significant reduction in anti-apoptotic genes (c-FLIP and Bcl-2). It was reported that overdose with APAP induced apoptosis in primary hepatocytes [51] and liver tissue of mice [52]. Moreover, hepatocyte apoptosis leads to elevated serum transaminases levels and further reduction of the liver function and preventing apoptosis inhibits liver failure [53]. The precise mechanisms by which APAP-induced apoptosis are linked to the ability of APAP to induce mitochondrial permeability transitions that are subsequently associated with the release of cytochrome c, second mitochondria-derived activator of caspase (Smac) and apoptosis inducing factor (AIF) [54]. Once in cytosol, cytochrome c binds to the adaptor protein Apaf-1 thus triggering the formation of apoptosome by activating caspases-9. Meanwhile, AIF translocates to the nucleus causing DNA damage due to its involvement in DNA fragmentation. Smac, meanwhile, plays a vital role in caspase-3 activation [55]. This induction of apoptosis in the liver was alleviated in the rats treated with MFKE, however, with the liver tissue of these rats showing marked improvement compared to the rats treated just with APAP. To understand this anti-apoptotic effect of MFKE, it can be noted that, in the current study, MFKE administration significantly upregulated c-FLIP. The cellular Fas-associated death domain-like IL-1β-converting enzyme-inhibitory protein is known as c-FLIP, and this is a master anti-apoptotic regulator with a multifunctional role in several signalling pathways, as well as in inducing several cytoprotective and pro-survival signalling pathways [56]. The anti-apoptotic effect of MFKE might, therefore, be related to the ability of myristicin and nectandrin B to inhibit endoplasmic reticulum stress or to promote c-FLIP expression and thus enhance the antioxidant defence system [40,57].

In conclusion, our findings highlight a body of evidence for the hepatoprotective effect of *Myristica fragrans* kernels in respect to the liver of rats exposed to APAP-induced hepatotoxicity. These favourable effects were mediated via suppressing oxidative stress, inflammation, and apoptosis, and this hepatoprotection effect could be linked to their ability to promote the Nrf2/ARE pathway. Interestingly, the hepatoprotective effect of MFKE was comparable to that exerted by the reference hepatoprotector SLY.

4. Materials and Methods

4.1. Chemicals

APAP and silymarin were obtained from Sigma-Aldrich (St. Louis, MO, USA). All other chemicals and reagents used in the experiment were of analytical grade.

4.2. Plant Materials and Extraction Procedure

The kernels of *Myristica fragrans* (MF) were obtained from a local market in West-Cairo, Egypt. The obtained kernels were identified by an expert taxonomist from the Botany Department, Faculty of Science, Helwan University, Egypt. The kernels were ground into powder using an electrical blender and extracted three times by maceration with 70% methanol. The ratio of nutmeg kernels to the solvent was 1:10 (w/v). The solvent was concentrated under a vacuum evaporator and subsequently lyophilized. The *Myristica fragrans* kernel extract (MFKE) was kept at $-20\,^{\circ}$C until used in the current experiment. Total phenolic and flavonoid compounds were determined as described previously [58]. The amount of the total phenolic compounds was 21.3 mg gallic acid equivalent/g extract, while the amount of flavonoid compounds was 20.8 mg quercetin equivalent/g extract.

4.3. HPLC Analysis

High performance liquid chromatography (HPLC) analysis was performed using a Perkin Elmer Series 200 liquid chromatography (PerkinElmer, Akron, OH, USA) to determine the polyphenol and flavonoid constituents of MFKE. The HPLC column was an AQUA column 150 mm 5 μ C18

(Phenomenex), with a detection wavelength of 280 nm. Elution was carried out using acetic acid (2%; A) and acetonitrile (B). The flow rate was set at 1 mL/min throughout the elution.

4.4. Animals

Adult male Wistar albino rats (10 weeks old, weighing 200–220 g) obtained from VACSERA (Cairo, Egypt) were used in the current experiments. The animal housing conditions was as described previously [59]. Briefly, the rats were housed in wire cages made from polypropylene in a room under standard laboratory conditions (12 h light-dark cycle; 25 ± 2 °C). A standard rodent diet (El Gomhorya Company, Ismailia, Egypt) and water were available *ad libitum*. The rats were acclimated to the environment for one week before the beginning of the experiment. The study was approved by the Committee of Research Ethics for Laboratory Animal Care, Department of Zoology, Faculty of Science, Helwan University (24/09/2018, Cairo; approval no, HU2017/Z/08) and was conducted according to the European Community Directive (86/609/EEC), the national rules on animal care and in accordance with the NIH Guidelines for the Care and Use of Laboratory Animals, eighth edition.

4.5. Experimental Design

The animals were fasted for 24 h prior to the experiment under standard laboratory conditions but allowed access to distilled water (dH$_2$O) *ad libitum*. Subsequently, the animals were randomly divided into five experimental groups of seven rats each and administered with test solutions orally once daily for seven consecutive days, as follows. Group I, serving as the normal control, received 10 mL/kg physiological saline (0.9% NaCl). Group II, serving as the MFKE control, received 300 mg/kg of MFKE. Group III, serving as the APAP (paracetamol) control group received 2 g/kg. Groups IV and V, pre-treatment groups, received 300 mg/kg MFKE and 50 mg/kg SLY (silymarin), respectively. All the treatments were suspended or dissolved in a physiological saline.

An acute toxicity study was performed using a maximum dose of 2000 mg/kg MFKE administered orally and observed in order to assess changes in skin and fur, eyes, and mucous membranes, and also respiratory, circulatory, autonomic and central nervous systems, and somatomotor activity and behavior pattern. This dose, after 14 days, did not lead to any signs of toxicity in the rats, and the subsequent oral dose of MFKE was selected based on a preliminary study using three doses of 100, 200, and 300 mg/kg, which showed that the oral administration of MFKE at a dose of 300 mg/kg effectively prevented APAP-induced hepatotoxicity. While, APAP dose was based on the previous studies to induce liver toxicity in rats by using APAP at 2 g/kg [60,61].

The oral administration of APAP (2 g/kg) was performed three hours prior to the last administration on the seventh day except for groups I and II, which received only 10 mL/kg physiological saline. Forty-eight hours after the induction of hepatic injury by these means, the animals were lightly anesthetized using an appropriated anaesthetic agent and blood was collected by cardiac puncture in sterilized centrifuged tubes which were then centrifuged at 3000 rpm for 10 min to get serum for the study of biochemical parameters. The animals were then killed by cervical dislocation and the liver was quickly removed, weighed and washed clean of blood with ice-cold saline. Subsequently, a small piece was homogenized in cold phosphate buffer (0.05 M, pH 7.4), and then centrifuged at 3000 rpm for 10 min at 4 °C. The resulting supernatant was used for determination of biochemical markers. The remaining parts of the liver were used for the histopathological and molecular studies.

4.6. Biochemical Parameters

4.6.1. Liver Functions Tests

A colorimetric method was used to determine the activity of alanine transaminase (ALT) and aspartate transaminase (AST) enzymes in the collected serum [62] using kits obtained from Bio-Diagnostic (Giza, Egypt) the results were recorded and analysed using an UV-visible spectrophotometer (V630; JASCO, Tokyo, Japan) at 505 nm. Alkaline phosphatase and total bilirubin

(TB) in serum were estimated according to the methods described by Shephard and Peake [63] and Schmidt and Eisenburg [64] using kits obtained from Bio-Diagnostic (Giza, Egypt) and Randox (Crumlin, UK). The results were recorded and analyzed using an UV-visible spectrophotometer (V630; JASCO, Japan) at 510 and 535 nm, respectively.

4.6.2. Determination of Malondialdehyde and Nitric Oxide

Hepatic malondialdehyde (MDA) was estimated using 1 mL of 0.67% thiobarbituric acid according to the method of Ohkawa et al. [65]. Meanwhile, the level of nitrite/nitrate [nitric oxide (NO)] was determined according to the method described by Green et al. [66].

4.6.3. Nonenzymatic and Enzymatic Antioxidant Molecules

Hepatic glutathione (GSH) was assayed spectrophotometrically using the Ellman [67] method. Determination of superoxide dismutase (SOD) and catalase (CAT) activities were assayed according to the methods described by Sun et al. [68] and Luck [69], respectively. In addition, the activity of glutathione related enzymes, namely, glutathione peroxidase (GSH-Px) and glutathione reductase (GSH-R), were assayed according to the techniques described by Paglia and Valentine [70] and Factor et al. [71], respectively.

4.6.4. Determination of Proinflammation Markers

The hepatic proinflammatory cytokines, namely, interleukin-1 beta (IL-1β; Cat. no. ERIL1B, ThermoFisher Scientific, Waltham, MA, USA) and tumour necrosis factor alpha (TNF-α; Cat. no. CSB-E11987r, CUSABIO Life Sciences, Wuhan, China), were assayed using enzyme-linked immunosorbent assay (ELISA) kits according to the manufacturer's instructions. Each liver homogenate sample was measured in duplicate.

4.6.5. Quantitative Reverse Transcription-Polymerase Chain Reaction (qRT-PCR) Analysis

Total RNA was isolated from liver samples using TRIzol reagent (Invitrogen, Carlsbad, CA, USA) and reverse transcribed to complementary DNA (cDNA) using the Script™ cDNA synthesis kit (Bio-Rad, California, CA, USA). Quantitative real-time polymerase chain reaction (qRT-PCR) was performed using Power SYBR® Green (Invitrogen) on an Applied Biosystems 7500 Instrument. The cycling conditions for amplification were 95 °C for 2 min, followed by 40 cycles at 94 °C for 60 s and 60 °C for 90 s. Each assay was performed in duplicate to determine delta CT. Specific primers were purchased from Jena Bioscience GmbH (Jena, Germany). The housekeeping gene was glyceraldehyde-3-phosphate dehydrogenase (GAPDH). The primer sequences for *Gapdh, Nos2, Nfe2l2, Nqo1, Gclc, Utg1a1, Hmox1, Cflar, Bcl2, Bax,* and *Casp3* were listed previously [72] (see Supplementary data: Table S1).

4.7. Histopathological Examination

The liver tissues were fixed in 10% neutral-buffered formalin for 24 h at room temperature, dehydrated in gradual ethanol, embedded in paraffin wax, and sectioned (4–5 μm). The deparaffinized sections were stained consistently with haematoxylin and eosin dye for microscopic examination. Images were recorded with 400× magnification (Nikon Eclipse E200-LED, Tokyo, Japan).

4.8. Immunohistochemistry Analysis

To investigate NF-κB, iNOS and αSMA, the paraffinized sections were blocked with hydrogen peroxide (0.1%) containing methanol for 15 min to damage the endogenous peroxidase. After blocking, the samples were incubated with the primary antibodies at 4 °C overnight. Thereafter, the samples were washed twice with phosphate-buffered saline and incubated with biotinylated secondary antibody labelled with horseradish peroxidase (HRP). The reactions were developed via an HRP-catalysed reaction with diaminobenzidine (DAB), followed by counterstaining with haematoxylin. Images were recorded at $400\times$ magnification (Nikon Eclipse E200-LED, Tokyo, Japan). Afterward, the color intensity of each examined protein was semi-quantitatively scored by a blinded pathologist. The intensity was expressed as + (weak immunoreaction), ++ (moderate immunoreaction), +++ (strong immunoreaction), or ++++ (very strong immunoreaction).

4.9. Statistical Analysis

Data obtained are expressed as the mean $\pm$ standard deviation. Data from various evaluations were analysed using one-way analysis of variance (ANOVA) and the differences between the groups were determined using Tukey's post hoc test; *p* values < 0.05 were taken to be statistically significant.

Supplementary Materials: Supplementary materials can be found at http://www.mdpi.com/1422-0067/20/4/993/s1.

Author Contributions: Conceptualization, M.A.D. and A.E.A.; Methodology, A.E.A.M.; Validation, M.A.D. and A.E.A.M.; Formal Analysis, A.E.A.M.; Investigation, A.E.A.M.; Data Curation, M.A.D. and A.E.A.M.; Writing-Original Draft Preparation, M.A.D. and A.E.A.M.; Writing-Review & Editing, M.A.D. and A.E.A.M.; Supervision, A.E.A.M. and S.A.; Project Administration, M.A.D. and S.A.; Funding Acquisition, T.A.H., M.A.M., W.F.M., F.A.T. and S.A.

Acknowledgments: The authors extend their appreciation to the Deanship of Scientific Research at King Saud University for funding the work through the research group project No. RG-198.

Conflicts of Interest: The authors declare no conflict of interest.

References

1. Almeer, R.S.; Alarifi, S.; Alkahtani, S.; Ibrahim, S.R.; Ali, D.; Moneim, A. The potential hepatoprotective effect of royal jelly against cadmium chloride-induced hepatotoxicity in mice is mediated by suppression of oxidative stress and upregulation of nrf2 expression. *Biomed. Pharmacother.* **2018**, *106*, 1490–1498. [CrossRef] [PubMed]

2. Bunchorntavakul, C.; Reddy, K.R. Acetaminophen-related hepatotoxicity. *Clin. Liver Dis.* **2013**, *17*, 587–607. [CrossRef] [PubMed]

3. Herndon, C.M.; Dankenbring, D.M. Patient perception and knowledge of acetaminophen in a large family medicine service. *J. Pain Palliat. Care Pharmacother.* **2014**, *28*, 109–116. [CrossRef] [PubMed]

4. Blieden, M.; Paramore, L.C.; Shah, D.; Ben-Joseph, R. A perspective on the epidemiology of acetaminophen exposure and toxicity in the united states. *Expert Rev. Clin. Pharmacol.* **2014**, *7*, 341–348. [CrossRef] [PubMed]

5. Yoon, E.; Babar, A.; Choudhary, M.; Kutner, M.; Pyrsopoulos, N. Acetaminophen-induced hepatotoxicity: A comprehensive update. *J. Clin. Transl. Hepatol.* **2016**, *4*, 131–142. [PubMed]

6. Mahmood, N.D.; Mamat, S.S.; Kamisan, F.H.; Yahya, F.; Kamarolzaman, M.F.F.; Nasir, N.; Mohtarrudin, N.; Tohid, S.F.M.; Zakaria, Z.A. Amelioration of paracetamol-induced hepatotoxicity in rat by the administration of methanol extract of muntingia calabura l. Leaves. *BioMed Res. Int.* **2014**, *2014*, 10. [CrossRef] [PubMed]

7. El-Kott, A.F.; Bin-Meferij, M.M. Use of arctium lappa extract against acetaminophen-induced hepatotoxicity in rats. *Curr. Ther. Res.* **2015**, *77*, 73–78. [CrossRef] [PubMed]

8. Abirami, A.; Nagarani, G.; Siddhuraju, P. Hepatoprotective effect of leaf extracts from citrus hystrix and c. Maxima against paracetamol induced liver injury in rats. *Food Sci. Hum. Wellness* **2015**, *4*, 35–41. [CrossRef]

9. Kareem, M.A.; Gadhamsetty, S.K.; Shaik, A.H.; Prasad, E.M.; Kodidhela, L.D. Protective effect of nutmeg aqueous extract against experimentally-induced hepatotoxicity and oxidative stress in rats. *J. Ayurveda Integr. Med.* **2013**, *4*, 216–223. [CrossRef] [PubMed]

10. Francis, S.K.; James, B.; Varughese, S.; Nair, M.S. Phytochemical investigation on myristica fragrans stem bark. *Nat. Prod. Res.* **2018**. [CrossRef] [PubMed]

11. Zhang, W.K.; Tao, S.S.; Li, T.T.; Li, Y.S.; Li, X.J.; Tang, H.B.; Cong, R.H.; Ma, F.L.; Wan, C.J. Nutmeg oil alleviates chronic inflammatory pain through inhibition of cox-2 expression and substance p release in vivo. *Food Nutr. Res.* **2016**, *60*, 30849. [CrossRef] [PubMed]

12. Broadhurst, C.L.; Polansky, M.M.; Anderson, R.A. Insulin-like biological activity of culinary and medicinal plant aqueous extracts in vitro. *J. Agric. Food Chem.* **2000**, *48*, 849–852. [CrossRef] [PubMed]

13. Kimura, Y.; Ito, H.; Hatano, T. Effects of mace and nutmeg on human cytochrome p450 3a4 and 2c9 activity. *Biol. Pharm. Bull.* **2010**, *33*, 1977–1982. [CrossRef] [PubMed]
14. Stec, D.E.; John, K.; Trabbic, C.J.; Luniwal, A.; Hankins, M.W.; Baum, J.; Hinds, T.D., Jr. Bilirubin binding to pparalpha inhibits lipid accumulation. *PLoS ONE* **2016**, *11*, e0153427. [CrossRef] [PubMed]
15. Hinds, T.D., Jr.; Sodhi, K.; Meadows, C.; Fedorova, L.; Puri, N.; Kim, D.H.; Peterson, S.J.; Shapiro, J.; Abraham, N.G.; Kappas, A. Increased ho-1 levels ameliorate fatty liver development through a reduction of heme and recruitment of fgf21. *Obesity* **2014**, *22*, 705–712. [CrossRef] [PubMed]
16. Morita, T.; Jinno, K.; Kawagishi, H.; Arimoto, Y.; Suganuma, H.; Inakuma, T.; Sugiyama, K. Hepatoprotective effect of myristicin from nutmeg (myristica fragrans) on lipopolysaccharide/d-galactosamine-induced liver injury. *J. Agric. Food Chem.* **2003**, *51*, 1560–1565. [CrossRef] [PubMed]
17. Sharma, M.; Kumar, M. Radioprotection of swiss albino mice by myristica fragrans houtt. *J. Radiat. Res.* **2007**, *48*, 135–141. [CrossRef] [PubMed]
18. Du, K.; Ramachandran, A.; Jaeschke, H. Oxidative stress during acetaminophen hepatotoxicity: Sources, pathophysiological role and therapeutic potential. *Redox Biol.* **2016**, *10*, 148–156. [CrossRef] [PubMed]
19. Du, K.; McGill, M.R.; Xie, Y.; Bajt, M.L.; Jaeschke, H. Resveratrol prevents protein nitration and release of endonucleases from mitochondria during acetaminophen hepatotoxicity. *Food Chem. Toxicol.* **2015**, *81*, 62–70. [CrossRef] [PubMed]
20. Papackova, Z.; Heczkova, M.; Dankova, H.; Sticova, E.; Lodererova, A.; Bartonova, L.; Poruba, M.; Cahova, M. Silymarin prevents acetaminophen-induced hepatotoxicity in mice. *PLoS ONE* **2018**, *13*, e0191353. [CrossRef] [PubMed]
21. Shanmugam, S.; Thangaraj, P.; Lima, B.D.S.; Chandran, R.; de Souza Araujo, A.A.; Narain, N.; Serafini, M.R.; Junior, L.J.Q. Effects of luteolin and quercetin 3-beta-d-glucoside identified from passiflora subpeltata leaves against acetaminophen induced hepatotoxicity in rats. *Biomed. Pharmacother.* **2016**, *83*, 1278–1285. [CrossRef] [PubMed]
22. Biswas, S.K. Does the interdependence between oxidative stress and inflammation explain the antioxidant paradox? *Oxid. Med. Cell. Longev.* **2016**, *2016*, 9. [CrossRef] [PubMed]
23. Posadas, I.; Santos, P.; Cena, V. Acetaminophen induces human neuroblastoma cell death through nfkb activation. *PLoS ONE* **2012**, *7*, e50160. [CrossRef] [PubMed]
24. Yoshioka, H.; Usuda, H.; Fujii, H.; Nonogaki, T. Sasa veitchii extracts suppress acetaminophen-induced hepatotoxicity in mice. *Environ. Health Prev. Med.* **2017**, *22*, 54. [CrossRef] [PubMed]
25. Akinboro, A.; Mohamed, K.B.; Asmawi, M.Z.; Sulaiman, S.F.; Sofiman, O.A. Antioxidants in aqueous extract of myristica fragrans (houtt.) suppress mitosis and cyclophosphamide-induced chromosomal aberrations in allium cepa l. Cells. *J. Zhejiang Univ. Sci. B* **2011**, *12*, 915–922. [CrossRef] [PubMed]
26. Calliste, C.A.; Kozlowski, D.; Duroux, J.L.; Champavier, Y.; Chulia, A.J.; Trouillas, P. A new antioxidant from wild nutmeg. *Food Chem.* **2010**, *118*, 489–496. [CrossRef]
27. Kapoor, I.P.S.; Singh, B.; Singh, G.; De Heluani, C.S.; De Lampasona, M.P.; Catalan, C.A.N. Chemical composition and antioxidant activity of essential oil and oleoresins of nutmeg (myristica fragrans houtt.) fruits. *Int. J. Food Prop.* **2013**, *16*, 1059–1070. [CrossRef]
28. Erukainure, O.L.; Ajiboye, J.A.; Abbah, U.A.; Asieba, G.O.; Mamuru, S.; Zaruwa, M.Z.; Manhas, N.; Singh, P.; Islam, M.S. *Monodora myristica* (African nutmeg) modulates redox homeostasis and alters functional chemistry in sickled erythrocytes. *Hum. Exp. Toxicol.* **2018**, *37*, 458–467. [CrossRef] [PubMed]
29. Tisserand, R.; Young, R. 14—Constituent profiles. In *Essential Oil Safety*, 2nd ed.; Tisserand, R., Young, R., Eds.; Churchill Livingstone: St. Louis, MO, USA, 2014; pp. 483–647.
30. Jeong, H.G.; Yun, C.H. Induction of rat hepatic cytochrome p450 enzymes by myristicin. *Biochem. Biophys. Res. Commun.* **1995**, *217*, 966–971. [CrossRef] [PubMed]
31. Lee, J.Y.; Park, W. Anti-inflammatory effect of myristicin on raw 264.7 macrophages stimulated with polyinosinic-polycytidylic acid. *Molecules* **2011**, *16*, 7132–7142. [CrossRef] [PubMed]
32. Serhan, C.N.; Chiang, N.; Van Dyke, T.E. Resolving inflammation: Dual anti-inflammatory and pro-resolution lipid mediators. *Nat. Rev. Immunol.* **2008**, *8*, 349–361. [CrossRef] [PubMed]
33. Acuña, U.M.; Blanco Carcache, P.J.; Matthew, S.; Carcache de Blanco, E.J. New acyclic bis phenylpropanoid and neolignans, from myristica fragrans houtt., exhibiting parp-1 and nf-κb inhibitory effects. *Food Chem.* **2016**, *202*, 269–275. [CrossRef] [PubMed]

34. Watelet, J.; Laurent, V.; Bressenot, A.; Bronowicki, J.P.; Larrey, D.; Peyrin-Biroulet, L. Toxicity of chronic paracetamol ingestion. *Aliment. Pharmacol. Ther.* **2007**, *26*, 1543–1544. [CrossRef] [PubMed]

35. Bai, Q.; Yan, H.; Sheng, Y.; Jin, Y.; Shi, L.; Ji, L.; Wang, Z. Long-term acetaminophen treatment induced liver fibrosis in mice and the involvement of egr-1. *Toxicology* **2017**, *382*, 47–58. [CrossRef] [PubMed]

36. Almeer, R.S.; El-Khadragy, M.F.; Abdelhabib, S.; Abdel Moneim, A.E. Ziziphus spina-christi leaf extract ameliorates schistosomiasis liver granuloma, fibrosis, and oxidative stress through downregulation of fibrinogenic signaling in mice. *PLoS ONE* **2018**, *13*, e0204923. [CrossRef] [PubMed]

37. Shen, K.; Chang, W.; Gao, X.; Wang, H.; Niu, W.; Song, L.; Qin, X. Depletion of activated hepatic stellate cell correlates with severe liver damage and abnormal liver regeneration in acetaminophen-induced liver injury. *Acta Biochim. Biophys. Sin.* **2011**, *43*, 307–315. [CrossRef] [PubMed]

38. Park, E.Y.; Shin, S.M.; Ma, C.J.; Kim, Y.C.; Kim, S.G. Meso-dihydroguaiaretic acid from machilus thunbergii down-regulates tgf-beta1 gene expression in activated hepatic stellate cells via inhibition of ap-1 activity. *Planta Med.* **2005**, *71*, 393–398. [CrossRef] [PubMed]

39. Chan, K.; Han, X.D.; Kan, Y.W. An important function of nrf2 in combating oxidative stress: Detoxification of acetaminophen. *Proc. Natl. Acad. Sci. USA* **2001**, *98*, 4611–4616. [CrossRef] [PubMed]

40. Song, J.S.; Kim, E.K.; Choi, Y.W.; Oh, W.K.; Kim, Y.M. Hepatocyte-protective effect of nectandrin b, a nutmeg lignan, against oxidative stress: Role of nrf2 activation through erk phosphorylation and ampk-dependent inhibition of gsk-3beta. *Toxicol. Appl. Pharmacol.* **2016**, *307*, 138–149. [CrossRef] [PubMed]

41. Kim, E.K.; Song, J.S.; Choi, D.G.; Choi, Y.W.; Oh, W.K.; Kim, Y.-M. Therapeutic potential of nectandrin b, a nutmeg lignan, in nonalcoholic fatty liver disease: Anti-lipogenic and hepatocyte-protective effects through amp-activated protein kinase and nrf2 activation. *FASEB J.* **2017**, *31*. [CrossRef]

42. Roubalová, L.; Dinkova-Kostova, A.T.; Biedermann, D.; Křen, V.; Ulrichová, J.; Vrba, J. Flavonolignan 2,3-dehydrosilydianin activates nrf2 and upregulates nad(p)h:Quinone oxidoreductase 1 in hepa1c1c7 cells. *Fitoterapia* **2017**, *119*, 115–120. [CrossRef] [PubMed]

43. Johnson, J.A.; Johnson, D.A.; Kraft, A.D.; Calkins, M.J.; Jakel, R.J.; Vargas, M.R.; Chen, P.C. The nrf2-are pathway: An indicator and modulator of oxidative stress in neurodegeneration. *Ann. N. Y. Acad. Sci.* **2008**, *1147*, 61–69. [CrossRef] [PubMed]

44. Lv, X.; Xia, Y.; Finel, M.; Wu, J.; Ge, G.; Yang, L. Recent progress and challenges in screening and characterization of ugt1a1 inhibitors. *Acta Pharm. Sin. B* **2018**. [CrossRef]

45. Kostrubsky, S.E.; Sinclair, J.F.; Strom, S.C.; Wood, S.; Urda, E.; Stolz, D.B.; Wen, Y.H.; Kulkarni, S.; Mutlib, A. Phenobarbital and phenytoin increased acetaminophen hepatotoxicity due to inhibition of udp-glucuronosyltransferases in cultured human hepatocytes. *Toxicol. Sci.* **2005**, *87*, 146–155. [CrossRef] [PubMed]

46. Li, S.; Fujino, M.; Takahara, T.; Li, X.K. Protective role of heme oxygenase-1 in fatty liver ischemia-reperfusion injury. *Med. Mol. Morphol.* **2018**. [CrossRef] [PubMed]

47. Bauer, M.; Bauer, I. Heme oxygenase-1: Redox regulation and role in the hepatic response to oxidative stress. *Antioxid. Redox Signal.* **2002**, *4*, 749–758. [CrossRef] [PubMed]

48. Chen, Y.H.; Chau, L.Y.; Chen, J.W.; Lin, S.J. Serum bilirubin and ferritin levels link heme oxygenase-1 gene promoter polymorphism and susceptibility to coronary artery disease in diabetic patients. *Diabetes Care* **2008**, *31*, 1615–1620. [CrossRef] [PubMed]

49. Bauer, I.; Vollmar, B.; Jaeschke, H.; Rensing, H.; Kraemer, T.; Larsen, R.; Bauer, M. Transcriptional activation of heme oxygenase-1 and its functional significance in acetaminophen-induced hepatitis and hepatocellular injury in the rat. *J. Hepatol.* **2000**, *33*, 395–406. [CrossRef]

50. Gao, Y.; Cao, Z.; Yang, X.; Abdelmegeed, M.A.; Sun, J.; Chen, S.; Beger, R.D.; Davis, K.; Salminen, W.F.; Song, B.J.; et al. Proteomic analysis of acetaminophen-induced hepatotoxicity and identification of heme oxygenase 1 as a potential plasma biomarker of liver injury. *Proteom. Clin. Appl.* **2017**, *11*, 1600123. [CrossRef] [PubMed]

51. Sharma, S.; Singh, R.L.; Kakkar, P. Modulation of bax/bcl-2 and caspases by probiotics during acetaminophen induced apoptosis in primary hepatocytes. *Food Chem. Toxicol.* **2011**, *49*, 770–779. [CrossRef] [PubMed]

52. Hu, J.; Yan, D.; Gao, J.; Xu, C.; Yuan, Y.; Zhu, R.; Xiang, D.; Weng, S.; Han, W.; Zang, G.; et al. Rhil-1ra reduces hepatocellular apoptosis in mice with acetaminophen-induced acute liver failure. *Lab. Investig.* **2010**, *90*, 1737–1746. [CrossRef] [PubMed]

53. Li, G.; Chen, J.B.; Wang, C.; Xu, Z.; Nie, H.; Qin, X.Y.; Chen, X.M.; Gong, Q. Curcumin protects against acetaminophen-induced apoptosis in hepatic injury. *World J. Gastroenterol.* **2013**, *19*, 7440–7446. [CrossRef] [PubMed]

54. Ramachandran, A.; Jaeschke, H. Mechanisms of acetaminophen hepatotoxicity and their translation to the human pathophysiology. *J. Clin. Transl. Res.* **2017**, *3*, 157–169. [CrossRef] [PubMed]

55. Vaux, D.L. Apoptogenic factors released from mitochondria. *Biochim. Biophys. Acta-Mol. Cell Res.* **2011**, *1813*, 546–550. [CrossRef] [PubMed]

56. Safa, A.R. C-flip, a master anti-apoptotic regulator. *Exp. Oncol.* **2012**, *34*, 176–184. [PubMed]

57. Zhao, Q.; Liu, C.; Shen, X.; Xiao, L.; Wang, H.; Liu, P.; Wang, L.; Xu, H. Cytoprotective effects of myristicin against hypoxiainduced apoptosis and endoplasmic reticulum stress in rat dorsal root ganglion neurons. *Mol. Med. Rep.* **2017**, *15*, 2280–2288. [CrossRef] [PubMed]

58. Abdel Moneim, A.E. The neuroprotective effects of purslane (portulaca oleracea) on rotenone-induced biochemical changes and apoptosis in brain of rat. *CNS Neurol. Disord. Drug Targets* **2013**, *12*, 830–841. [CrossRef] [PubMed]

59. Abdel Moneim, A.E. Indigofera oblongifolia prevents lead acetate-induced hepatotoxicity, oxidative stress, fibrosis and apoptosis in rats. *PLoS ONE* **2016**, *11*, e0158965. [CrossRef] [PubMed]

60. Sehitoglu, M.H.; Yayla, M.; Kiraz, A.; Oztopuz, R.O.; Bayir, Y.; Karaca, T.; Khalid, S.; Akpinar, E. The effects of apomorphine on paracetamol-induced hepatotoxicity in rats. *Cell. Mol. Biol.* **2017**, *63*, 40–44. [CrossRef] [PubMed]

61. El-Maddawy, Z.K.; El-Sayed, Y.S. Comparative analysis of the protective effects of curcumin and n-acetyl cysteine against paracetamol-induced hepatic, renal, and testicular toxicity in wistar rats. *Environ. Sci. Pollut. Res.* **2018**, *25*, 3468–3479. [CrossRef] [PubMed]

62. Reitman, S.; Frankel, S. A colorimetric method for the determination of serum glutamic oxalacetic and glutamic pyruvic transaminases. *Am. J. Clin. Pathol.* **1957**, *28*, 56–63. [CrossRef] [PubMed]

63. Shephard, M.D.; Peake, M.J. Quantitative method for determining serum alkaline phosphatase isoenzyme activity i. Guanidine hydrochloride: New reagent for selectively inhibiting major serum isoenzymes of alkaline phosphatase. *J. Clin. Pathol.* **1986**, *39*, 1025–1030. [CrossRef] [PubMed]

64. Schmidt, M.; Eisenburg, J. Serum bilirubin determination in newborn infants. A new micromethod for the determination of serum of plasma bilirubin in newborn infants. *Fortschr. Med.* **1975**, *93*, 1461–1466. [PubMed]

65. Ohkawa, H.; Ohishi, N.; Yagi, K. Assay for lipid peroxides in animal tissues by thiobarbituric acid reaction. *Anal. Biochem.* **1979**, *95*, 351–358. [CrossRef]

66. Green, L.C.; Wagner, D.A.; Glogowski, J.; Skipper, P.L.; Wishnok, J.S.; Tannenbaum, S.R. Analysis of nitrate, nitrite, and [15n]nitrate in biological fluids. *Anal. Biochem.* **1982**, *126*, 131–138. [CrossRef]

67. Ellman, G.L. Tissue sulfhydryl groups. *Arch. Biochem. Biophys.* **1959**, *82*, 70–77. [CrossRef]

68. Sun, Y.; Oberley, L.W.; Li, Y. A simple method for clinical assay of superoxide dismutase. *Clin. Chem.* **1988**, *34*, 497–500. [PubMed]

69. Luck, H. Catalase. In *Methods of Enzymatic Analysis*; Bergmeyer, H.U., Ed.; Academic Press: New York, NY, USA, 1965; pp. 855–888.

70. Paglia, D.E.; Valentine, W.N. Studies on the quantitative and qualitative characterization of erythrocyte glutathione peroxidase. *J. Lab. Clin. Med.* **1967**, *70*, 158–169. [PubMed]

71. Factor, V.M.; Kiss, A.; Woitach, J.T.; Wirth, P.J.; Thorgeirsson, S.S. Disruption of redox homeostasis in the transforming growth factor-alpha/c-myc transgenic mouse model of accelerated hepatocarcinogenesis. *J. Biol. Chem.* **1998**, *273*, 15846–15853. [CrossRef] [PubMed]

72. Almeer, R.S.; Abdel Moneim, A.E. Evaluation of the protective effect of olive leaf extract on cisplatin-induced testicular damage in rats. *Oxid. Med. Cell. Longev.* **2018**, *2018*, 11. [CrossRef] [PubMed]

International Journal of
Molecular Sciences

MDPI

Article

Heme Oxygenase 1 Impairs Glucocorticoid Receptor Activity in Prostate Cancer

Daiana B. Leonardi [1,2], Nicolás Anselmino [1,2], Javier N. Brandani [1,2], Felipe M. Jaworski [1,2], Alejandra V. Páez [1,2], Gisela Mazaira [2,3], Roberto P. Meiss [4], Myriam Nuñez [5], Sergio I. Nemirovsky [2,6], Jimena Giudice [7,8], Mario Galigniana [3,9], Adalí Pecci [6,10], Geraldine Gueron [1,2], Elba Vazquez [1,2,*] and Javier Cotignola [1,2,*]

[1] Laboratorio de Inflamación y Cáncer, Departamento de Química Biológica, Facultad de Ciencias Exactas y Naturales, Universidad de Buenos Aires, Buenos Aires C1428EGA, Argentina; daialeonardi@yahoo.com.ar (D.B.L.); nicoanselmino@hotmail.com (N.A.); jnbrandani@gmail.com (J.N.B.); felipejaw@gmail.com (F.M.J.); alejandravpaez@gmail.com (A.V.P.); ggueron@gmail.com (G.G.)
[2] Instituto de Química Biológica de la Facultad de Ciencias Exactas y Naturales (IQUIBICEN), CONICET Universidad de Buenos Aires, Buenos Aires C1428EGA, Argentina; gisela.mazaira@gmail.com (G.M.); sergio.nemirovsky@gmail.com (S.I.N.)
[3] Laboratorio de Biología Celular y Molecular, Departamento de Química Biológica, Facultad de Ciencias Exactas y Naturales, Universidad de Buenos Aires, Buenos Aires C1428EGA, Argentina; mgali@qb.fcen.uba.ar
[4] Departamento de Patología, Instituto de Estudios Oncológicos, Academia Nacional de Medicina, Buenos Aires C1425AUM, Argentina; rpmeiss@gmail.com
[5] Departamento de Físico-Matemática, Facultad de Farmacia y Bioquímica, Universidad de Buenos Aires, Cátedra de Matemática, Buenos Aires C1113AAD, Argentina; myr1710@yahoo.com
[6] Departamento de Química Biológica, Facultad de Ciencias Exactas y Naturales, Universidad de Buenos Aires, Buenos Aires C1428EGA, Argentina; pecciadali@gmail.com
[7] Department of Cell Biology and Physiology, University of North Carolina, Chapel Hill, NC 27599-7545, USA; jimena.giudice@gmail.com
[8] McAllister Heart Institute, University of North Carolina, Chapel Hill, NC 27599-7126, USA
[9] Laboratorio de Receptores Nucleares, Instituto de Biología y Medicina Experimental (IBYME), CONICET, Buenos Aires C1428ADN, Argentina
[10] Biología Molecular y Neurociencias (IFIBYNE), Instituto de Fisiología, CONICET—Universidad de Buenos Aires, Buenos Aires C1428EGA, Argentina
[*] Correspondence: elba@qb.fcen.uba.ar (E.V.); jcotignola@qb.fcen.uba.ar (J.C.)

Received: 30 January 2019; Accepted: 21 February 2019; Published: 26 February 2019

Abstract: Glucocorticoids are used during prostate cancer (PCa) treatment. However, they may also have the potential to drive castration resistant prostate cancer (CRPC) growth via the glucocorticoid receptor (GR). Given the association between inflammation and PCa, and the anti-inflammatory role of heme oxygenase 1 (HO-1), we aimed at identifying the molecular processes governed by the interaction between HO-1 and GR. PCa-derived cell lines were treated with Hemin, Dexamethasone (Dex), or both. We studied GR gene expression by RTqPCR, protein expression by Western Blot, transcriptional activity using reporter assays, and nuclear translocation by confocal microscopy. We also evaluated the expression of HO-1, FKBP51, and FKBP52 by Western Blot. Hemin pre-treatment reduced Dex-induced GR activity in PC3 cells. Protein levels of FKBP51, a cytoplasmic GR-binding immunophilin, were significantly increased in Hemin+Dex treated cells, possibly accounting for lower GR activity. We also evaluated these treatments in vivo using PC3 tumors growing as xenografts. We found non-significant differences in tumor growth among treatments. Immunohistochemistry analyses revealed strong nuclear GR staining in almost all groups. We did not observe HO-1 staining in tumor cells, but high HO-1 reactivity was detected in tumor infiltrating macrophages. Our results suggest an association and crossed modulation between HO-1 and GR pathways.

Int. J. Mol. Sci. **2019**, *20*, 1006

Keywords: glucocorticoid receptor; GR; heme oxygenase 1; HO-1; prostate cancer

1. Introduction

Glucocorticoid receptor (GR) activation, mediated by the binding of glucocorticoids, regulates the expression of inflammatory factors. Many chaperones are involved in the trafficking and turnover of this receptor. Cytoplasmic GR forms heterocomplexes with the Hsp90-binding immunophilin FKBP51 [1]. Upon steroid binding, FKBP51 is released from GR and replaced by FKBP52 which, in turn, recruits dynein/dynactin [1]. Therefore, this immunophilin exchange assemblies the molecular machinery for the efficient and fast nuclear transport of GR.

In prostate cancer (PCa), glucocorticoids are usually used to counteract the pain associated with bone metastasis and the toxic effects triggered by therapy. On the other hand, GR is upregulated in castration-resistant tumors and its over-expression is involved in the development of abiraterone and enzalutamide resistance [2], bypassing androgen receptor (AR) blockage and regulating a subset of distinguishable AR target genes [3]. The GR upregulation can be explained by the lack of AR binding to a negative androgen response element in the GR promoter [4].

Heme oxygenase 1 (HO-1), the rate limiting enzyme of heme degradation, is of vital importance in the inflammatory response [5]. Although in many tumors its expression has been associated with progression and increased aggressiveness, our group showed that high HO-1 expression inhibits PCa cell proliferation, migration, and invasion in vitro [6], and prevents tumor growth and angiogenesis in vivo [6,7]. Similar results were reported for other tumors [5]. We also showed that HO-1 can negatively modulate AR transcriptional activity by interfering with STAT3 signaling [8]. Additionally, HO-1 exhibits nuclear localization in different tumors [9–12]; however, the role of nuclear HO-1 is still unknown.

The current literature on HO-1 and GR association is controversial and non-conclusive. Lutton et al. showed that HO-1 was under-expressed when hepatocarcinoma cells were treated with Dexamethasone [13]. Additionally, Lavrovsky et al. revealed the existence of a negative glucocorticoid response element in the HO-1 promoter [14] and that glucocorticoid-treated endothelial cells had lower HO-1 expression [15]. On the other hand, an upregulation of HO-1 by Dexamethasone was shown in monocytes [16] and larynx carcinoma cells [17].

The aim of this study was to analyse the interaction between GR and HO-1 and to identify the molecular processes governed by the association between these two proteins in PCa.

2. Results

2.1. NR3C1 Expression and Signaling is Modulated by Hemin Treatment

We first analyzed whether glucocorticoids or hemin treatments affected PC3 (GR+/AR−) and C4-2B (GR+/AR+) viability. Hemin, a potent inducer of HO-1, significantly decreased the viability of both cell lines, either in the presence or absence of Dexamethasone (Figure 1A). Hemin-induced HO-1 over-expression was confirmed by Western Blot and RTqPCR (Figure 1B and Supplemental Figure S1, respectively). *NR3C1* (GR gene) mRNA expression was significantly decreased under Hemin and Hemin+Dexamethasone treatments in both cell lines, while it did not change under Dexamethasone treatment (Supplemental Figure S1). However, we observed that GR protein levels were significantly elevated by more than 3-fold in cells exposed to Dexamethasone (Figure 1B). The higher protein levels detected, even when *NR3C1* mRNA levels were similar or lower to control, might be due to lower GR proteasome degradation, as was previously demonstrated [18].

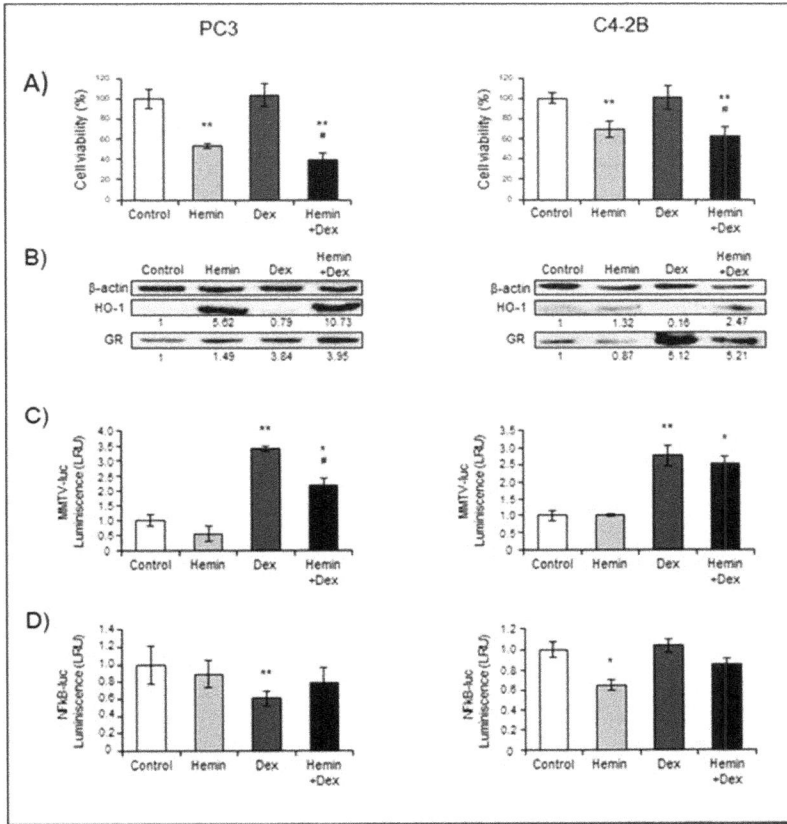

Figure 1. Hemin treatment modulates Dexamethasone-induced GR expression and signaling. PC3 and C4-2B cells were treated with Hemin (80 μM for 24 h), Dexamethasone (Dex; 10 nM for 6 h post Hemin/PBS 24-h treatment), the combination of both drugs, or PBS as control. (**A**) MTS viability assay was performed and results are presented as percentage of viable cells compared to control (100%). (**B**) Western blot analysis showing HO-1, GR, and β-actin as loading control. Protein quantification was performed by densitometry analysis using ImageJ software. The numbers under the bands indicate the quantitation normalized to β-actin and control lane. One representative experiment is shown. Panels C and D depict reporter assays. Cell lines were transiently transfected with the MMTV-luc (**C**) or NFkB-luc (**D**) reporter plasmids, and after treatments, cells were lysed and luciferase activity assay was performed. Data were normalized to total protein values. Results are shown as mean ± SEM from at least three independent experiments; * $p < 0.05$ and ** $p < 0.01$ versus control cells, # $p < 0.05$ versus Dex treated cells.

We further analyzed the expression levels of *IKBA* and *BCLXL*, two GR downstream targets, to infer GR activity. Hemin significantly downregulated the transcription of both genes in PC3 independently of Dexamethasone treatment (Supplemental Figure S1), while in C4-2B this repressive effect was observed only on the Dexamethasone-induced condition (Supplemental Figure S1).

In order to further investigate the transcriptional activity of GR, we used a reporter plasmid (MMTV-luc) with GR response elements. Dexamethasone-induced luciferase activity was confirmed in both cell lines (Figure 1C), but hormone-activation was partially abrogated in PC3 cells pre-treated with Hemin (Figure 1C, left panel), albeit the protein levels were similar to those in Hemin-alone treated cells. Additionally, we studied the GR-mediated transrepression activity on NFkB pathway, using a luciferase reporter plasmid (NFkB-luc) that responds to this pro-inflammatory factor [7]. Dexamethasone inhibited luciferase activity in PC3 cells, thus reflecting the higher GR-transrepressive effect provoked by the glucocorticoid (Figure 1D), which was reversed when cells were pre-treated with Hemin. This effect was not observed in C4-2B (Figure 1D). However, in this cell line, Hemin reduced NFkB-luc activity under the experimental conditions assayed (Figure 1D).

Given our previous data showing that HO-1 interacts with and modulates STAT3 [8], a critical transcription factor in PCa, we sought to test whether a direct protein association exists between GR and HO-1. Co-immunoprecipitation assays in PC3 cells suggested that HO-1 and GR interact (Figure 2).

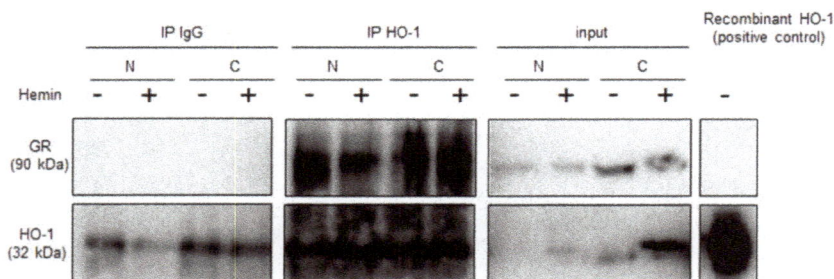

Figure 2. HO-1 associates to GR in PC3 cells. PC3 cells were treated with Hemin (80 μM, 24 h) or PBS as control. Nuclear and cytoplasmic compartments were isolated. Cell extracts were immunoprecipitated using an anti-HO-1 polyclonal antibody or IgG as negative control. Complexes were analysed by SDS-polyacrylamide gel electrophoresis and immunoblot assay with anti-GR and anti-HO-1 antibodies.

We previously reported that Hemin treatment retained STAT3 in the cytoplasm [8], therefore, here we analyzed GR subcellular localization under HO-1 induced expression. Analysis of confocal immunofluorescence microscopy confirmed HO-1 over-expression under Hemin treatment and a higher nuclear localization of this protein (Figure 3). GR expression and its nuclear localization increased under Dexamethasone treatment, and this effect was not impaired by Hemin pre-treatment (Figure 3).

Altogether, these results suggest that HO-1 modulates GR signaling in PCa cells without interfering with GR nuclear translocation.

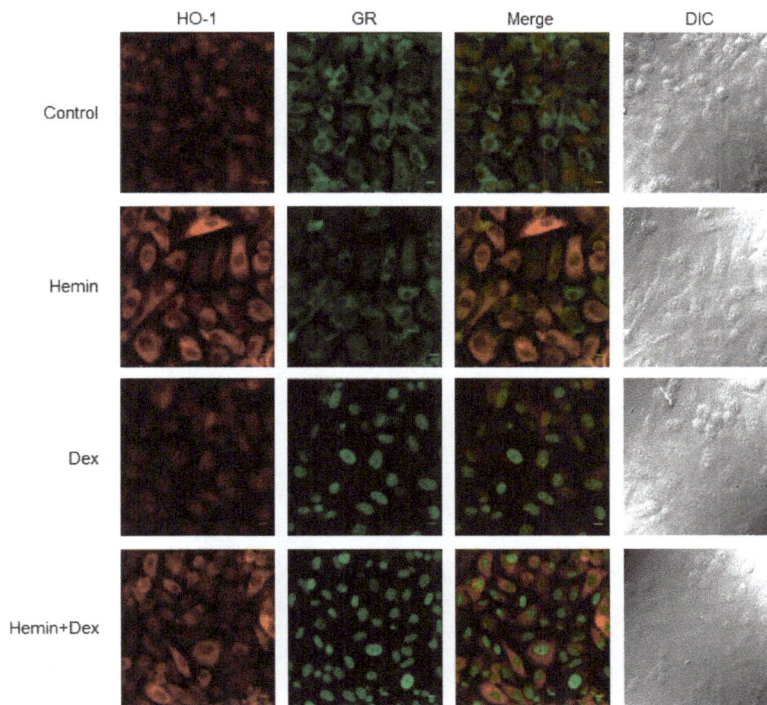

Figure 3. Analysis of HO-1 and GR subcellular localization. Immunofluorescence analysis of HO-1 and GR expression and localization in PC3 cells treated with Hemin (80 µM for 24 h), Dex (10 nM for 6 h post Hemin/PBS 24-h treatment), and the combination of both drugs or PBS as control. Cells were fixed, stained with anti-HO1, and anti-GR primary antibodies and secondary antibodies conjugated with Alexa Fluor 488 (red, HO-1) and 555 (green, GR) antibodies. Cells were imaged by confocal microscopy using the same parameters for all the treatments. A representative image for each condition is shown. Final magnification: ×60.

2.2. Identification of Glucocorticoid Response Elements in HO-1 Proximal Promoter

Confocal Immunofluorescent microscopy revealed altered HO-1 expression in Hemin+Dexamethasone treated cells compared to Hemin alone treatment (Figure 3). We performed an *in-silico* analysis of the HO-1 promoter region (estimated at Chr22:35379360–35380560) to identify glucocorticoid response elements (GRE). As shown in Supplemental Table S1, HO-1 proximal promoter does not contain consensus GRE sequences. However, we cannot rule out the presence of other GRE in distant regions, such as enhancers.

2.3. Hemin Treatment Increases FKBP51 Expression in the Presence of Dexamethasone

Strong evidence suggest that FKBP51 and FKBP52 have a role in the modulation of GR activity and glucocorticoid-dependent translocation of GR from the cytosol to the nucleus [1]. Western blot analysis revealed a significant increase of FKBP51 in cells under HO-1 induction and GR stimulation with respect to cells that received only single treatments or vehicle (Figure 4A). Furthermore, Hemin+Dexamethasone treatment triggered a higher FKBP51/52 expression ratio (Figure 4B).

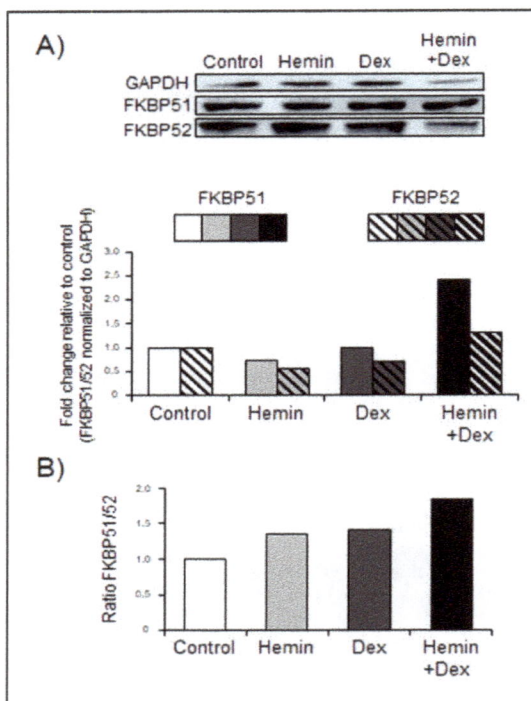

Figure 4. Hemin increases FKBP51 expression under Dexamethasone stimulation. (**A**) Western blot analysis showing FKBP51 and FKBP52 expression in PC3 cells treated with Hemin (80 μM for 24 h), Dex (10 nM for 6 h post Hemin/PBS 24-h treatment), the combination of both drugs, or PBS as control. Total protein was extracted and protein expression was analyzed by western blotting using specific antibodies. GAPDH levels are shown as control for equal loading. Protein quantification was performed by densitometry analysis using ImageJ software and bands were normalized to GAPDH and control. (**B**) FKBP51/FKBP52 ratio was calculated for each condition. One representative from at least three independent experiments is shown.

2.4. Study of Hemin and/or Dexamethasone Treatment in PC3 Xenografts

Given that Dexamethasone is frequently used as a palliative treatment in patients with PCa, and the relevance of inflammation in this disease, we used a human PCa xenograft model to investigate the effect of Hemin, Dexamethasone, or both on tumor growth. No significant alterations were observed in the body weight of the animals in the different groups (Supplemental Figure S2). For each treatment ($n = 7$), the exponential regression of tumor volume curve was plotted and duplication time for each condition was calculated (Figure 5A). Non-significant differences were observed among treatments in the tumor growth nor in *HMOX1*, *BCLXL*, *IKBA*, and *MKI67* gene expression (Supplemental Figure S2). As previously reported in vitro [19], a significant down-regulation of GR mRNA levels was detected in animals of the Dexamethasone group (Supplemental Figure S2).

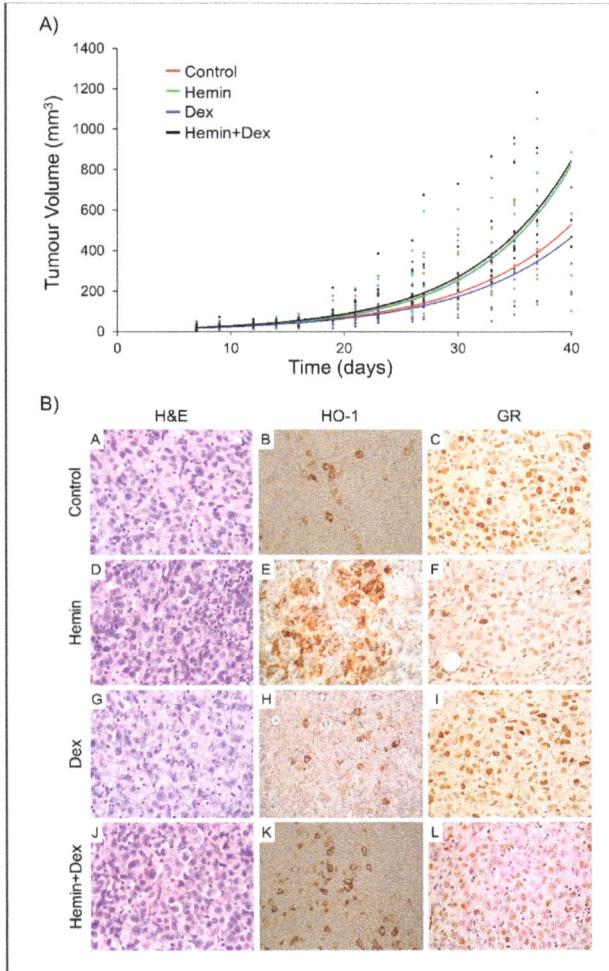

Figure 5. In vivo effect of Hemin and/or Dexamethasone on PC3 xenografts growth. Six- to eight-week-old male athymic nude (nu/nu) mice were randomized into four groups (*n* = 7 per group). PC3 cells (3.6 × 10^6) were injected *s.c.* in the right flank. When tumors reached a volume of around 150 mm^3, animals were *i.p.* injected every 48 h with 6 doses of Hemin (25 mg/kg), Dexamethasone (0.2 mg/kg), Hemin+Dexamethasone (same doses of individual treatments), or PBS (control). (**A**) Exponential regression of tumor volume was calculated for each treatment according to the volume measured, as described in Materials and Methods along the experimental procedure. (**B**) Histological (H&E, left panel) and immunohistochemical analysis of paraffin-embedded tumor sections obtained from treated or control mice at the experimental end point. HO-1 and GR immunohistochemical analysis were performed using specific antibodies: HO-1 negative immunostaining in tumor cells and positive HO-1 reactivity in macrophages (central panel) and GR-positive nuclear immunostain (right panel). Final magnification: H&E × 250, HO-1 × 250, GR × 100. One representative image of each condition is shown.

Histological analysis at the time of euthanasia revealed the presence of poorly differentiated carcinomas, with irregular nuclei, prominent nucleoli, and aberrant mitotic images (Figure 5B). Immunohistochemical staining showed negative HO-1 expression in the tumor cells in all the groups. However, positive HO-1 expression was detected in macrophages in all the animals with higher immunoreactivity in Hemin, Dexamethasone, and Hemin+Dexamethasone groups (Figure 5B). Nuclear GR staining was detected in all cases; nevertheless, different intensities were observed—strong in control and Dexamethasone treated animals, weak to moderate in the Hemin group and weak in the co-treated group (Figure 5B).

2.5. High NR3C1 and HMOX1 Expression Reduces PCa-Patient Disease-Free Survival

To evaluate whether *NR3C1* and *HMOX1* tumoral expression have clinical implications, we analyzed disease-free survival and relapse-free survival using data from public repositories. RNAseq data showed non-significant differences in survival when both genes were examined individually (Figure 6A,B). However, when the combined expression of *NR3C1* and *HMOX1* was considered, we found that patients with high expression of both genes have significantly worse disease-free survival (HR = 9.3, 95% CI = 2.0–42.9, $p = 0.004$; Figure 6C).

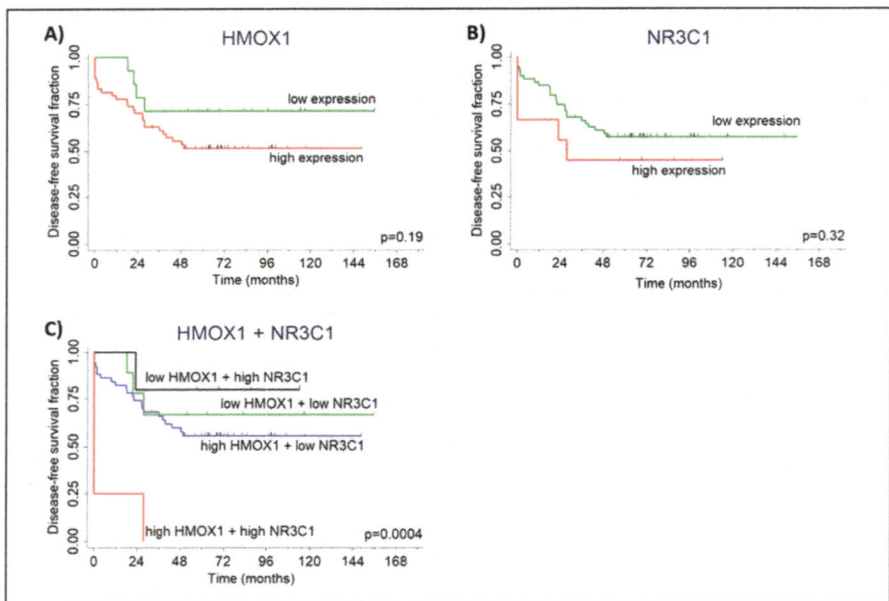

Figure 6. *NR3C1* and *HMOX1* high expression reduces PCa-patient disease-free survival. Kaplan Meier plot showing groups with low and high *HMOX1* (**A**) and *NR3C1* (**B**) expression according to ROC curve threshold. (**C**) Kaplan Meier plot showing groups with low and high *HMOX1* and *NR3C1* combined expression according to ROC curve threshold. Patients with high expression of both genes have shorter disease-free survival compared to other groups (log-rank $p = 0.0004$). Vertical marks show censored patients.

3. Discussion

In this study, we report for the first time that Hemin negatively modulates GR activity induced by Dexamethasone in PCa cells (PC3, GR+/AR−) (Figure 1). The reduced GR activity observed under Hemin+Dexamethasone treatment could be explained by the increased expression of FKBP51

(Figure 4). These findings suggest a crosstalk between HO-1 and GR pathways in PC3 cells. Furthermore, co-immunoprecipitation analysis suggested a physical interaction between these two proteins. Overall, we demonstrated that HO-1 alters GR signaling in PCa cells without impairing GR nuclear translocation.

We have previously shown the direct interaction between HO-1 and STAT3, the cytoplasmic retention of this transcription factor, and the consequent down-regulation of AR transcriptional activity [8]. HO-1 was also shown to interact with other proteins [20,21]. In this study, we demonstrated that Hemin treatment is able to impair GR activity in vitro, even though GR is able to translocate to the nucleus in the presence of Dexamethasone. This observation agrees with results already reported showing a FKBP51-dose-dependent inhibition of GR activity [22] partially mediated by the reduction of hormone binding [23]. Furthermore, it was demonstrated that in WCL2 cells that constitutively over-expresses GR, this receptor was localized in the nucleus, even without steroid stimulation [24]. Since GR activity and subcellular localization were performed at 6 h post Dexamethasone treatment, we cannot rule out that HO-1 interferes with GR nuclear translocation kinetics at shorter times (<2 h). In agreement with this, it was previously reported that geldanamycin (Hsp90 inhibitor) reduced the mineralocorticoid receptor nuclear translocation [24]. At longer times, hormone receptors could translocate to the nucleus in a cytoskeleton-independent manner.

HO-1 is one of the NRF2 downstream targets [25] and NRF2 activity is modulated by nuclear receptors, including GR [26]. It was reported that GR signaling represses NRF2-dependent transcriptional activation [27]. Accordingly, we observed that HO-1 mRNA levels were lower under Dexamethasone treatment (Figure 1B). However, Dexamethasone-mediated repression was not detected when cells were pretreated with hemin, probably due to the high levels of HO-1 reached before the glucocorticoid treatment.

Although GR and HO-1 expression profile were similar in C4-2B (GR+/AR+) and PC3 (GR+/AR−) in all treatments (Supplemental Figure S1), GR activity induced by Dexamethasone was not significantly altered in Hemin-treated C4-2B cells (Figure 1). The lack of effect could be explained by the presence of a mutant AR, as it is well known that MMTV-luc reporter plasmid can also be activated by AR [28]. Moreover, it was recently documented that PCa cells with high AR expression have lower GR dependency [29].

Dexamethasone is frequently used as a palliative treatment in patients with PCa [30]. Considering that Hemin modulates GR activity in vitro, we assessed the effect of this protoporphyrin and Dexamethasone co-treatment using PC3-derived tumors growing as xenografts. In accordance with a previous report [31], no significant differences in tumor growth was observed when animals received Dexamethasone. On the other hand, DU145 xenografts treated with peritumoral injections of Dexamethasone showed a significantly lower growth rate compared to the control group [32]. Furthermore, Hemin treatment had no effect on tumor growth when administered, either alone or in combination with the glucocorticoid. It is worth mentioning that we recently reported that Hemin conditioning prior to tumor challenge resulted in a significant increase in tumor latency compromising the tumor vascularization and the immune response [33]. However, these results were obtained using a pre-clinical model of PCa in immunocompetent mice, another PCa cell line was used, and drugs were administered with different protocols [32,33].

Immunohistochemistry analysis of PC3 xenografts revealed negative HO-1 staining in tumor cells demonstrating that Hemin *i.p.* treatment did not induce HO-1 expression in tumor epithelial cells. However, we observed strong positive HO-1 immunoreactivity was detected in tumor infiltrating macrophages (Figure 5). In agreement with these results, analyses of human prostate carcinomas and benign prostatic hyperplasia samples showed HO-1 positive staining in stromal and infiltrating immune cells [11,34]. HO-1 expression was also reported in extra tumoral macrophages and associated with tumor aggressiveness [34], metastatic behavior of PC3 xenografts [35], and tumor development and progression [34]. In addition, HO-1 positivity was almost exclusively seen in macrophages at the tumor invasive front in high-grade tumors from human samples [34].

Strong nuclear heterogeneous GR immunostaining was observed in Dexamethasone-treated animals and its intensity decreased in the group receiving both agents (Figure 5). Recently, other authors reported that GR is down-regulated in PCa tissue from patients sensitive to enzalutamide and abiraterone treatment compared to normal prostate [2,3]. They also described that GR expression was higher in enzalutamide/abiraterone-resistant CRPC samples [2,3].

Finally, we analyzed RNAseq data from a public repository and found that patients with high expression of *NR3C1* and *HMOX1* have shorter disease-free survival than patients with low expression of both genes. Similarly, Puhr et al. showed that patients expressing GR high levels have shorter biochemical relapse free survival [29]. However, we cannot rule out that this result was confounded by inflammation.

In summary, we demonstrated the importance of GR signaling in PCa and provided evidence about the association between GR and HO-1. Further studies will elucidate the therapeutic potential of targeting GR/HO-1 pathways in PCa therapy.

4. Materials and Methods

4.1. In Vitro Experiments

4.1.1. Cell Culture and Treatments

C4-2B and PC3 cells were cultured in RPMI 1640 (Invitrogen, Buenos Aires, Argentina) supplemented with 10% FBS (fetal bovine serum) and antibiotics (penicillin 100 U/mL, streptomycin 100 µg/mL, and amphotericin 0.5 µg/mL). Cultures were maintained at 37 °C in a humidified incubator with a 5% CO_2.

Cells were incubated 24 h in RPMI media containing 10% charcoaled FBS before they were exposed to either Hemin (80 µM, 24 h), Dexamethasone (10 nM, 6 h post Hemin/PBS treatment), the combination of both drugs, or PBS as control. Dexamethasone and Hemin were purchased from Sigma-Aldrich (St. Louis, MO, USA).

Cell viability was measured using CellTiter 96® AQueous One Solution Cell Proliferation Assay (Promega, Madison, WI, USA).

4.1.2. Transfections and Luciferase Reporter Assay

C4-2B and PC3 cells were seeded on 12-well plates (90% confluence) and transiently transfected with luciferase plasmid (MMTV-luc or NFkB-luc; 2 µg) using PEI (Polyethyilenimine, Polysciences Inc., Warrington, PA, USA; 4 µg). Luciferase activity was determined by the Luciferase Assay System (Promega, Madison, WI, USA) in a Glomax luminometer (Promega, Madison, WI, USA). Transfections were performed in triplicate and each experiment was repeated at least three times. Data were normalized to total protein determined by Bradford assay.

4.1.3. RNA Isolation and Reverse Transcription–Quantitative PCR (RTqPCR)

Total RNA was isolated with Quick-Zol (Kalium technologies, Argentina) according to the manufacturer's protocol. The cDNAs were synthesized with RevertAid Premium First Strand cDNA Synthesis Kit (Fermentas, Waltham, MA, USA) and used for real-time PCR amplification with Taq DNA Polymerase (Invitrogen, Waltham, MA, USA) in a Stratagene MX3000P (Agilent Technologies, Santa Clara, CA, USA). Primers sequences (5'->3') used were: *NR3C1* (GR): TAT CTC GGC TGC GGC GGG AA and AGC GAC AGC CAG TGA GGG TGA; *HMOX1* (HO-1): ACT GCG TTC CTG CTC AAC AT and GGG GCA GAA TCT TGC ACT TT; *IkB*: ACC ATG GAA GTG ATC CGC CAG G and AGC TCC CAG AAG TGC CTC AGC AA; *PPIA*: CCC ATT TGC TCG CAG TAT CCT AGA and GGC ATG GGA GGG AAC AAG GAA AAC; *BCLXL*: GGT ATT GGT GAG TCG GAT CG and TTC CAC AAA AGT ATC CCA GC; *MKI67*: GCC AGC ACG TCG TGT CTC AAG AT and ACA CTG TCT TTT GAG TCA TCT GCG G.

Data were analyzed by Mx3000P software and normalized to the reference gene *PPIA* and control group. Data were analyzed using de $2^{-\Delta\Delta Ct}$ method [36].

4.1.4. Immunoblot Analysis and Antibodies

Total cell lysates and immunoblot analysis were carried out as previously described [6]. Briefly, cells were lysed with RIPA buffer (Tris HCl 50 mM pH 7.4; NaCl 150 mM; EDTA 20 mM pH 8; sodium deoxycholate 1%; SDS 0.1%; Triton X-100 1%, 1 mM Na_3VO_4, 20 mM NaF and 1 mM $Na_4P_2O_7$, pH 7.9) and homogenized. After 20 min of incubation at 4 °C, the lysates were centrifuged at $12,000 \times g$ for 20 min at 4 °C and the supernatant kept at -80 °C. Lysates containing equal amounts of proteins (50 μg) were resolved on 7.5–12.5% SDS–PAGE depending on the molecular weight of the proteins under study. PageRuler Plus Prestained Protein Ladder (Fermentas, Waltham, MA, USA) was used for the estimation of molecular weight. The proteins were blotted to a Hybond-ECL nitrocellulose membrane (GE Healthcare, Little Chalfont, UK). Membranes were blocked with 5% dry non-fat milk in TBS containing 0.1% Tween 20 (TBST) for 1 h at room temperature, and incubated with primary antibodies diluted in TBST for 1 h at room temperature. Membranes were then incubated with horseradish peroxidase-labelled secondary antibody for 1 h at room temperature.

The following antibodies were used: monoclonal anti–HO-1 (catalogue 13248. Abcam, UK), monoclonal anti-GR (catalogue 12041. Cell Signaling, Danvers, MA, USA), monoclonal anti-FKBP51 (catalogue PA1-020, Affinity BioReagents, Golden, CO, USA), polyclonal anti-FKBP52 (catalogue UP30. Pharmacia and Upjohn, Inc., Pfizer, New York, NY, USA), anti–β-actin antibody (catalogue 3700. Cell Signaling, Danvers, MA, USA), and anti-mouse and anti-rabbit secondary antibodies (catalogue 7076S and catalogue 7074, respectively. Cell Signaling, Danvers, MA, USA).

4.1.5. Co-Immunoprecipitation

PC3 cells were treated as described above and harvested in lysis buffer (20 mM Tris HCl, pH 8; 137 mM NaCl; 10% glycerol; 1% Triton X-100; 2 mL de EDTA 0.5 M, 1x protease inhibitor mixture (Sigma-Aldrich, St. Louis, MO, USA), 100 μg/mL PMSF, 20 mM NaF, and 1 mM Na_3VO_4). Proteins (500 μg) in lysis buffer were incubated overnight at 4 °C with 8 μg of anti–HO-1 antibody. Protein G Agarose beads (Invitrogen, Waltham, MA, USA) were added to each tube for 3 h at 4 °C. Beads were washed with ice-cold lysis buffer. Fifty micrograms of the lysate was used as input. Immune complexes were analyzed by immunoblot with anti-GR and anti–HO-1 antibodies.

4.1.6. Immunofluorescence and Microscopy

Cells were seeded in 12-well plates at a density of 1×10^5 cells per well on coverslips overnight. Cells were treated as described above and were fixed in ice-cold methanol and permeabilized for 10 min with 0.5% Triton X-100/PBS, washed with PBS, and then blocked with 5% BSA/PBS. Cells were incubated overnight with primary antibodies diluted in 4% BSA and 0.1% Tween 20 in PBS. Cells were then washed with PBS and incubated with fluorescent secondary antibodies Alexa Fluor 488 goat anti-mouse and Alexa Fluor 555 goat anti-rabbit antibodies were from Molecular Probes (Invitrogen, Waltham, MA, USA). Negative controls were carried out using PBS instead of primary antibodies. Cells were washed, mounted, and imaged by confocal laser scanning microscopy, which was performed with an Olympus Fluo view FV 1000 microscope, using an Olympus 60x/1.20 NA UPLAN APO water immersion objective.

4.1.7. Statistical Analysis

GraphPad Prism software was used for statistical analysis and results are shown as mean $\pm$ standard error (SEM) of at least three independent experiments, unless stated otherwise. Student's *t*-test or Mann-Whitney U-test were used to compare two experimental groups. For multiple comparisons, one-way ANOVA tests were performed.

4.2. In Vivo Experiments

4.2.1. Human PCa Xenograft Model

Six- to eight-week-old male athymic nude (nu/nu) mice, each weighing at least 20 g, were purchased from Cátedra de Animales de Laboratorio y Bioterio of Facultad de Ciencias Veterinarias, UNLP, La Plata, Argentina. Mice were used in accordance with the Guidelines for the Welfare of Animals in Experimental Neoplasia [37]. The protocol was approved by the Ethical Committee (CICUAL N°46). Mice were randomly assigned to four groups of seven animals each: Hemin, Dexamethasone, Hemin+Dexamethasone, or control (PBS). PC3 cells (3.6×10^6 in 200 µL of RPMI) were injected *s.c.* (subcutaneously) in the right flank of mice using a 22G needle. Tumors were measured with a caliper starting at day 8 after inoculation, when the tumors became detectable under the skin. Tumor volumes were calculated using the formula (length $\times$ width^2)/2. When tumors reached a volume around 150 mm^3, the animals were *i.p.* (intraperitoneal) injected every 48 h with 6 doses of the following treatments: Hemin (25 mg/kg), Dexamethasone (0.2 mg/kg), Hemin+Dexamethasone (same doses that individual treatments), or PBS (control). Animals were sacrificed 24 h after the last dose. Resected tumors were divided, one piece was immediately placed in Quick-Zol (Kalium Technologies SRL, Bernal, Buenos Aires, Argentina) for RNA isolation, and the remaining piece was fixed in PFA 10% for immunohistochemistry staining.

4.2.2. Immunohistochemical Analyses

Immunohistochemical techniques were performed as previously described [6,11]. Briefly, immunohistochemistry was done using the streptavidin-biotin-peroxidase complex system LSAB + kit, horseradish peroxidase (DAKO, Santa Clara, CA, USA). Endogenous peroxide activity was quenched using hydrogen peroxide in distilled water (3%). Antigen retrieval was done by microwaving. Tissue slides were incubated overnight with GR and HO-1 primary antibodies. This was followed by sequential incubations with biotinylated antibody and peroxidase-labelled streptavidin complex. The peroxidase reaction was conducted, under microscope, using 3,3'-diaminobenzidine. Slides were counterstained with Mayer's hematoxylin and analyzed by standard light microscopy. Negative control slides were prepared by substituting primary antiserum with PBS. For semiquantitative analysis, the degree of staining was scored as high, moderate, low, or not detectable (3+, 2+, 1+, and 0, respectively); staining localization was also recorded. We considered positive expression when more than 10% cells exhibited positive staining.

4.2.3. Statistical Analysis

To analyse the difference between treatments and experiments, we used a three-factor analysis of variance with repeated measures in the time factor. The considered factors were: treatment, time, and inter-experiment variation. Tumor volume was used as dependent variable. Wilcoxon test was employed to determine if there were differences in tumor growth among treatments.

4.3. In Silico Analyses

4.3.1. HMOX1 Promoter Analysis

Promoter sequence was estimated between positions -1000 and $+200$ of *HMOX1*. Gene location and sequence belonging to the GRCh38 version of the human genome were obtained from Ensembl (http://www.ensembl.org) (access on august 2015). Possible binding of GR to the promoter sequence was assessed with the command-line version of the FIMO tool from the MEME Suit [38], using DNA binding motifs from the Cistrome platform [39] (IDs: EN0252, M00205 and MC00033) and the JASPAR database [40] (IDs: MA0113.1 and MA0113.2).

4.3.2. Analysis of Human Tumor Samples

The public repository from the European Bioinformatics Institute (EBI, EMBL), Wellcome Genome Campus, Hinxton, Cambridge, UK (www.ebi.ac.uk) (access on august 2016), was browsed for studies analyzing PCa tissues. One study that included transcriptome data from 100 tumors obtained by surgery was selected (E-GEOD-54460—RNAseq Analysis of Formalin-Fixed Paraffin-Embedded Prostate Cancer Tissues" [41]). The clinico-pathological variables available were: biochemical relapse, time to relapse, pre-surgical PSA levels, pathological T stage, Gleason score, and surgical margin involvement.

Kaplan-Meier curves to study disease-free survival and biochemical relapse-free survival were performed. For these analysis, we dichotomized the patients according to high or low *NR3C1* and *HMOX1* expression. We performed Receiver Operating Characteristics (ROC) curves to determine gene expression cutoffs with the best sensitivity and specificity to predict biochemical relapse. The comparison between the groups was done with the Log-rank test. We estimated the hazard ratios (HR), 95 % confidence intervals (95% IC), and *p*-values using Cox proportional hazard models. Differences between the groups were considered significant if *p*-value ≤ 0.05. Analysis were performed using Stata v14 (StataCorp, College Station, TX, USA).

Supplementary Materials: Supplementary materials can be found at http://www.mdpi.com/1422-0067/20/5/1006/s1. Figure S1. Gene expression profile in PCa cells treated with Hemin or Dexamethasone. Figure S2. In vivo experiments. Table S1. Predicted GR binding sites in *HMOX1* promoter region.

Author Contributions: D.B.L., E.V., and J.C., conceptualization, original draft preparation. and writing. D.B.L., N.A., J.N.B., F.M.J., A.P., and G.M., methodology. D.B.L., investigation. R.P.M., histopathological and immunohistochemistry formal analysis. M.N., statistical formal analysis of in vivo experiments. S.I.N., bioinformatic formal analysis. J.G., formal analysis of immunofluorescence images. D.B.L., M.G., A.P., G.G., E.V., and J.C., formal analysis. G.G., E.V., and J.C., funding acquisition. E.V. and J.C., supervision. All authors participated in reviewing the manuscript.

Acknowledgments: This study was supported by Agencia Nacional de Promoción Científica y Tecnológica (ANPCyT), Consejo Nacional de Investigaciones Científicas y Técnicas (CONICET), Instituto Nacional del Cáncer (INC), and Universidad de Buenos Aires (UBA).

Conflicts of Interest: The authors declare no conflict of interest.

References

1. Galigniana, M.D.; Echeverria, P.C.; Erlejman, A.G.; Piwien-Pilipuk, G. Role of molecular chaperones and TPR-domain proteins in the cytoplasmic transport of steroid receptors and their passage through the nuclear pore. *Nucleus* **2010**, *1*, 299–308. [CrossRef] [PubMed]
2. Arora, V.K.; Schenkein, E.; Murali, R.; Subudhi, S.K.; Wongvipat, J.; Balbas, M.D.; Shah, N.; Cai, L.; Efstathiou, E.; Logothetis, C.; et al. Glucocorticoid receptor confers resistance to antiandrogens by bypassing androgen receptor blockade. *Cell* **2013**, *155*, 1309–1322. [CrossRef] [PubMed]
3. Shah, N.; Wang, P.; Wongvipat, J.; Karthaus, W.R.; Abida, W.; Armenia, J.; Rockowitz, S.; Drier, Y.; Bernstein, B.E.; Long, H.W.; et al. Regulation of the glucocorticoid receptor via a BET-dependent enhancer drives antiandrogen resistance in prostate cancer. *eLife* **2017**, *6*. [CrossRef] [PubMed]
4. Xie, N.; Cheng, H.; Lin, D.; Liu, L.; Yang, O.; Jia, L.; Fazli, L.; Gleave, M.E.; Wang, Y.; Rennie, P.; et al. The expression of glucocorticoid receptor is negatively regulated by active androgen receptor signaling in prostate tumors. *Int. J. Cancer* **2015**, *136*, E27–E38. [CrossRef] [PubMed]
5. Jozkowicz, A.; Was, H.; Dulak, J. Heme oxygenase-1 in tumors: Is it a false friend? *Antioxid. Redox Signal.* **2007**, *9*, 2099–2117. [CrossRef] [PubMed]
6. Gueron, G.; De Siervi, A.; Ferrando, M.; Salierno, M.; De Luca, P.; Elguero, B.; Meiss, R.; Navone, N.; Vazquez, E.S. Critical role of endogenous heme oxygenase 1 as a tuner of the invasive potential of prostate cancer cells. *Mol. Cancer Res.* **2009**, *7*, 1745–1755. [CrossRef] [PubMed]
7. Ferrando, M.; Gueron, G.; Elguero, B.; Giudice, J.; Salles, A.; Leskow, F.C.; Jares-Erijman, E.A.; Colombo, L.; Meiss, R.; Navone, N.; et al. Heme oxygenase 1 (HO-1) challenges the angiogenic switch in prostate cancer. *Angiogenesis* **2011**, *14*, 467–479. [CrossRef] [PubMed]

8. Elguero, B.; Gueron, G.; Giudice, J.; Toscani, M.A.; De Luca, P.; Zalazar, F.; Coluccio-Leskow, F.; Meiss, R.; Navone, N.; De Siervi, A.; et al. Unveiling the association of STAT3 and HO-1 in prostate cancer: Role beyond heme degradation. *Neoplasia* **2012**, *14*, 1043–1056. [CrossRef] [PubMed]

9. Gandini, N.A.; Fermento, M.E.; Salomon, D.G.; Blasco, J.; Patel, V.; Gutkind, J.S.; Molinolo, A.A.; Facchinetti, M.M.; Curino, A.C. Nuclear localization of heme oxygenase-1 is associated with tumor progression of head and neck squamous cell carcinomas. *Exp. Mol. Pathol.* **2012**, *93*, 237–245. [CrossRef] [PubMed]

10. Hsu, F.F.; Yeh, C.T.; Sun, Y.J.; Chiang, M.T.; Lan, W.M.; Li, F.A.; Lee, W.H.; Chau, L.Y. Signal peptide peptidase-mediated nuclear localization of heme oxygenase-1 promotes cancer cell proliferation and invasion independent of its enzymatic activity. *Oncogene* **2015**, *34*, 2360–2370. [CrossRef] [PubMed]

11. Sacca, P.; Meiss, R.; Casas, G.; Mazza, O.; Calvo, J.C.; Navone, N.; Vazquez, E. Nuclear translocation of haeme oxygenase-1 is associated to prostate cancer. *Br. J. Cancer* **2007**, *97*, 1683–1689. [CrossRef] [PubMed]

12. Tibullo, D.; Barbagallo, I.; Giallongo, C.; La Cava, P.; Parrinello, N.; Vanella, L.; Stagno, F.; Palumbo, G.A.; Li Volti, G.; Di Raimondo, F. Nuclear translocation of heme oxygenase-1 confers resistance to imatinib in chronic myeloid leukemia cells. *Curr. Pharm. Des.* **2013**, *19*, 2765–2770. [CrossRef] [PubMed]

13. Lutton, J.D.; da Silva, J.L.; Moqattash, S.; Brown, A.C.; Levere, R.D.; Abraham, N.G. Differential induction of heme oxygenase in the hepatocarcinoma cell line (Hep3B) by environmental agents. *J. Cell. Biochem.* **1992**, *49*, 259–265. [CrossRef] [PubMed]

14. Lavrovsky, Y.; Drummond, G.S.; Abraham, N.G. Downregulation of the human heme oxygenase gene by glucocorticoids and identification of 56b regulatory elements. *Biochem. Biophys. Res. Commun.* **1996**, *218*, 759–765. [CrossRef] [PubMed]

15. Deramaudt, T.B.; da Silva, J.L.; Remy, P.; Kappas, A.; Abraham, N.G. Negative regulation of human heme oxygenase in microvessel endothelial cells by dexamethasone. *Proc. Soc. Exp. Biol. Med.* **1999**, *222*, 185–193. [CrossRef] [PubMed]

16. Vallelian, F.; Schaer, C.A.; Kaempfer, T.; Gehrig, P.; Duerst, E.; Schoedon, G.; Schaer, D.J. Glucocorticoid treatment skews human monocyte differentiation into a hemoglobin-clearance phenotype with enhanced heme-iron recycling and antioxidant capacity. *Blood* **2010**, *116*, 5347–5356. [CrossRef] [PubMed]

17. Duzgun, A.; Bedir, A.; Ozdemir, T.; Nar, R.; Kilinc, V.; Salis, O.; Alacam, H.; Gulten, S. Effect of dexamethasone on unfolded protein response genes (MTJ1, Grp78, Grp94, CHOP, HMOX-1) in HEp2 cell line. *Indian J. Biochem. Biophys.* **2013**, *50*, 505–510. [PubMed]

18. Deroo, B.J.; Rentsch, C.; Sampath, S.; Young, J.; DeFranco, D.B.; Archer, T.K. Proteasomal inhibition enhances glucocorticoid receptor transactivation and alters its subnuclear trafficking. *Mol. Cell. Biol.* **2002**, *22*, 4113–4123. [CrossRef] [PubMed]

19. Silva, C.M.; Powell-Oliver, F.E.; Jewell, C.M.; Sar, M.; Allgood, V.E.; Cidlowski, J.A. Regulation of the human glucocorticoid receptor by long-term and chronic treatment with glucocorticoid. *Steroids* **1994**, *59*, 436–442. [CrossRef]

20. Gueron, G.; Giudice, J.; Valacco, P.; Paez, A.; Elguero, B.; Toscani, M.; Jaworski, F.; Leskow, F.C.; Cotignola, J.; Marti, M.; et al. Heme-oxygenase-1 implications in cell morphology and the adhesive behavior of prostate cancer cells. *Oncotarget* **2014**, *5*, 4087–4102. [CrossRef] [PubMed]

21. Paez, A.V.; Pallavicini, C.; Schuster, F.; Valacco, M.P.; Giudice, J.; Ortiz, E.G.; Anselmino, N.; Labanca, E.; Binaghi, M.; Salierno, M.; et al. Heme oxygenase-1 in the forefront of a multi-molecular network that governs cell-cell contacts and filopodia-induced zippering in prostate cancer. *Cell Death Dis.* **2016**, *7*, e2570. [CrossRef] [PubMed]

22. Wochnik, G.M.; Ruegg, J.; Abel, G.A.; Schmidt, U.; Holsboer, F.; Rein, T. FK506-binding proteins 51 and 52 differentially regulate dynein interaction and nuclear translocation of the glucocorticoid receptor in mammalian cells. *J. Biol. Chem.* **2005**, *280*, 4609–4616. [CrossRef] [PubMed]

23. Denny, W.B.; Valentine, D.L.; Reynolds, P.D.; Smith, D.F.; Scammell, J.G. Squirrel monkey immunophilin FKBP51 is a potent inhibitor of glucocorticoid receptor binding. *Endocrinology* **2000**, *141*, 4107–4113. [CrossRef] [PubMed]

24. Galigniana, M.D.; Erlejman, A.G.; Monte, M.; Gomez-Sanchez, C.; Piwien-Pilipuk, G. The hsp90-FKBP52 complex links the mineralocorticoid receptor to motor proteins and persists bound to the receptor in early nuclear events. *Mol. Cell. Biol.* **2010**, *30*, 1285–1298. [CrossRef] [PubMed]

25. Loboda, A.; Damulewicz, M.; Pyza, E.; Jozkowicz, A.; Dulak, J. Role of Nrf2/HO-1 system in development, oxidative stress response and diseases: An evolutionarily conserved mechanism. *Cell. Mol. Life Sci.* **2016**, *73*, 3221–3247. [CrossRef] [PubMed]

26. Namani, A.; Li, Y.; Wang, X.J.; Tang, X. Modulation of NRF2 signaling pathway by nuclear receptors: Implications for cancer. *Biochim. Biophys. Acta* **2014**, *1843*, 1875–1885. [CrossRef] [PubMed]

27. Kratschmar, D.V.; Calabrese, D.; Walsh, J.; Lister, A.; Birk, J.; Appenzeller-Herzog, C.; Moulin, P.; Goldring, C.E.; Odermatt, A. Suppression of the Nrf2-dependent antioxidant response by glucocorticoids and 11beta-HSD1-mediated glucocorticoid activation in hepatic cells. *PLoS ONE* **2012**, *7*, e36774. [CrossRef] [PubMed]

28. Iguchi, K.; Toyama, T.; Ito, T.; Shakui, T.; Usui, S.; Oyama, M.; Iinuma, M.; Hirano, K. Antiandrogenic activity of resveratrol analogs in prostate cancer LNCaP cells. *J. Androl.* **2012**, *33*, 1208–1215. [CrossRef] [PubMed]

29. Puhr, M.; Hoefer, J.; Eigentler, A.; Ploner, C.; Handle, F.; Schaefer, G.; Kroon, J.; Leo, A.; Heidegger, I.; Eder, I.; et al. The Glucocorticoid Receptor Is a Key Player for Prostate Cancer Cell Survival and a Target for Improved Antiandrogen Therapy. *Clin. Cancer Res.* **2018**, *24*, 927–938. [CrossRef] [PubMed]

30. Dorff, T.B.; Crawford, E.D. Management and challenges of corticosteroid therapy in men with metastatic castrate-resistant prostate cancer. *Ann. Oncol.* **2013**, *24*, 31–38. [CrossRef] [PubMed]

31. Tuttle, R.M.; Loop, S.; Jones, R.E.; Meikle, A.W.; Ostenson, R.C.; Plymate, S.R. Effect of 5-alpha-reductase inhibition and dexamethasone administration on the growth characteristics and intratumor androgen levels of the human prostate cancer cell line PC-3. *Prostate* **1994**, *24*, 229–236. [CrossRef] [PubMed]

32. Yano, A.; Fujii, Y.; Iwai, A.; Kageyama, Y.; Kihara, K. Glucocorticoids suppress tumor angiogenesis and in vivo growth of prostate cancer cells. *Clin. Cancer Res.* **2006**, *12*, 3003–3009. [CrossRef] [PubMed]

33. Jaworski, F.M.; Gentilini, L.D.; Gueron, G.; Meiss, R.P.; Ortiz, E.G.; Berguer, P.M.; Ahmed, A.; Navone, N.; Rabinovich, G.A.; Compagno, D.; et al. In Vivo Hemin Conditioning Targets the Vascular and Immunologic Compartments and Restrains Prostate Tumor Development. *Clin. Cancer Res.* **2017**, *23*, 5135–5148. [CrossRef] [PubMed]

34. Halin Bergstrom, S.; Nilsson, M.; Adamo, H.; Thysell, E.; Jernberg, E.; Stattin, P.; Widmark, A.; Wikstrom, P.; Bergh, A. Extratumoral Heme Oxygenase-1 (HO-1) Expressing Macrophages Likely Promote Primary and Metastatic Prostate Tumor Growth. *PLoS ONE* **2016**, *11*, e0157280. [CrossRef] [PubMed]

35. Nemeth, Z.; Li, M.; Csizmadia, E.; Dome, B.; Johansson, M.; Persson, J.L.; Seth, P.; Otterbein, L.; Wegiel, B. Heme oxygenase-1 in macrophages controls prostate cancer progression. *Oncotarget* **2015**, *6*, 33675–33688. [CrossRef] [PubMed]

36. Livak, K.J.; Schmittgen, T.D. Analysis of relative gene expression data using real-time quantitative PCR and the 2(-Delta Delta C(T)) Method. *Methods* **2001**, *25*, 402–408. [CrossRef] [PubMed]

37. Workman, P.; Balmain, A.; Hickman, J.A.; McNally, N.J.; Rohas, A.M.; Mitchison, N.A.; Pierrepoint, C.G.; Raymond, R.; Rowlatt, C.; Stephens, T.C.; et al. UKCCCR guidelines for the welfare of animals in experimental neoplasia. *Lab. Anim.* **1988**, *22*, 195–201. [CrossRef] [PubMed]

38. Bailey, T.L.; Boden, M.; Buske, F.A.; Frith, M.; Grant, C.E.; Clementi, L.; Ren, J.; Li, W.W.; Noble, W.S. MEME SUITE: Tools for motif discovery and searching. *Nucleic Acids Res.* **2009**, *37*, W202–W208. [CrossRef] [PubMed]

39. Liu, T.; Ortiz, J.A.; Taing, L.; Meyer, C.A.; Lee, B.; Zhang, Y.; Shin, H.; Wong, S.S.; Ma, J.; Lei, Y.; et al. Cistrome: An integrative platform for transcriptional regulation studies. *Genome Biol.* **2011**, *12*, R83. [CrossRef] [PubMed]

40. Sandelin, A.; Alkema, W.; Engstrom, P.; Wasserman, W.W.; Lenhard, B. JASPAR: An open-access database for eukaryotic transcription factor binding profiles. *Nucleic Acids Res.* **2004**, *32*, D91–D94. [CrossRef] [PubMed]

41. Long, Q.; Xu, J.; Osunkoya, A.O.; Sannigrahi, S.; Johnson, B.A.; Zhou, W.; Gillespie, T.; Park, J.Y.; Nam, R.K.; Sugar, L.; et al. Global transcriptome analysis of formalin-fixed prostate cancer specimens identifies biomarkers of disease recurrence. *Cancer Res.* **2014**, *74*, 3228–3237. [CrossRef] [PubMed]

International Journal of
Molecular Sciences

MDPI

Article

Genetic Restoration of Heme Oxygenase-1 Expression Protects from Type 1 Diabetes in NOD Mice

Julien Pogu [1], Sotiria Tzima [2], Georges Kollias [2], Ignacio Anegon [1], Philippe Blancou [1,3] and Thomas Simon [1,3,*]

[1] Centre de Recherche en Transplantation et Immunologie, Institut National de la Santé Et de la Recherche Médicale (INSERM), Université de Nantes, 44000 Nantes, France; pogujulien@gmail.com (J.P.); Ignacio.Anegon@univ-nantes.fr (I.A.); blancou@unice.fr (P.B.)
[2] Institute of Immunology, Biomedical Sciences Research Centre "Alexander Fleming", Vari, 210 Attica, Greece; sotiria.tzima@roche.com (S.T.); kollias@fleming.gr (G.K.)
[3] Université Côte d'Azur, Centre National de la Recherche Scientifique (CNRS), Institut de Pharmacologie Moléculaire et Cellulaire, 06560 Valbonne, France
* Correspondence: simon@ipmc.cnrs.fr; Tel.: +33-493-957-782

Received: 5 March 2019; Accepted: 29 March 2019; Published: 3 April 2019

Abstract: Antigen-presenting cells (APCs) including dendritic cells (DCs) play a critical role in the development of autoimmune diseases by presenting self-antigen to T-cells. Different signals modulate the ability of APCs to activate or tolerize autoreactive T-cells. Since the expression of heme oxygenase-1 (HO-1) by APCs has been associated with the tolerization of autoreactive T-cells, we hypothesized that HO-1 expression might be altered in APCs from autoimmune-prone non-obese diabetic (NOD) mice. We found that, compared to control mice, NOD mice exhibited a lower percentage of HO-1-expressing cells among the splenic DCs, suggesting an impairment of their tolerogenic functions. To investigate whether restored expression of HO-1 in APCs could alter the development of diabetes in NOD mice, we generated a transgenic mouse strain in which HO-1 expression can be specifically induced in DCs using a tetracycline-controlled transcriptional activation system. Mice in which HO-1 expression was induced in DCs exhibited a lower Type 1 Diabetes (T1D) incidence and a reduced insulitis compared to non-induced mice. Upregulation of HO-1 in DCs also prevented further increase of glycemia in recently diabetic NOD mice. Altogether, our data demonstrated the potential of induction of HO-1 expression in DCs as a preventative treatment, and potential as a curative approach for T1D.

Keywords: ANTIGEN presenting cell; tolerance; Tet-ON system; antigen presentation

---

1. Introduction

Heme oxygenase-1 (HO-1) is one of the three isoforms of the heme oxygenase enzymes that catabolizes the degradation of heme into biliverdin, free iron, and carbon monoxide (CO). In contrast to the two other isoforms, HO-1 is the only one which is induced by oxidative stress or pro-inflammatory cytokines. Under stressful conditions such as increased level of reactive oxygen species (ROS), the nuclear factor erythroid 2-related factor 2 (Nrf2) is released from the inhibitor protein Kelch-like ECH-associated protein 1 (Keap1). After translocation in the nucleus, Nrf2 binds to the antioxidant response element (ARE) sequence and increases the transcription of anti-oxidant genes including HO-1 (*Hmox-1*) and NAD(P)H quinone dehydrogenase 1 (*nqo-1*) [1]. Heme oxygenase-1 is also induced by its own substrates, whether they are natural (heme or hemin) or synthetic (Cobalt-protoporphyrin, CoPP). Many lines of evidence have pointed towards an anti-inflammatory role for HO-1. A chronic inflammation state was evidenced in both HO-1 knock-out mice [2] and in a patient exhibiting a mutation in the HO-1 coding *Hmox-1* gene [3]. Induction of experimental autoimmune encephalomyelitis (EAE) in HO-1 knock-out mice led to enhanced neurological symptoms as compared to wild-type (wt) mice [4]. In agreement with this latter result, both genetic and pharmacological manipulations aimed at inducing HO-1 expression protected mice against EAE [4], autoimmune type 1 diabetes (T1D) [5,6], and allergic asthma [7]. Furthermore, we and others have shown that HO-1 and CO treatment improved graft survival in both mice and rats [8–12]. Although the anti-inflammatory properties of HO-1 rely on heme degradation products, i.e., biliverdin, free iron, and CO, the cellular targets of these latter molecules remain to be identified. In this respect, studies in both mice and humans have ruled out the possibility that HO-1 degradation products act directly on regulatory T-cells (Tregs) [13–15]. In contrast, other studies have suggested that HO-1 degradation products could promote the development of tolerogenic dendritic cells (DCs) [16–18], that could eventually inhibit T-cell responses either by inhibiting the migration of autoreactive CD8+ T-cells to the target organ [19] or by inducing Tregs cells [20].

Type I diabetes (T1D) is a chronic autoimmune disease that results from the killing of pancreatic β-cells by autoreactive T-lymphocytes. Clinical diagnosis is preceded by a prediabetic asymptomatic phase characterized by the infiltration of several immune cell types including CD4+ and CD8+ T-cells in pancreatic islets. The diabetes-prone non-obese diabetic (NOD) mouse strain has been extensively studied as a clinically relevant model of T1D [21]. Non-obese diabetic mice, and particularly females, spontaneously develop insulitis starting at 4–5 weeks of age. A significant proportion of these animals (between 20 and 80% depending on housing conditions) eventually progress to diabetes as the result of pancreatic β-cells destruction by CD4$^+$ and CD8$^+$ T-cells. Dendritic cells (DCs) play a critical role in the initiation of TD1 by capturing and processing β-cells antigens. These DCs then migrate to the pancreatic lymph node where they present antigenic peptides to diabetogenic naïve T-cells leading to their activation and differentiation into effector T-cells [22].

Given the anti-inflammatory properties of HO-1 and the critical role of DCs in the initiation of T1D, we hypothesized 1) that DCs from NOD mice could be deficient in HO-1 expression, and ii) that the selective upregulation of HO-1 in DCs could inhibit the development of T1D. Here we demonstrated for the first time that genetic induction of HO-1 limited to DCs was sufficient both to prevent T1D in non-diabetic NOD mice and to stabilize glycemia in some recently diabetic NOD mice.

2. Results

2.1. DCs from NOD Mice are Deficient in HO-1 Expression

In contrast to NOD mice that spontaneously develop insulitis and eventually T1D, Major Histocompatibility Complex (MHC)-matched non-obese diabetes resistant (NOR) mice only developed peri-insulitis and never progressed to insulitis and T1D [23]. We found that the percentage of HO-1-expressing cells among spleen CD11c$^+$ cells was two-fold lower in NOD mice compared to NOR mice (1.8% versus 3.8%) demonstrating a strain-specific difference (Figure 1A). To investigate whether the lower percentage of HO-1-positive DCs in NOD mice resulted from an intrinsic defect, we generated bone marrow-derived DCs (BMDCs) from both NOD and NOR mice and assessed these cells for HO-1 expression before and after exposure to the HO-1 inducer CoPP. In the absence of CoPP, the percentage of HO-1-positive DCs was lower in BMDCs from NOD mice compared to those from NOR mice. While CoPP increased the percentage of BMDCs expressing HO-1 in both NOD and NOR mice in a dose-dependent manner, the percentage of HO-1-positive BMDCs remained lower in NOD mice whatever the concentration of CoPP (Figure 1B). Altogether, these data demonstrated that DCs from NOD mice exhibited an intrinsic defect in HO-1 expression.

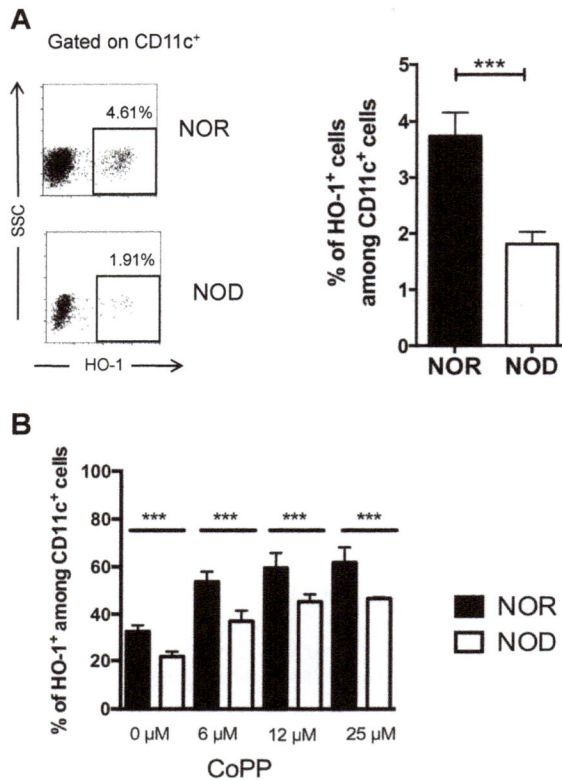

Figure 1. Heme oxygenase-1 expression deficiency in NOD DCs. (**A**) HO-1 expression in splenic CD11c+ cells was evaluated in one-month-old female NOD and NOR mice by flow cytometry. Representative dot plots (left panel) and mean frequencies of splenic HO-1+ CD11c+ cells ± s.e.m. ($n \geq 7$ mice/group) (right panel) are reported. (**B**) Flow cytometry analysis showing HO-1 expression in CD11c+ of bone marrow-derived DCs (BMDC) from NOD and NOR mice treated or not with CoPP for 24 h. Mean frequencies of splenic HO-1+ CD11c+ cells ± s.e.m. ($n \geq 8$ mice/group) from 3 independent experiments are reported. An unpaired *t*-test (**A**) and a one-way ANOVA followed by Tukey's post-hoc test (**B**) were performed. ***, $p < 0.001$.

2.2. Generation of a HO-1 Inducible Transgenic Mouse Strain on the NOD Background

To investigate whether the upregulation of HO-1 in DCs could impact the development of T1D in NOD mice, we used a pIi-tTA+ transgenic mouse strain in which the tetracycline transactivator (*tTA*) gene was under the control of the MHC-II invariant chain (Eα-Ii) promoter [16]. We generated a new transgenic mouse strain, termed TetO-HO-1+, in which the *Homx-1* gene was cloned downstream a hybrid cytomegalovirus-Tet operator (Figure 2A). We then crossed pIi-tTA+ to TetO-HO-1+ transgenic mice, identified double transgenic animals, and further crossed these mice onto the NOD background for eight generations. We expected the resultant TetO-HO-1+ pIi-tTA+ double-transgenic mice to show a doxycycline (DOX)-driven expression of HO-1 in MHC-II+ cells. To investigate whether this was the case, we purified splenic CD11c+ from double transgenic TetO-HO-1+ pIi-tTA+ and single transgenic TetO-HO-1+ pIi-tTA- mice and incubated these cells with different doses of DOX (Figure 2B). As expected, DOX did not increase the percentage of HO-1-positive DCs in single transgenic TetO-HO-1+ pIi-tTA- mice, but it did in TetO-HO-1+ pIi-tTA+ double transgenic mice.

An increase in the percentage of HO-1-positive DCs was observed as early as two hours after DOX treatment in a dose- and time-dependent manner.

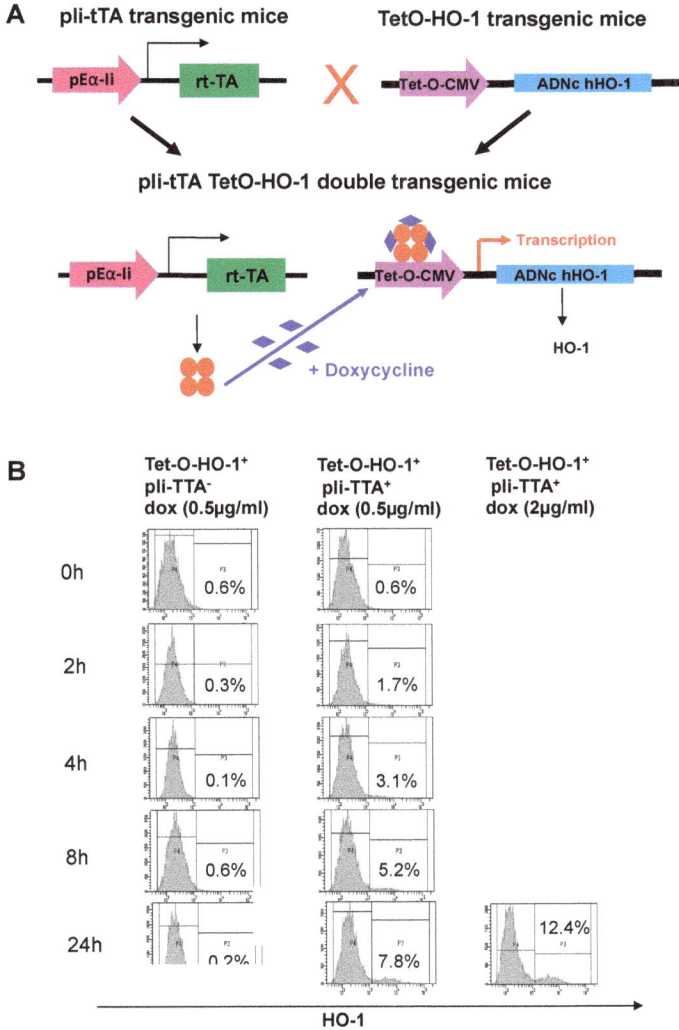

Figure 2. In vitro doxycycline treatment of splenic CD11c[+] DCs from double transgenic TetO-HO-1[+] pli-tTA[+] NOD mice induces HO-1 expression. (**A**) Schematic representation of the transgenic mouse strains used in the study. pIi-tTA are transgenic mice in which the tetracycline transactivator (*tTA*) gene is under the control of the Major Histocompatibility Complex II (MHC-II) invariant chain (Eα-Ii) promoter [24]. TetO-HO-1 are transgenic mice in which the human *homx-1* cDNA was cloned downstream a hybrid cytomegalovirus (CMV)-Tet operator rtTA. Double transgenic mouse strain show induction of HO-1 expression in MHC-II[+] cells in the presence of doxycycline (DOX). (**B**) Splenic CD11c[+] DCs from single transgenic Tet-O-HO-1[+]pli-tTA[−] or double transgenic TetO-HO-1[+] pIi-tTA[+] NOD mice were cultured with doxycycline for 24 h and analyzed by flow cytometry for HO-1 expression. One representative experiment out of two is shown.

To investigate whether DOX could upregulate HO-1 expression in vivo and to further identify in which cell types we treated or not TetO-HO-1$^+$ pIi-tTA$^+$ double-transgenic mice with DOX for 24 h, we then analyzed splenic cells for HO-1 expression by flow cytometry after intercellular staining with anti-HO-1 monoclonal antibody (mAb) and surface staining with mAbs directed to the identify APCs (CD11c to identify DCs; B220 to identify B cells; F4/80 and CD11b to identify macrophages, Supplementary Materials Figure S1). While DOX increased the percentage of DCs that expressed HO-1, it did not have any impact on the percentage of macrophages or B lymphocytes expressing HO-1 (Figure 3A). As expected, DOX increased the percentage of HO-1-expressing DCs TetO-HO-1$^+$ pIi-tTA$^+$ double-transgenic mice, but not in their TetO-HO-1$^+$ pIi-tTA$^-$ single transgenic littermates (Figure 3B). Of note, the percentage of HO-1-positive DCs in DOX-treated TetO-HO-1$^+$ pIi-tTA$^+$ double transgenic mice was comparable to the percentage of HO-1-positive DCs in NOR mice.

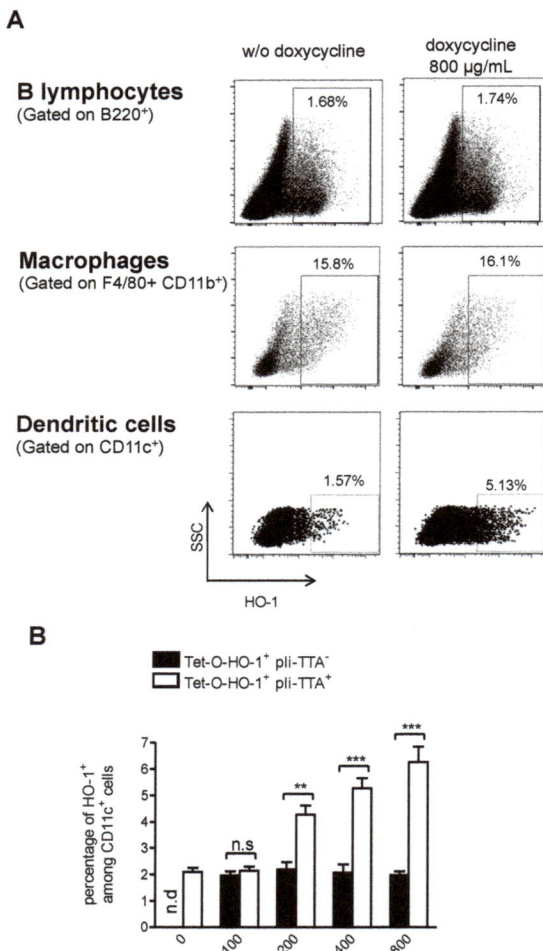

Figure 3. Induction of HO-1 expression by addition of doxycycline in the drinking water of double transgenic TetO-HO-1$^+$ pIi-tTA$^+$ NOD mice. (**A,B**) Different doses of doxycycline (DOX) were added to the drinking water in single transgenic Tet-O-HO-1$^+$pIi-tTA$^-$ or double transgenic TetO-HO-1$^+$ pIi-tTA$^+$ NOD mice to induce HO-1 expression. (**A**) Splenic cell populations (DCs cells, B lymphocyte

or macrophages) were analyzed for HO-1 expression by intracellular HO-1 staining. (**B**) Expression of HO-1 in splenic DCs are presented as mean frequencies $\pm$ s.e.m. ($n \geq 4$ mice/group) for two independent experiments. One-way ANOVA followed by Tukey's post-hoc tests were performed. *** $p < 0.001$. n.s.: not significant; n.d.: not determined.

2.3. Upregulation of HO-1 in DCs Prevents T1D in NOD Mice

We next investigated whether the selective upregulation of HO-1 in DCs could prevent T1D in NOD mice. We treated TetO-HO-1$^+$ pIi-tTA$^+$ double transgenic mice and their TetO-HO-1$^+$ pIi-tTA$^-$ single transgenic littermates with DOX starting at 4 weeks of age. Control single and double transgenic mice were left untreated. Doxycycline treatment did not have any impact on the incidence of T1D in TetO-HO-1$^+$ pIi-tTA$^-$ single transgenic mice (Figure 4A). Likewise, TetO-HO-1$^+$ pIi-tTA$^-$ single transgenic mice and TetO-HO-1$^+$ pIi-tTA$^+$ double transgenic mice developed T1D with a similar incidence in the absence of DOX treatment. In striking contrast, only 10% of DOX-treated TetO-HO-1$^+$ pIi-tTA$^+$ double transgenic mice developed T1D before 9 months of age, compared to 70% of DOX-treated TetO-HO-1$^+$ pIi-tTA$^-$ single transgenic mice (Figure 4A). In agreement with this latter result, the percentage of islets exhibiting immune cell infiltration was strongly reduced in DOX-treated TetO-HO-1$^+$ pIi-tTA$^+$ double transgenic mice compared to single transgenic controls (Figure 4B).

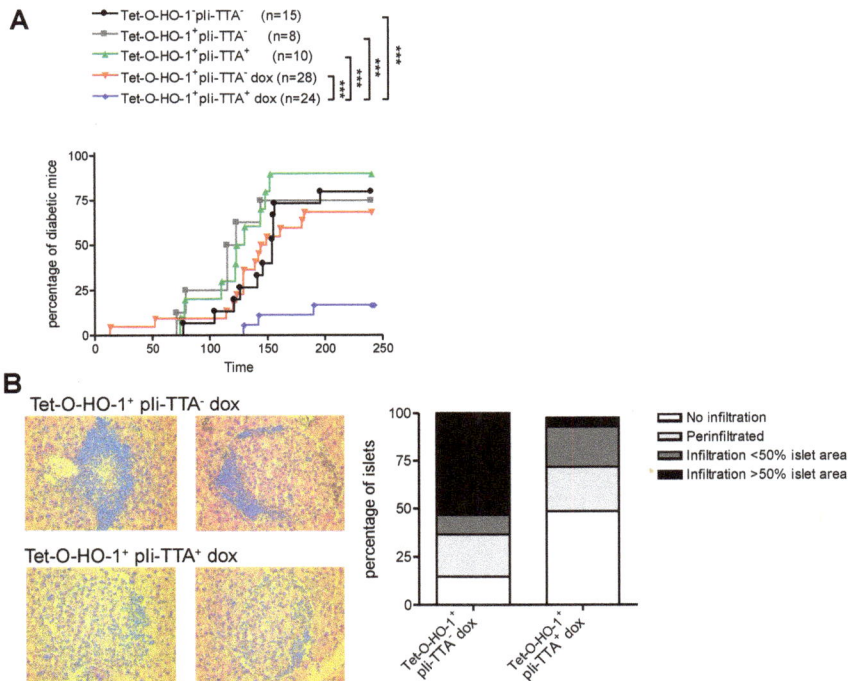

Figure 4. HO-1 induction in DCs decreased diabetes incidence in NOD mice. (**A**,**B**) Starting at one month of age, non-transgenic, single transgenic Tet-O-HO-1$^+$pIi-tTA$^-$ or double transgenic TetO-HO-1$^+$ pIi-tTA$^+$ female NOD mice were provided or not with 800 µg/mL of doxycycline (DOX) in their drinking water. (**A**) Cohorts were monitored for the development of diabetes by measurement of glycemia. Mice were considered diabetic when glycemia was over 200 mg/dL for two consecutive days. (**B**) In three-month-old female single transgenic Tet-O-HO-1$^+$pIi-tTA$^-$ or double transgenic TetO-HO-1$^+$ pIi-tTA$^+$ NOD mice, insulitis was evaluated by hematoxylin and eosin (H&E) staining of pancreatic

sections. Representative H&E staining of pancreatic sections are shown (left panels). The extension of insulitis is reported (right panel) as the percentage of non-infiltrated, peri-infiltrated, slightly infiltrated (less than 50% of islet area), or highly infiltrated (more than 50% of islet area) islets (*n* > 4 mice/group). The total numbers of analyzed islets for single transgenic Tet-O-HO-1+pIi-tTA⁻ or double transgenic TetO-HO-1+ pIi-tTA+ female transgenic NOD mice were 52 and 47, respectively. (**A**) Log-rank test was performed, and only significant differences were reported. *** *p* < 0.001.

2.4. Upregulation of HO-1 in DCs Prevent A Further Increase in Glycemia in Recently Diabetic NOD Mice

We next investigated whether the selective upregulation of HO-1 in DCs could impact glycemia in recently diabetic NOD mice. To this aim, we monitored glycemia in TetO-HO-1+ pIi-tTA+ double transgenic mice and TetO-HO-1+ pIi-tTA⁻ single transgenic mice every day. Mice exhibiting a glycemia higher than 200 mg/dL for two consecutive days were immediately given DOX in their drinking water and further followed for glycemia. While all animals in the TetO-HO-1+ pIi-tTA⁻ single transgenic mice group had to be sacrificed because they reached humane end-points, three animals out of eight in the TetO-HO-1+ pIi-tTA+ double transgenic mice group remained normoglycemic. Glycemia progressively increased in TetO-HO-1+ pIi-tTA⁻ single transgenic mice following DOX treatment, whereas it remained constant in DOX-treated TetO-HO-1+ pIi-tTA+ double transgenic mice (Figure 5). Therefore, the selective upregulation of HO-1 in NOD DCs could inhibit T1D progression in recently diabetic animals.

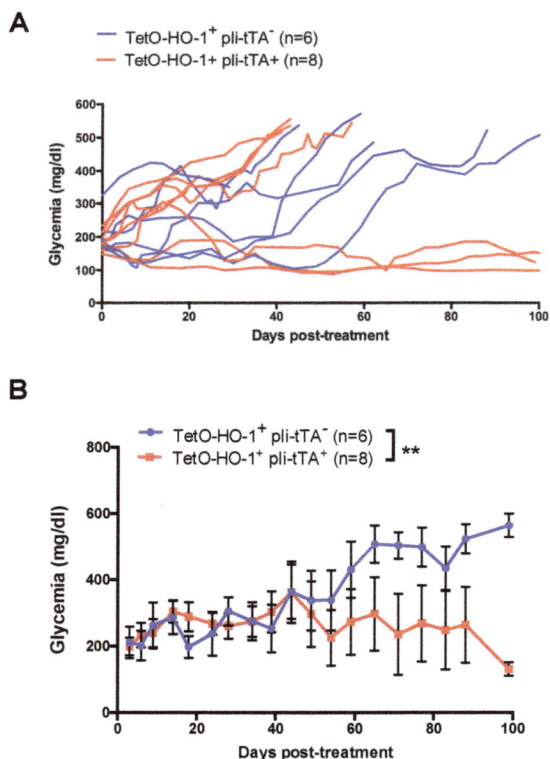

Figure 5. HO-1 induction in DCs inhibited diabetes in spontaneously diabetic mice. Diabetic mice with a glycemia between 200 mg/dL and 250 mg/dL for two consecutive days were identified in single

transgenic Tet-O-HO-1⁺pli-tTA⁻ or double transgenic TetO-HO-1⁺ pIi-tTA⁺ female NOD mice cohorts. These diabetic mice were treated with doxycycline (DOX) in the drinking water (800 µg/mL). Mice then were monitored for glycemia twice a week. Graphs show glycemia over time for each animal (**A**) or as a mean ± s.e.m. ($n \geq 6$ mice/group) (**B**). Log-rank test was performed in (**B**). ** $p < 0.01$.

3. Discussion

Type 1 diabetes is a complex multifactorial autoimmune disease in which genetic factors play a critical role. As many as 60 genetic susceptibility loci, termed insulin-dependent diabetes (Idd) loci, have been identified in both mice and humans. Here, we have shown that DCs from NOD mice are deficient in HO-1 expression compared to those from NOR mice that do not develop T1D, although these two strains share 88% of their genome [23]. To which extent this defect contributes to the susceptibility of NOD mice to T1D remains a matter of speculation. While in humans the gene coding for HO-1, *Hmox1*, does not map to any identified Insulin dependent diabetes (Idd) locus, the gene coding interleukin-10 (IL-10), which regulates HO-1 expression [25], colocalizes with Idd3 [26]. It has been previously demonstrated that IL-10 could induce HO-1 in antigen presenting cells (APCs) through Signal transducer and activator of transcription 3 (STAT-3)- and Phosphoinositide 3-kinase (PI3K)-dependent mechanisms [25,27]. Given the fact that decreased IL-10 levels have been associated with T1D in both NOD mice and humans [28,29], this may explain the low level of HO-1 observed in NOD DCs. This defect of HO-1 expression in DCs may also extend to other HO-1 expressing cells including macrophages and B lymphocytes.

On a different but related topic, several defects associated to APCs' phenotype and functions have been identified in NOD mice that may explain why this strain is prone to T1D. As an example, hyperactivation of nuclear factor kappa-light-chain-enhancer of activated B cells (NF-κB) has been detected in DCs of NOD mice, and was directly correlated to elevated levels of IL-12 secretion relative to DCs from control animals [30]. Elevated NF-κB activation in DCs resulted in an increased capacity of these cells to stimulate naïve CD8⁺ T cells and promote their differentiation into cytotoxic effector T-cells [31]. Decreased expression of molecules associated with tolerance induction (CD103, Langerin, C-type lectin domain family 9, member A (CLEC9A), C-C chemokine receptor type 5 (CCR5) or increased expression of co-stimulatory molecules (CD80, CD86) [32] in the DC population of NOD mice also suggested that abnormal differentiation of pancreatic DCs contributes to the loss of tolerance. Moreover, reduced numbers of tolerogenic DCs have been observed in NOD mice [31]. Altogether, these studies demonstrated that NOD DCs possess intrinsic characteristics that may contribute to the autoimmune phenotype. Whether there is a causal relationship between these defects and HO-1 partial deficiency remains to be investigated.

We have generated transgenic NOD mice in which HO-1 expression is regulated by the Tet ON system in MHC class II-positive cells. Surprisingly, HO-1 induction seems to be restricted to DCs among MHC-II⁺ cells. This could be due to the Tet ON system which was under the control of the MHC-II invariant chain (Eα-Ii) promoter. Indeed, RNA sequencing data from the Immgen Consortium showed that mRNA levels of the MHC-II invariant chain (CD74) was higher in splenic DCs compared to splenic B lymphocytes and macrophages, respectively (Supplementary Materials Figure S2). Thus, a higher expression of Eα-Ii in DCs could be associated with a higher sensitivity to DOX resulting in HO-1 upregulation in DCs at a similar level as observed in the NOR control strain. Recovery of HO-1 expression in DCs to levels comparable to NOR mice in DOX-treated pIi-tTA-tHO-1 NOD mice dramatically lowered T1D incidence. While the proportion of DCs expressing HO-1 in non-diabetic mice is low, their capacity to induce tolerogenic immune response may be very powerful, even with a small proportion of cells expressing HO-1 as previously shown [19]. Concerning the role of HO-1 negative DCs, we can hypothesize that these cells might be of physiological importance to induce inflammatory response upon detection of immune danger signals.

The low level of protection observed in DOX-treated diabetic NOD mice (3 animals out of 8) may be due to either the limited induction of HO-1 in DCs or the extent of beta cells' destruction of time of intervention. The efficacy of protection might be improved by combining HO-1 induction in DCs with immunosuppression treatments.

We [19] and others [5] have previously shown the protective effect of upregulating HO-1 in T1D with chemical inducers or systemic HO-1 transduction. Here, we demonstrated for the first time that genetic induction of HO-1 limited to DCs was sufficient both to prevent T1D in non-diabetic NOD mice and to stabilize glycemia in some recently diabetic NOD mice. As an anti-inflammatory protein, HO-1 exerts its immunomodulatory effects through multiple mechanisms by modifying T-cell responses, either directly at the level of the T-cell, or, most likely, by indirectly influencing APCs [33]. HO-1 induction inhibits DCs' maturation and secretion of pro-inflammatory cytokines [8]. Moreover, HO-1+ DCs or DCs treated by the HO-1 end-product CO, showed impaired antigen presentation abilities [4,34] possibly resulting in fewer islet-specific pathogenic T-cells. The absence of HO-1 expression in APCs also impaired suppressive functions of Tregs [35]. The low proportion of HO-1+ DCs in NOD mice could in part explain the Treg defect observed in T1D. We also demonstrated previously that ex vivo treatment with CO conferred tolerogenic properties to DCs and inhibited the pathogenicity of naïve T-cells stimulated by those DCs in a T1D model [17]. Moreover, intradermal injection of HO-1 inducers promoted the accumulation of DCs overexpressing HO-1 in draining lymph nodes (LNs) [19]. These HO-1^{high} DCs exhibited antigen-specific tolerogenic properties as demonstrated by their ability to inhibit the diabetogenic potential of autoreactive cytotoxic T-cells. The same protective mechanisms could also be at play following the restoration of HO-1 level in DCs of DOX-treated pIi-tTA-tHO-1 NOD mice.

Another mechanism of protection could rely on the antioxidant properties of HO-1. As an example, the induction of HO-1 by hemin treatment has been showed to reduce hyperglycemia and to improved glucose metabolism in streptozotocin-treated rats [36] [. The protective effect and the reduction of lesions in the pancreas were due to the inhibition of oxidative stress mediated by HO-1 activity. The oxidative stress mediated by hyperglycemia is a major pathophysiological factor in T1D [37] that could be reduced by the genetic induction of HO-1 limited to DCs in our model.

Altogether, our results showed that restoration of HO-1 expression levels in DCs in NOD mice prevent the development of T1D but also highlight the therapeutic beneficial effect of inducing HO-1 in APCs as a treatment for T1D.

4. Materials and Methods

4.1. Animals

The NOD/LtJ and NOR/Lt mice were originally purchased from Charles River. The pIi-tTA mice were a kind gift from Christophe Benoist [38]. For generation of the TetO-HO-1 mice, the human α-globin intron located upstream of the cDNA sequence, the human HO-1 cDNA, and the bovine growth hormone polyA located downstream of the cDNA were cloned at the Not-I/Xho-I sites into pBluKSM-tet-O-CMV vector containing the Tet-Responsive-Element (TRE) downstream of the minimal CMV promoter followed by the human α-globin intron and the bovine growth hormone polyA. Transgenic mice were generated by pronuclear microinjection of CBA/C57BL6 eggs with the XhoI-NotI fragment of the vector described. Seventeen different founders were carrying the transgene as tested by PCR and Southern blot. Of the 17 lines, 3 founders contained high copies of the hHO-1 cDNA. One of these was further analyzed by crossing with actin-rtTA mice. The hHO-1 expression was confirmed by Western blotting. Finally, both strains pIi-tTA and TetO-HO-1 were backcrossed to NOD/LtJ mice for at least eight generations. Only females were used in experiments. All animal breeding and experiments were performed under conditions in accordance with the European Union Guidelines.

4.2. Genotyping of rTA-HO-1

Mice were bled from the cheek into 100 μL of heparin (125 UI/mL). For DNA purification, we used the DNeasy Blood and Tissue Kit following the manufacturer's instructions. Consequently, we performed a PCR for each transgene inserted in the transgenic mice: rtTA and TetO-HO-1using GoTaq G2 Flexi DNA Polymerase (Promega, Madison, WI, USA). Then, samples were loaded in an agarose gel electrophoresis. The mice that only had the *TetO-HO-1* gene but lacked the *tTA* gene were named littermates and used as controls.

4.3. Doxycycline Treatment

Doxycycline hyclate powder (Sigma–Aldrich, St. Louis, MO, USA) was diluted in drinking water at different concentrations (100 µg/mL up to 800 µg/mL), protected from light, and were changed every 2 to 3 days.

4.4. Diabetes Follow-Up

Diabetes monitoring was done by daily glycemia measurement to identify diabetic mice. Mice were considered diabetic when glycemia was superior to 200 mg/dL for two consecutive days. For incidence follow-up, glycosuria was measured twice a week and diabetic animals were confirmed by glycemia measurements (superior to 200 mg/dL for two consecutive days).

4.5. Flow Cytometry

Spleen was digested in collagenase D (Roche, Basel, Switzerland) during 30 min at 37 °C then crushed and filtered on 70 µm tamis before staining. Single-cell suspensions were stained with mouse F4.80 (BM8, eBioscience), mouse CD11c (HL3, BD), and mouse B220 (RA3-6B2) antibodies from BD Biosciences. The HO-1 Intracellular stainings were performed using the BD cytofix/cytoperm kit with anti-HO-1 antibody (clone HO-1-1, abcam) followed by anti-mouse IgG1 (clone A85-1) conjugated to FITC. We used IgG1 (Immunotech) as negative controls. Stained cells were acquired on a FACSAria™ (Becton Dickinson, Franklin Lakes, NJ, USA) flow cytometer and analyzed using Flow Jo v10.0.7 software (FlowJo LLC, Ashland, OR, USA).

4.6. Histological Analysis.

Insulitis was evaluated on snap-frozen acetone-fixed cryosections (8-µm thick). Hematoxylin and eosin staining (Thermo Electron Corp, Waltham MA, USA) were performed, and we evaluated the degree of inflammation microscopically. The percentage of non-infiltrated, peri-infiltrated, slightly infiltrated (less than 50% of islet area), or highly infiltrated (more than 50% of islet area) islets were evaluated.

4.7. Bone-Marrow Derived DCs

DCs were derived from bone marrow culture (BM-derived DCs) from NOD and NOR mice, as described [10]. At day 6, non-adherent cells were harvested and a CD11c$^+$ sorting was performed using magnetic beads (purity > 95%) (Miltenyi Biotech, Bergisch Gladbach, Germany). CD11c$^+$ cells were treated with different concentrations of CoPP (Frontier Scientific, Carnforth, UK), as previously described [39]

4.8. Isolation of Splenic DCs

Single-cell suspensions were prepared by enzymatic spleen disaggregation with collagenase D (Sigma–Aldrich) and CD11c$^+$ population was enriched by cell sorting using magnetic beads (purity > 95%) (Miltenyi Biotech). The DCs from conditional transgenic mice were incubated for 24 h with 0.5 or 2 µg/mL of doxycycline.

4.9. Statistics

For diabetes incidence, significance was calculated using the log-rank test. For all other parameters, significance was calculated by *t*-paired test, Mann–Whitney non-parametric *t*-test. One-way ANOVA

or two-way ANOVA using Prism software (GraphPad Software, San Diego, CA, USA): * $p < 0.05$,
** $p < 0.01$, and *** $p < 0.001$.

Supplementary Materials: Supplementary materials can be found at http://www.mdpi.com/1422-0067/20/7/1676/s1.

Author Contributions: I.A., P.B. and IA conceived the study. T.S. designed the experiments. J.P. and T.S. performed the experiments. S.T. and G.K. generated the TetO-HO-1+ transgenic mice. P.B. and T.S. interpreted the data. T.S. wrote the manuscript. T.S, P.B. and I.A. edited the manuscript. All authors read and approved the final manuscript.

Funding: This project was funded by JDRF (to P.B., grant number 5-2010-640).

Conflicts of Interest: The authors declare no conflict of interest.

Abbreviations

APC	Antigen Presenting Cells
BMDC	Bone Marrow-Derived Dendritic Cells
CO	Carbon Monoxide
CoPP	Cobalt-Protoporphyrin
DCs	Dendritic Cells
DOX	Doxycycline
HO-1	Heme Oxygenase 1
idd	Insulin-Dependent Diabetes
MHC-II	Major Histocompatibility Complex II
NOD	Non-Obese Diabetic
NOR	Non-Obese Resistant
T1D	Type 1 Diabetes
Tregs	Regulatory T-cells
tTA	Tetracycline TransActivator

References

1. Kobayashi, M.; Yamamoto, M. Nrf2-Keap1 regulation of cellular defense mechanisms against electrophiles and reactive oxygen species. *Adv. Enzyme Regul.* **2006**, *46*, 113–140. [CrossRef] [PubMed]
2. Poss, K.D.; Tonegawa, S. Reduced stress defense in heme oxygenase 1-deficient cells. *Proc. Natl. Acad. Sci. USA* **1997**, *94*, 10925–10930. [CrossRef] [PubMed]
3. Yachie, A.; Niida, Y.; Wada, T.; Igarashi, N.; Kaneda, H.; Toma, T.; Ohta, K.; Kasahara, Y.; Koizumi, S. Oxidative stress causes enhanced endothelial cell injury in human heme oxygenase-1 deficiency. *J. Clin. Investig.* **1999**, *103*, 129–135. [CrossRef]
4. Chora, A.A.; Fontoura, P.; Cunha, A.; Pais, T.F.; Cardoso, S.; Ho, P.P.; Lee, L.Y.; Sobel, R.A.; Steinman, L.; Soares, M.P. Heme oxygenase-1 and carbon monoxide suppress autoimmune neuroinflammation. *J. Clin. Investig.* **2007**, *117*, 438–447. [CrossRef]
5. Li, M.; Peterson, S.; Husney, D.; Inaba, M.; Guo, K.; Kappas, A.; Ikehara, S.; Abraham, N.G. Long-lasting expression of HO-1 delays progression of type I diabetes in NOD mice. *Cell Cycle* **2007**, *6*, 567–571. [CrossRef] [PubMed]
6. Hu, C.-M.; Lin, H.-H.; Chiang, M.-T.; Chang, P.-F.; Chau, L.-Y. Systemic expression of heme oxygenase-1 ameliorates type 1 diabetes in NOD mice. *Diabetes* **2007**, *56*, 1240–1247. [CrossRef]
7. Xia, Z.-W.; Zhong, W.-W.; Xu, L.-Q.; Sun, J.-L.; Shen, Q.-X.; Wang, J.-G.; Shao, J.; Li, Y.-Z.; Yu, S.-C. Heme oxygenase-1-mediated CD4+CD25high regulatory T cells suppress allergic airway inflammation. *J. Immunol.* **2006**, *177*, 5936–5945. [CrossRef]
8. Chauveau, C.; Bouchet, D.; Roussel, J.-C.; Mathieu, P.; Braudeau, C.; Renaudin, K.; Tesson, L.; Soulillou, J.-P.; Iyer, S.; Buelow, R.; et al. Gene transfer of heme oxygenase-1 and carbon monoxide delivery inhibit chronic rejection. *Am. J. Transplant.* **2002**, *2*, 581–592. [CrossRef]
9. Braudeau, C.; Bouchet, D.; Tesson, L.; Iyer, S.; Rémy, S.; Buelow, R.; Anegon, I.; Chauveau, C. Induction of long-term cardiac allograft survival by heme oxygenase-1 gene transfer. *Gene Ther.* **2004**, *11*, 701–710. [CrossRef]

10. Soares, M.P.; Lin, Y.; Anrather, J.; Csizmadia, E.; Takigami, K.; Sato, K.; Grey, S.T.; Colvin, R.B.; Choi, A.M.; Poss, K.D.; et al. Expression of heme oxygenase-1 can determine cardiac xenograft survival. *Nat. Med.* **1998**, *4*, 1073–1077. [CrossRef] [PubMed]

11. Hancock, W.W.; Buelow, R.; Sayegh, M.H.; Turka, L.A. Antibody-induced transplant arteriosclerosis is prevented by graft expression of anti-oxidant and anti-apoptotic genes. *Nat. Med.* **1998**, *4*, 1392–1396. [CrossRef]

12. Zhao, Y.; Jia, Y.; Wang, L.; Chen, S.; Huang, X.; Xu, B.; Zhao, G.; Xiang, Y.; Yang, J.; Chen, G. Upregulation of Heme Oxygenase-1 Endues Immature Dendritic Cells With More Potent and Durable Immunoregulatory Properties and Promotes Engraftment in a Stringent Mouse Cardiac Allotransplant Model. *Front. Immunol.* **2018**, *9*, 1515. [CrossRef] [PubMed]

13. Biburger, M.; Theiner, G.; Schädle, M.; Schuler, G.; Tiegs, G. Pivotal Advance: Heme oxygenase 1 expression by human CD4+ T cells is not sufficient for their development of immunoregulatory capacity. *J. Leukoc. Biol.* **2010**, *87*, 193–202. [CrossRef]

14. Zelenay, S.; Chora, A.; Soares, M.P.; Demengeot, J. Heme oxygenase-1 is not required for mouse regulatory T cell development and function. *Int. Immunol.* **2007**, *19*, 11–18. [CrossRef]

15. Blancou, P.; Anegon, I. Editorial: Heme oxygenase-1 and dendritic cells: What else? *J. Leukoc. Biol.* **2010**, *87*, 185–187. [CrossRef] [PubMed]

16. Rémy, S.; Blancou, P.; Tesson, L.; Tardif, V.; Brion, R.; Royer, P.J.; Motterlini, R.; Foresti, R.; Painchaut, M.; Pogu, S.; et al. Carbon monoxide inhibits TLR-induced dendritic cell immunogenicity. *J. Immunol.* **2009**, *182*, 1877–1884. [CrossRef] [PubMed]

17. Simon, T.; Pogu, S.; Tardif, V.; Rigaud, K.; Rémy, S.; Piaggio, E.; Bach, J.-M.; Anegon, I.; Blancou, P. Carbon monoxide-treated dendritic cells decrease β1-integrin induction on CD8$^+$ T cells and protect from type 1 diabetes. *Eur. J. Immunol.* **2013**, *43*, 209–218. [CrossRef]

18. Moreau, A.; Hill, M.; Thébault, P.; Deschamps, J.Y.; Chiffoleau, E.; Chauveau, C.; Moullier, P.; Anegon, I.; Alliot-Licht, B.; Cuturi, M.C. Tolerogenic dendritic cells actively inhibit T cells through heme oxygenase-1 in rodents and in nonhuman primates. *FASEB J.* **2009**, *23*, 3070–3077. [CrossRef]

19. Simon, T.; Pogu, J.; Rémy, S.; Brau, F.; Pogu, S.; Maquigneau, M.; Fonteneau, J.-F.; Poirier, N.; Vanhove, B.; Blancho, G.; et al. Inhibition of effector antigen-specific T cells by intradermal administration of heme oxygenase-1 inducers. *J. Autoimmun.* **2017**, *81*, 44–55. [CrossRef]

20. Wong, T.-H.; Chen, H.-A.; Gau, R.-J.; Yen, J.-H.; Suen, J.-L. Heme Oxygenase-1-Expressing Dendritic Cells Promote Foxp3+ Regulatory T Cell Differentiation and Induce Less Severe Airway Inflammation in Murine Models. *PLoS ONE* **2016**, *11*, e0168919. [CrossRef]

21. Mullen, Y. Development of the Nonobese Diabetic Mouse and Contribution of Animal Models for Understanding Type 1 Diabetes. *Pancreas* **2017**, *46*, 455–466. [CrossRef]

22. Herold, K.C.; Vignali, D.A.A.; Cooke, A.; Bluestone, J.A. Type 1 diabetes: Translating mechanistic observations into effective clinical outcomes. *Nat. Rev. Immunol.* **2013**, *13*, 243–256. [CrossRef] [PubMed]

23. Prochazka, M.; Serreze, D.V.; Frankel, W.N.; Leiter, E.H. NOR/Lt Mice: MHC-Matched Diabetes-Resistant Control Strain for NOD Mice. *Diabetes* **1992**, *41*, 98–106. [CrossRef]

24. Van Santen, H.; Benoist, C.; Mathis, D. A cassette vector for high-level reporter expression driven by a hybrid invariant chain promoter in transgenic mice. *J. Immunol. Methods* **2000**, *245*, 133–137. [CrossRef]

25. Lee, T.-S.; Chau, L.-Y. Heme oxygenase-1 mediates the anti-inflammatory effect of interleukin-10 in mice. *Nat. Med.* **2002**, *8*, 240–246. [CrossRef] [PubMed]

26. Barrett, J.C.; Clayton, D.G.; Concannon, P.; Akolkar, B.; Cooper, J.D.; Erlich, H.A.; Julier, C.; Morahan, G.; Nerup, J.; Nierras, C.; et al. Genome-wide association study and meta-analysis find that over 40 loci affect risk of type 1 diabetes. *Nat. Genet.* **2009**, *41*, 703–707. [CrossRef] [PubMed]

27. Ricchetti, G.A.; Williams, L.M.; Foxwell, B.M.J. Heme oxygenase 1 expression induced by IL-10 requires STAT-3 and phosphoinositol-3 kinase and is inhibited by lipopolysaccharide. *J. Leukoc. Biol.* **2004**, *76*, 719–726. [CrossRef] [PubMed]

28. Alleva, D.G.; Pavlovich, R.P.; Grant, C.; Kaser, S.B.; Beller, D.I. Aberrant macrophage cytokine production is a conserved feature among autoimmune-prone mouse strains: Elevated interleukin (IL)-12 and an imbalance in tumor necrosis factor-alpha and IL-10 define a unique cytokine profile in macrophages from young nonobese diabetic mice. *Diabetes* **2000**, *49*, 1106–1115. [PubMed]

29. Szelachowska, M.; Kretowski, A.; Kinalska, I. Decreased in vitro IL-4 [corrected] and IL-10 production by peripheral blood in first degree relatives at high risk of diabetes type-I. *Horm. Metab. Res.* **1998**, *30*, 526–530. [CrossRef] [PubMed]

30. Weaver, D.J.; Poligone, B.; Bui, T.; Abdel-Motal, U.M.; Baldwin, A.S.; Tisch, R. Dendritic cells from nonobese diabetic mice exhibit a defect in NF-kappa B regulation due to a hyperactive I kappa B kinase. *J. Immunol.* **2001**, *167*, 1461–1468. [CrossRef]

31. Welzen-Coppens, J.M.C.; van Helden-Meeuwsen, C.G.; Leenen, P.J.M.; Drexhage, H.A.; Versnel, M.A. Reduced numbers of dendritic cells with a tolerogenic phenotype in the prediabetic pancreas of NOD mice. *J. Leukoc. Biol.* **2012**, *92*, 1207–1213. [CrossRef]

32. Marleau, A.M.; Singh, B. Myeloid dendritic cells in non-obese diabetic mice have elevated costimulatory and T helper-1-inducing abilities. *J. Autoimmun.* **2002**, *19*, 23–35. [CrossRef] [PubMed]

33. Blancou, P.; Tardif, V.; Simon, T.; Rémy, S.; Carreño, L.; Kalergis, A.; Anegon, I. Immunoregulatory properties of heme oxygenase-1. *Methods Mol. Biol.* **2011**, *677*, 247–268.

34. Tardif, V.; Riquelme, S.A.; Remy, S.; Carreño, L.J.; Cortés, C.M.; Simon, T.; Hill, M.; Louvet, C.; Riedel, C.A.; Blancou, P.; et al. Carbon monoxide decreases endosome-lysosome fusion and inhibits soluble antigen presentation by dendritic cells to T cells. *Eur. J. Immunol.* **2013**, *43*, 2832–2844. [CrossRef]

35. George, J.F.; Braun, A.; Brusko, T.M.; Joseph, R.; Bolisetty, S.; Wasserfall, C.H.; Atkinson, M.A.; Agarwal, A.; Kapturczak, M.H. Suppression by CD4+CD25+ regulatory T cells is dependent on expression of heme oxygenase-1 in antigen-presenting cells. *Am. J. Pathol.* **2008**, *173*, 154–160. [CrossRef] [PubMed]

36. Ndisang, J.F.; Jadhav, A. Heme oxygenase system enhances insulin sensitivity and glucose metabolism in streptozotocin-induced diabetes. *Am. J. Physiol. Endocrinol. Metab.* **2009**, *296*, E829–E841. [CrossRef] [PubMed]

37. Mabley, J.G.; Southan, G.J.; Salzman, A.L.; Szabó, C. The combined inducible nitric oxide synthase inhibitor and free radical scavenger guanidinoethyldisulfide prevents multiple low-dose streptozotocin-induced diabetes in vivo and interleukin-1beta-induced suppression of islet insulin secretion in vitro. *Pancreas* **2004**, *28*, E39–E44. [CrossRef]

38. Witherden, D.; van Oers, N.; Waltzinger, C.; Weiss, A.; Benoist, C.; Mathis, D. Tetracycline-controllable selection of CD4(+) T cells: Half-life and survival signals in the absence of major histocompatibility complex class II molecules. *J. Exp. Med.* **2000**, *191*, 355–364. [CrossRef]

39. Chauveau, C.; Rémy, S.; Royer, P.J.; Hill, M.; Tanguy-Royer, S.; Hubert, F.-X.; Tesson, L.; Brion, R.; Beriou, G.; Gregoire, M.; et al. Heme oxygenase-1 expression inhibits dendritic cell maturation and proinflammatory function but conserves IL-10 expression. *Blood* **2005**, *106*, 1694–1702. [CrossRef]

International Journal of
Molecular Sciences

MDPI

Article

The Effects of Cobalt Protoporphyrin IX and Tricarbonyldichlororuthenium (II) Dimer Treatments and Its Interaction with Nitric Oxide in the Locus Coeruleus of Mice with Peripheral Inflammation

Patricia Moreno [1,2,†], Rafael Alves Cazuza [3,†], Joyce Mendes-Gomes [3,†], Andrés Felipe Díaz [1,2], Sara Polo [1,2], Sergi Leánez [1,2], Christie Ramos Andrade Leite-Panissi [3] and Olga Pol [1,2,*]

[1]   Grup de Neurofarmacologia Molecular, Institut d'Investigació Biomèdica Sant Pau, Hospital de la Santa Creu i Sant Pau, 08025 Barcelona, Spain; patricia.morenov@e-campus.uab.cat (P.M.); andresfelipe.diaz@e-campus.uab.cat (A.F.D.); sara.polor@e-campus.uab.cat (S.P.); sergi.leanez@uab.es (S.L.)
[2]   Institut de Neurociències, Universitat Autònoma de Barcelona, 08193 Barcelona, Spain
[3]   Department of Psychology, Faculty of Philosophy, Science and Letters, University of São Paulo, Ribeirão Preto 14040-901, SP, Brazil; cazuzarafaelalves@usp.br (R.A.C.); joypharm1@yahoo.com.br (J.M.-G.); christie@usp.br (C.R.A.L.-P.)
*   Correspondence: opol@santpau.es; Tel.: +34-619-757-054
†   These authors contributed equally to this work.

Received: 23 April 2019; Accepted: 2 May 2019; Published: 5 May 2019

Abstract: Heme oxygenase 1 (HO-1) and carbon monoxide were shown to normalize oxidative stress and inflammatory reactions induced by neuropathic pain in the central nervous system, but their effects in the locus coeruleus (LC) of animals with peripheral inflammation and their interaction with nitric oxide are unknown. In wild-type (WT) and knockout mice for neuronal (NOS1-KO) or inducible (NOS2-KO) nitric oxide synthases with inflammatory pain induced by complete Freund's adjuvant (CFA), we assessed: (1) antinociceptive actions of cobalt protoporphyrin IX (CoPP), an HO-1 inducer; (2) effects of CoPP and tricarbonyldichlororuthenium(II) dimer (CORM-2), a carbon monoxide-liberating compound, on the expression of HO-1, NOS1, NOS2, CD11b/c, GFAP, and mitogen-activated protein kinases (MAPK) in the LC. CoPP reduced inflammatory pain in different time-dependent manners in WT and KO mice. Peripheral inflammation activated astroglia in the LC of all genotypes and increased the levels of NOS1 and phosphorylated extracellular signal-regulated kinase 1/2 (p-ERK 1/2) in WT mice. CoPP and CORM-2 enhanced HO-1 and inhibited astroglial activation in all genotypes. Both treatments blocked NOS1 overexpression, and CoPP normalized ERK 1/2 activation. This study reveals an interaction between HO-1 and NOS1/NOS2 during peripheral inflammation and shows that CoPP and CORM-2 improved HO-1 expression and modulated the inflammatory and/or plasticity changes caused by peripheral inflammation in the LC.

Keywords: analgesia; carbon monoxide; heme oxygenase 1; inflammatory pain; locus coeruleus; nitric oxide

---

1. Introduction

The effects of carbon monoxide and nitric oxide on the regulation of the nociceptive responses induced by acute painful stimuli, chronic inflammation, or nerve injury and those associated with diabetic neuropathy have been investigated [1–4], but the possible interaction among them has been poorly evaluated.

Several studies demonstrated that, whereas carbon monoxide is a potent modulator of inflammation and nociception, nitric oxide has a more complex role in the development and maintenance of chronic pain. That is, while treatment with carbon monoxide inhaled and/or released by tricarbonyldichlororuthenium(II)dimer (CORM-2) exerted robust antiinflammatory [5,6] and antinociceptive actions during inflammatory and neuropathic pain [2,7,8], both types of pain were inhibited with the administration of selective nitric oxide synthase (NOS) inhibitors or in mice lacking neuronal (NOS1-KO) or inducible (NOS2-KO) nitric oxide synthases [9–13].

More interesting is the detail that, although the interaction among carbon monoxide and nitric oxide has been widely investigated at the vascular level [14–16], in the control of fewer, sepsis, hemorrhagic shock, etc. [17,18], only few studies examined this interaction in pain regulation. In this line, previous works revealed that carbon monoxide required the NOS pathway for its antinociceptive effects, whereas nitric oxide effects were produced independently of the carbon monoxide system [19,20]. Nonetheless, the interaction among heme oxygenase 1 (HO-1) enzyme, principally responsible for the antinociceptive effects induced by carbon monoxide [2], and NOS1 or NOS2 in inflammatory pain has not been evaluated. In this study, we assessed this interaction by testing the antinociceptive effects of cobalt protoporphyrin IX (CoPP), an HO-1 inducer, in both NOS1- and NOS2-deficient mice with chronic peripheral inflammation.

Recent works revealed that the systemic administration of CoPP and CORM-2, besides inhibiting neuropathic pain and blocking NOS1 and NOS2 over-expression in the spinal cords [2,21], activated powerful anti-inflammatory and antioxidant responses in several brain areas [22]. Considering that the locus coeruleus (LC) is implicated in the control of nociception [23] and both HO-1 [24,25] and NOS1/NOS2 enzymes [26,27] are well expressed in it, our objective was to evaluate the potential interaction between them in the LC of animals with inflammatory pain.

Several authors have demonstrated the relevant role played by glial cells in the development and maintenance of pain [28]. Therefore, and considering the inhibitory effects induced by CoPP and CORM-2 treatments on glial activation induced by sciatic nerve injury in the spinal cord and specific brain areas such as amygdala and hippocampus [2,22], in this study, we also evaluated the effects of these treatments on the expression of GFAP (an astroglial marker) and CD11b/c (a microglial marker) in LC of animals with complete Freund's adjuvant (CFA)-induced inflammatory pain.

It is well known that peripheral inflammation, in addition to induce the phosphorylation of several mitogen-activated protein kinases (MAPK)—especially, the extracellular signal-regulated kinase 1/2 (ERK 1/2) and c-Jun N-terminal kinase (JNK) in the spinal cord [29,30]—also activated ERK 1/2 in LC [31,32]. Considering that the systemic and peripheral administration of CoPP and other antiinflammatory agents, such as diclofenac, normalized the up regulation of p-ERK 1/2 and p-JNK induced by chronic pain in different brain areas [22,32], we examined the effects of CoPP and CORM-2 on the expression of p-ERK 1/2 and p-JNK in LC of animals with inflammatory pain.

Then, using wild-type (WT), NOS1-KO, and NOS2-KO mice with chronic peripheral inflammation, we assessed the antinociceptive actions of the repeated administration of CoPP and the effects of CoPP and CORM-2 treatments on the protein levels of HO-1, NOS1, NOS2, CD11b/c, GFAP, p-ERK $\frac{1}{2}$, and p-JNK in the LC of these animals.

2. Results

2.1. Antinociceptive Effects of CoPP in WT, NOS1-KO, and NOS2-KO Mice

Mechanical allodynia and thermal hyperalgesia were assessed after 1, 4, 7, and 11 days of repeated administration of CoPP or vehicle in WT, NOS1-KO, and NOS2-KO mice (Figure 1).

Our data showed that CFA injection caused mechanical allodynia in the ipsilateral paw of all genotypes from day 3 to day 14 since CFA injection ($p < 0.001$; one-way ANOVA vs. their respective contralateral paws). In NOS1-KO animals (Figure 1A), three-way ANOVA displayed significant effects of the genotype on day 11 ($p < 0.011$), of the treatment on days 4, 7, and 11 ($p < 0.020$), and with respect to the paw on days 0, 1, 4, 7, and 11 of CoPP treatment ($p < 0.001$). In addition, significant interactions among genotype and treatment at days 7 and 11 ($p < 0.042$), genotype and paw at day 14 ($p < 0.005$), treatment and paw at days 4, 7, and 11 ($p < 0.017$), and genotype, treatment, and paw at days 7 and 11 of CoPP treatment ($p < 0.019$) were also demonstrated. Our results, besides confirming that NOS1-KO mice had faster recovery of the mechanical allodynia than WT animals from days 10 to 14 after CFA injection ($p < 0.001$, one-way ANOVA), demonstrated that mechanical allodynia caused by CFA was further reduced in WT compared to NOS1-KO mice after 4 and 7 days of CoPP treatment ($p < 0.001$, one-way ANOVA; Figure 1A).

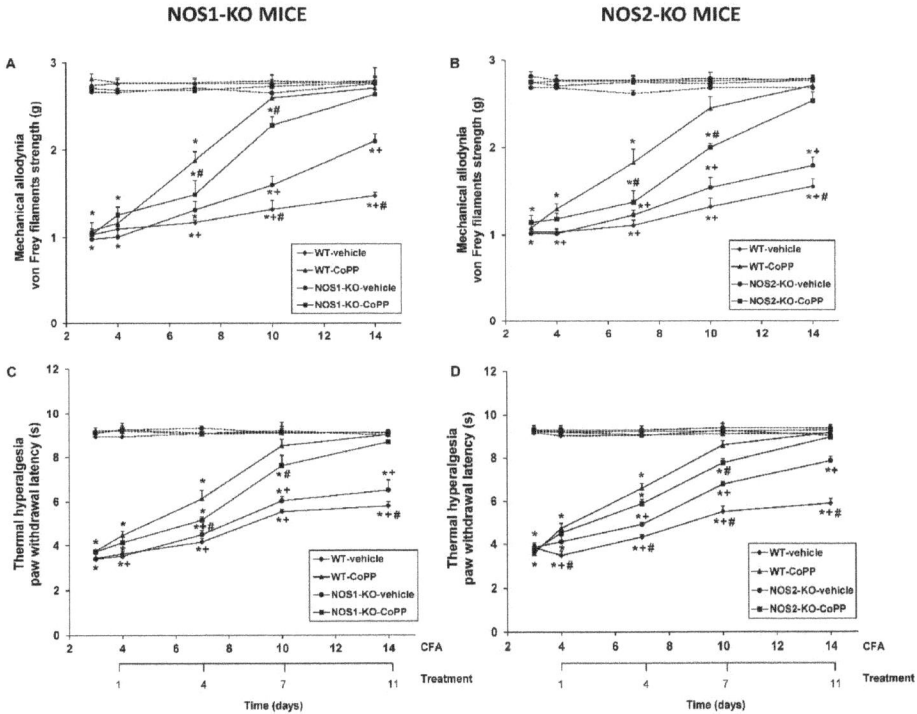

Figure 1. Antiallodynic and antihyperalgesic effects of cobalt protoporphyrin IX (CoPP) in wild-type (WT), neuronal nitric oxide synthase knock-out (NOS1-KO), and inducible nitric oxide synthase knock-out (NOS2-KO) mice with peripheral inflammation. Mechanical allodynia (**A,B**) and thermal hyperalgesia (**C,D**) in ipsilateral (continuous lines) and contralateral paws (discontinuous lines) of WT, NOS1-KO, and NOS2-KO mice treated for 11 days with vehicle or CoPP at 4, 7, 10, and 14 days after complete Freund's adjuvant (CFA) injection are shown. For each genotype, day and treatment were assessed; * denotes significant differences when compared with their respective contralateral paw, + denotes significant differences when compared with their respective ipsilateral paw treated with CoPP, and # denotes significant differences of the same treatment between genotypes ($p < 0.05$, one-way ANOVA, Student–Newman–Keuls test). Data are shown as mean values ± SEM; $n = 8$ animals per group.

In NOS2-KO animals (Figure 1B), three-way ANOVA also showed effects of treatment at days 1, 4, 7, and 11 ($p < 0.007$), and with respect to the paw at days 0, 1, 4, 7, and 11 of CoPP administration ($p < 0.001$). Moreover, significant interactions amongst genotype and treatment at day 7 ($p < 0.007$), treatment and paw at days 4, 7, and 11 ($p < 0.025$), as well as among genotype, treatment, and paw at days 4, 7, and 11 days of CoPP treatment ($p < 0.032$) were evident. Therefore, although similar mechanical allodynia caused by CFA was observed in WT and NOS2-KO animals, the inhibitory

Int. J. Mol. Sci. **2019**, *20*, 2211

effects of CoPP were stronger in WT that in NOS2-KO mice at 4 and 7 days of treatment ($p < 0.001$, one-way ANOVA, Figure 1B). In all genotypes, no treatment produced no effect on the respective contralateral paw.

In all genotypes, peripheral inflammation also reduced the threshold for evoking ipsilateral paw withdrawal to thermal stimulus from days 3 to 14 after CFA injection ($p < 0.001$; one-way ANOVA vs. the corresponding contralateral paw) (Figure 1C,D).

In NOS1-KO animals, three-way ANOVA proved significant effects of treatment at days 1, 4, 7, and 11 ($p < 0.002$) and of paw at days 0, 1, 4, 7, and 11 of CoPP administration ($p < 0.001$). Moreover, significant interactions among genotype and treatment at day 4 ($p < 0.008$), treatment and paw at days 1, 4, 7, and 11 of CoPP treatment ($p < 0.018$), and among genotype, treatment, and paw at days 4 and 7 of CoPP treatment in NOS1-KO mice were demonstrated ($p < 0.045$) (Figure 1C). Therefore, although similar thermal hyperalgesia induced by CFA was observed in both genotypes, its inhibition by CoPP was higher in WT mice than in NOS1-KO animals at days 4 and 7 of treatment ($p < 0.001$, one-way ANOVA).

In NOS2-KO mice, our data confirmed that the absence of this isoform improved the recovery from thermal hyperalgesia induced by CFA (Figure 1D). In addition, significant effects of genotype at days 7 and 11 ($p < 0.010$), treatment at days 1, 4, 7, and 11 ($p < 0.004$), and paw at days 0, 1, 4, 7, and 11 of CoPP treatment ($p < 0.001$) were demonstrated. Moreover, significant interactions among genotype and treatment at days 4, 7, and 11 ($p < 0.010$), treatment and paw on days 1, 4, 7, and 11 ($p < 0.005$), as well as among genotype, treatment, and paw at days 4, 7, and 11 of CoPP administration ($p < 0.002$) were shown. Thus, the reduced thermal hyperalgesia induced by CoPP in WT mice was similar to that induced by this compound in NOS2-KO animals during the overall experiment, except at day 7 of treatment, in which it was higher in WT than in NOS2-KO mice ($p < 0.05$; one-way ANOVA). In all genotypes, the absence of treatment did not produced any effect on the respective contralateral paw.

2.2. Effects of CoPP and CORM-2 on HO-1, NOS1, NOS2, CD11b/c, GFAP, p-ERK 1/2, and p-JNK Expression in the LC of WT Mice with Peripheral Inflammation

Our data showed that CoPP and CORM-2 treatments augmented the expression of HO-1 in the LC of CFA-injected WT mice ($p < 0.001$; one-way ANOVA vs. naïve and CFA-injected mice treated with vehicle) (Figure 2A), and the enhanced expression of NOS1 (Figure 2B) and GFAP (Figure 2F) caused by peripheral inflammation ($p < 0.024$; one-way ANOVA compared to the corresponding naïve vehicle-treated animals) was normalized by both CoPP and CORM-2 treatments. Moreover, the enhanced expression of p-ERK 1/2 caused by peripheral inflammation ($p < 0.005$; one-way ANOVA vs. naïve vehicle-treated mice) was normalized by CoPP treatment (Figure 2G). None of these treatments changed the unaltered expression of NOS2 (Figure 2C), CD11b/c (Figure 2D), or p-JNK (Figure 2H) in the LC of CFA-injected mice.

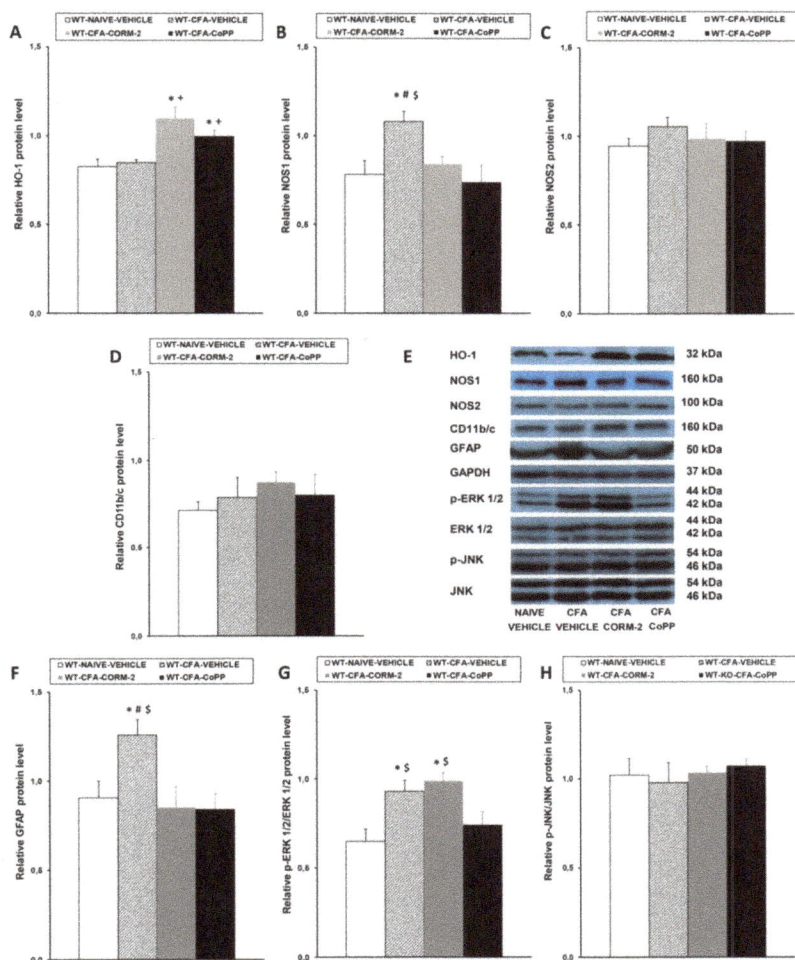

Figure 2. Effects of CoPP and tricarbonyldichlororuthenium(II) dimer (CORM-2) on the protein levels of HO-1, NOS1, NOS2, CD11b/c, GFAP, p-ERK 1/2, and p-JNK in the LC of WT mice. Protein levels of HO-1 (**A**), NOS1 (**B**), NOS2 (**C**), CD11b/c (**D**), GFAP (**F**), p-ERK 1/2 (**G**), and p-JNK (**H**) in the LC from CFA-injected WT mice treated with vehicle, CORM-2, or CoPP are represented. These levels in naïve mice treated with vehicle are also represented as controls. In all panels, * denotes significant differences vs. naïve vehicle-treated mice, + denotes significant differences vs. CFA-injected mice treated with vehicle, # denotes significant differences vs. CFA-injected mice treated with CORM-2, and $ denotes significant differences vs. CFA-injected mice treated with CoPP ($p < 0.05$, one-way ANOVA, Student–Newman–Keuls test). Examples of western blots for HO-1 (32 kDa), NOS1 (160 kDa), NOS2 (100 kDa), CD11b/c (160 kDa), GFAP (50 kDa), GAPDH (37 kDa), p-ERK 1/2/total ERK 1/2 (42–44 kDa), and p-JNK/total JNK (46–54 kDA) are shown (**E**). The levels of phosphorylated proteins are indicated relative to the corresponding total protein levels, while the levels of the remaining proteins are relative to those of GAPDH. Results are expressed as mean ± SEM; $n = 4$ samples per group.

2.3. Effects of CoPP and CORM-2 on HO-1, NOS2, CD11b/c, GFAP, p-ERK 1/2, and p-JNK Expression in the LC of NOS1-KO Mice with Peripheral Inflammation

Similar to what observed in WT mice, while peripheral inflammation did not modify the protein levels of HO-1 in the LC of NOS1-KO animals (Figure 3A), HO-1 expression was significantly augmented in CoPP- and CORM-2-treated animals ($p < 0.001$; one-way ANOVA vs. naïve and CFA-injected mice treated with vehicle). None of these treatments altered the expression of NOS2 (Figure 3B), CD11b/c (Figure 3C), and p-JNK (Figure 3G) in the LC of CFA-injected NOS1-KO mice and, although peripheral inflammation did not activate ERK 1/2, an increased expression of p-ERK 1/2 was detected in NOS1-KO mice treated with CORM-2 ($p < 0.019$ vs. naïve and CFA-injected mice treated with vehicle or CoPP) (Figure 3F). In addition, the overexpression of GFAP induced by CFA was completely inhibited by CORM-2 and CoPP treatments (Figure 3E).

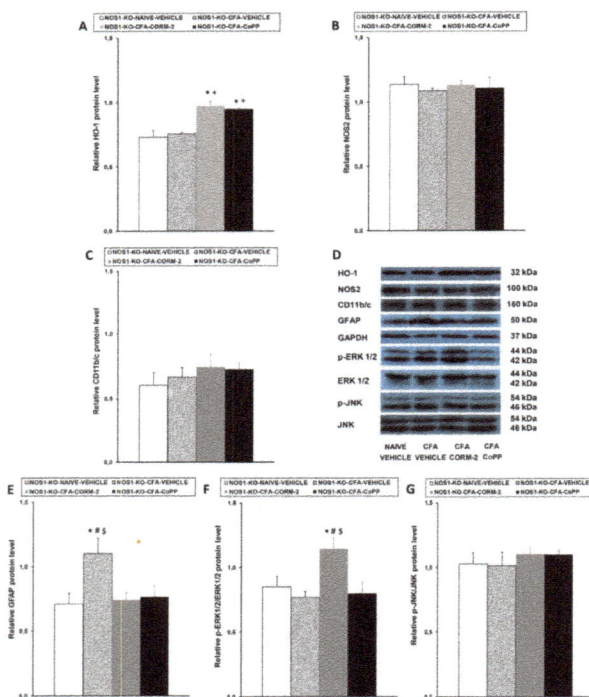

Figure 3. Effects of CoPP and CORM-2 on the protein levels of HO-1, NOS2, CD11b/c, GFAP, p-ERK 1/2, and p-JNK in the LC of NOS1-KO mice. Protein levels of HO-1 (**A**), NOS2 (**B**), CD11b/c (**C**), GFAP (**E**), p-ERK 1/2 (**F**), and p-JNK (**G**) in the LC of CFA-injected NOS1-KO mice treated with vehicle, CORM-2, or CoPP are represented. These levels from naïve mice treated with vehicle are also represented as controls. In all panels, * represents significant differences vs. naïve vehicle-treated mice, + represents significant differences vs. CFA-injected mice treated with vehicle, # represents significant differences vs. CFA-injected mice treated with CORM-2, and $ represents significant differences vs. CFA-injected mice treated with CoPP ($p < 0.05$, one-way ANOVA, Student–Newman–Keuls test). Examples of western blots for HO-1(32 kDa), NOS2 (100 kDa), CD11b/c (160 kDa), GFAP (50 kDa), GAPDH (37 kDa), p-ERK 1/2/total ERK 1/2 (42–44 kDa), and p-JNK/total JNK (46–54 kDA) are shown (**D**). The levels of phosphorylated proteins are relative to the total levels of the corresponding proteins, while the levels of the remaining proteins are relative to those of GAPDH. Results are expressed as mean ± SEM; $n = 4$ samples per group.

2.4. Effects of CoPP and CORM-2 on HO-1, NOS1, CD11b/c, GFAP, p-ERK 1/2, and p-JNK Expression in the LC from NOS2-KO Mice with Peripheral Inflammation

Similar to what observed in NOS1-KO mice, although peripheral inflammation did not change the expression of HO-1 in the LC of NOS2-KO mice (Figure 4A), CoPP and CORM-2 treatments significantly enhanced its expression ($p < 0.007$; one-way ANOVA vs. naïve and CFA-injected mice treated with vehicle). Moreover, repeated treatment with CORM-2 or CoPP significantly reduced the enhanced expression of GFAP induced by peripheral inflammation in the LC (Figure 4E). In contrast, neither CORM-2 nor CoPP treatments altered the unchanged levels of NOS1 (Figure 4B), CD11b/c (Figure 4C), p-ERK 1/2 (Figure 4F), or p-JNK (Figure 4G) in the LC of CFA-injected NOS2-KO mice.

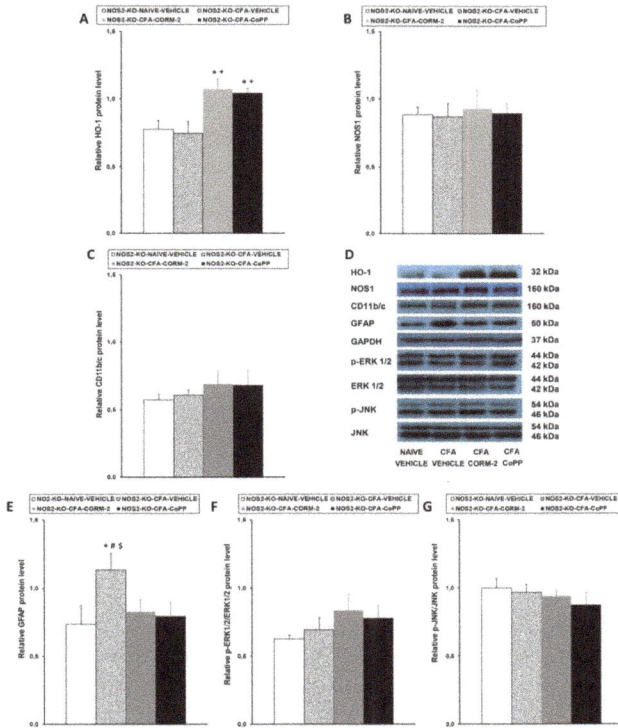

Figure 4. Effects of CoPP and CORM-2 on the protein levels of HO-1, NOS1, CD11b/c, GFAP, p-ERK 1/2, and p-JNK in the LC of NOS2-KO mice. Protein levels of HO-1 (**A**), NOS1 (**B**), CD11b/c (**C**), GFAP (**E**), p-ERK 1/2 (**F**), and p-JNK (**G**) in the LC of CFA-injected NOS2-KO mice treated with vehicle, CORM-2, or CoPP are represented. These levels from naïve mice treated with vehicle are also represented as controls. In all panels, * denotes significant differences vs. naïve vehicle-treated mice, + represents significant differences vs. CFA-injected mice treated with vehicle, # represents significant differences vs. CFA-injected mice treated with CORM-2, and $ represents significant differences vs. CFA-injected mice treated with CoPP ($p < 0.05$, one-way ANOVA, Student–Newman–Keuls test). Examples of western blots for HO-1 (32 kDa), NOS1 (160 kDa), CD11b/c (160 kDa), GFAP (50 kDa), GAPDH (37 kDa), p-ERK 1/2/total ERK 1/2 (42–44 kDa), and p-JNK/total JNK (46–54 kDA) are shown (**D**). The levels of phosphorylated proteins are relative to the total levels of the corresponding proteins, while the levels of the remaining proteins are relative to those of GAPDH. Each column represents the mean, and the vertical bars indicate the SEM; $n = 4$ samples per group.

3. Discussion

This study revealed that the systemic repeated administration of CoPP differentially inhibited allodynia and hyperalgesia caused by CFA in WT and NOS1-KO or NOS2-KO mice. Moreover, CoPP and CORM-2 treatments induced HO-1 overexpression and inhibited activated astroglia in the LC of all genotypes. Both treatments also normalized the upregulation of NOS1 caused by peripheral inflammation in WT mice. Moreover, peripheral inflammation activated ERK 1/2 in the LC of WT animals but not in NOS1-KO or NOS2-KO mice, and only CoPP treatment inhibited ERK 1/2 phosphorylation.

Our findings indicated that treatment with CoPP throughout 11 consecutive days inhibited the mechanical and thermal hypersensitivity triggered by peripheral inflammation in a different time-dependent manner in WT and NOS1- or NOS2-deficient mice. That is, the antiallodynic and antihyperalgesic effects of CoPP in WT mice were stronger than in NOS1-KO or NOS2-KO mice after 4 to 7 days of treatment. These data demonstrate the involvement of both NOS isoforms in the analgesic effects of CoPP during inflammatory pain and reveal an interaction between HO-1 and NOS1/NOS2 isoenzymes in chronic inflammatory pain conditions. Our results are in agreement with the analgesic effects induced by HO-1 inducers during acute [33] and neuropathic pain [2,4,21,34,35]. Our findings also support the improved antinociceptive actions of carbon monoxide, liberated by CORM-2, in WT versus both KO mice with inflammatory pain [20] and further demonstrate the crucial role played by HO-1 in the interaction between carbon monoxide and nitric oxide during the management of inflammatory pain.

It is well known that LC is a supraspinal structure implicated in the control of pain, but the effects induced by chronic treatment with CoPP or CORM-2 during inflammatory pain in this brain area had not been investigated previously. In previous works, we showed that the repetitive administration of CoPP or CORM-2 upregulated HO-1 levels in paws and dorsal root ganglia of animals with inflammatory pain [20,36] as well as in spinal cords and sciatic nerves from diabetic mice [4,35]. Recently, an augmented expression of HO-1 was also demonstrated in the prefrontal cortex and hippocampus of WT mice with neuropathic pain repeatedly treated with CORM-2 or CoPP [22]. In this study, we demonstrated for the first time that, under inflammatory pain circumstances, both HO-1 and carbon monoxide inducers improved HO-1 levels in the LC of WT, NOS1-KO, and NOS2-KO mice. These data agree with those reporting augmented expression of HO-1 induced by CoPP and CORM-2 treatments in the central nervous system of WT mice with neuropathic pain [22] and further support data showing the upregulation of this enzyme in the paw of CFA-injected NOS1- and NOS2-deficient mice systemically treated with CORM-2 [20]. Our results reveal the central antioxidant effects induced by these treatments during inflammatory pain in the presence or absence of NOS enzymes and suggest that the enhanced expression of HO-1 produced by CoPP and CORM-2 in the LC might be involved in the antinociceptive effects of these compounds during inflammatory pain.

In this study, an increased expression of NOS1, but not of NOS2, was also revealed in the LC of WT mice with peripheral inflammation. This is in agreement with observations in inflamed paws, showing the relevant role performed by NOS1 in the maintenance of inflammatory pain in the central and peripheral nervous system [37,38]. Moreover, the fact that the systemic administration of CoPP and CORM-2 inhibited NOS1-upregulation in the LC revealed the central anti-inflammatory effects induced by these compounds during inflammatory pain. In contrast to NOS1, no changes in the protein levels of NOS2 were observed in the LC of these animals. These data are in agreement with the observed lack of changes in the expression of this enzyme in the dorsal root ganglia and spinal cords of WT mice at day 14 after CFA injection [30]. All of these data suggest that, in chronic peripheral inflammatory pain conditions, the effects induced by CoPP and CORM-2 treatments in the LC are mainly produced via regulating NOS1 expression.

The implication of spinal glia in the progress of chronic pain is well recognized, but less is known about its activation in supraspinal structures after peripheral inflammatory pain. Thus, in this study, we examined if peripheral inflammation induced astroglial and/or microglial activation in the LC

14 days after CFA injection. Our results showed that CFA induced astroglial activation in the LC of WT, NOS1-KO, and NOS2-KO mice. No changes in the expression of CD11b/c (a microglial marker) were detected in the LC of any genotype, confirming the crucial role of astroglia in the maintenance of chronic inflammatory pain [39,40]. Interestingly, CORM-2 and CoPP inhibited the overexpression of GFAP (an astroglial marker) in the LC of all genotypes, showing that the systemic administration of these compounds has anti-inflammatory properties in the LC. Our results also suggest the participation of astroglia in the analgesic effects of CORM-2 and CoPP during inflammatory pain. These findings are supported by the antinociceptive actions produced by the treatment with selective astroglial cell inhibitors, such as fluorocitrate and α-aminoadipate, during chronic pain [41,42].

The participation of MAPK in the development and maintenance of chronic pain is well documented [29,43]. Thus, under inflammatory pain conditions, the expression of ERK1/2 and JNK are activated in the spinal cord [30]. In this work, we analyzed the actions of CoPP and CORM-2 on the expression of these MAPK in the LC of WT and NOS-deficient mice. As occurs in other animal pain models [32,44], CFA-induced peripheral inflammation activated ERK 1/2 in the LC of WT mice. Curiously, this effect was not detected in NOS1-KO or NOS2-KO mice, suggesting that nitric oxide generated by these isoforms is involved in the plasticity changes induced by chronic peripheral inflammation in LC. Moreover, while CoPP normalized ERK 1/2 activation in WT animals, CORM-2 did not alter or enhanced p-ERK 1/2 levels in WT and NOS1-KO mice, respectively, thus revealing that ERK 1/2 activation induced by peripheral inflammation in the LC might be modulated by the systemic treatment with the antioxidant enzyme HO-1. The lack of effects of CORM-2 on the expression of p-ERK 1/2 in WT mice confirmed similar results obtained in the central and peripheral nervous system of animals with neuropathic pain [22,45], suggesting that CORM-2 might act via inhibiting other pathways implicated in the regulation of inflammatory pain. Finally, although more studies are required to explain this phenomenon, the increased expression of p-ERK 1/2 induced by CORM-2 in NOS1-KO mice is an indication of the plausible involvement of NOS1 in ERK 1/2 activation induced by CORM-2 in WT mice.

In conclusion, this study reveals an interaction between HO-1 and NOS1/NOS2 enzymes during peripheral inflammation and shows that CoPP treatment inhibited inflammatory pain by improving HO-1 expression and decreasing NOS1 overexpression, which would restrict the activation of astroglia and subsequent ERK 1/2 activation in the LC.

4. Materials and Methods

4.1. Animals

Experiments were performed in male NOS1-KO (C57BL/6 J background) and NOS2-KO mice (C57BL/6 J background) acquired from Jackson Laboratories (Bar Harbor, ME, USA). WT mice with the same genetic background (C57BL/6J) were obtained from Envigo Laboratories (Barcelona, Spain). Animals weighing 21–25 g were housed under 12 h/12 h light/dark conditions and controlled temperature (22°C) and humidity (66%). Animals with free access to food and water were utilized after 7 days acclimatization to the housing conditions. Experiments were performed between 9:00 a.m. and 5:00 p.m., executed in accordance to the animals guidelines of the European Communities Council (86/609/ECC, 90/679/ECC; 98/81/CEE, 2003/65/EC, and Commission Recommendation 2007/526/EC), and approved by the Comitè d'Ètica en Experimentació Animal of Universitat Autònoma de Barcelona (number: 1325R5, 29 November 2013). Maximal efforts to minimize the quantity of mice employed and their suffering were made.

4.2. Chronic Inflammatory Pain Induction

Chronic inflammatory pain was provoked with the sub-plantar injection of CFA (30 µL) (Sigma-Aldrich, St. Louis, MO, USA) into the right hind paw under brief anesthetic conditions with isoflurane (3% induction, 2% maintenance) according to the method used by our group [38].

Mechanical allodynia and thermal hyperalgesia were assessed with the von Frey filaments and plantar tests, respectively.

4.3. Nociceptive Behavioral Tests

Mechanical allodynia was estimated by determining the hind paw withdrawal response to von Frey filament stimulation. Thus, mice were positioned in methacrylate cylinders (20 cm high, 9 cm diameter) with a wire grid bottom through which the von Frey filaments (North Coast Medical, Inc., San Jose, CA, USA) were applied according to the up–down paradigm [46]. A filament of 0.4 g was applied first, and one of 3.5 g as a cut-off. The strength of the next filament was reduced or augmented depending on the response. Withdrawal, shaking, or licking the paw were considered nociceptive-like reactions.

Thermal hyperalgesia was assessed according to previous methods [47]. Paw withdrawal latency in reply to a radiant heat was assessed using the plantar test device (Ugo Basile, Italy). Mice were placed in methacrylate cylinders, 20 cm high × 9 cm diameter, situated on a glass surface. The heat source was situated under the plantar surface of the hind paw and activated with a light beam. A cut-off time of 12 s was utilized to avoid tissue damage. The mean paw withdrawal latencies were calculated from the average of 2–3 separate trials, taken at 5 min intervals to prevent thermal sensitization.

In both tests, the animals were habituated to the environment for 1 h before the experiment. Both ipsilateral and contralateral paws were tested.

4.4. Western Blot Analysis

Mice were euthanized by cervical dislocation after 0 (naïve) or 14 days from CFA injection. LC were extracted and preserved at 80 °C until use. Samples from four animals were combined to have satisfactory proteins levels to analyze HO-1, NOS1, NOS2, CD11b/c, GFAP, pERK1/2/ERK1/2, and p-JNK/JNK in LC by western blot assay. The homogenization of tissues was made in cold lysis buffer (50 mM Tris·Base, 150 nM NaCl, 1% NP-40, 2 mM EDTA, 1mM phenylmethylsulfonyl fluoride, 0.5 Triton X-100, 0.1% sodium dodecyl sulfate, 1 mM Na3VO4, 25 mM NaF, 0.5% protease inhibitor cocktail, and 1% phosphatase inhibitor cocktail). All reagents were acquired from Sigma-Aldrich (St. Louis, MO, USA), except for NP-40 which was bought from Calbiochem (Darmstadt, Germany). After solubilization for 1 h at 4 °C, crude homogenates were sonicated for 10 s and centrifuged at 700× g for 15 min at 4 °C. The supernatant (60 µg of total protein) was mixed with 4X Laemmli loading buffer and loaded onto a 4% stacking/10% separating sodium dodecyl sulfate polyacrylamide gels. Proteins were electrophoretically transferred onto a polyvinylidene fluoride membrane for 120 min and successfully blocked with phosphate-buffered saline containing 5% nonfat dry milk or Tris-buffered saline with Tween 20 containing 5% bovine serum albumin for 75 min; they were then incubated with specific rabbit primary antibodies anti HO-1 (1:300; Abcam, Cambridge, United Kingdom), NOS1 (1:200; Abcam, Cambridge, United Kingdom), NOS2 (1:100; Abcam, Cambridge, United Kingdom), CD11b/c (1:160, Novus Biologic, Littleton, CO, USA), GFAP (1:3000, Novus Biologic, Littleton, CO, USA), phospho-ERK 1/2 and total ERK 1/2 (1:250; Cell Signaling Technology, Danvers, MA, USA), phospho-JNK and total JNK (1:250; Cell Signaling Technology, Danvers, MA, USA), or glyceraldehyde-3-phosphate dehydrogenase antibody (GAPDH) (1:5000, Merck, Billerica, MA, USA) overnight at 4 °C. The blots were then incubated with anti-rabbit secondary polyclonal antibodies conjugated to horseradish peroxidase (GE Healthcare, Little Chalfont, Buckinghamshire, UK) for 1 h at r.t. Proteins were detected by using chemiluminescence reagents provided in the ECL kit (GE, Healthcare, Little Chalfont, Buckinghamshire, UK) and exposure onto hyperfilms (GE, Healthcare, Little Chalfont, Buckinghamshire, UK). Blots' intensity was quantified by densitometry.

4.5. Experimental Procedure

In WT, NOS1-KO, and NOS2-KO mice baseline responses were established in von Frey filaments and plantar tests. All mice were tested at days 3, 4, 7, 10, and 14 after CFA injection.

The animals were intraperitoneally injected with vehicle, 2.5 mg/kg CoPP, or 5 mg/kg CORM-2, two times a day for 11 days, from day 4 to day 14 after CFA injection, according to earlier studies [2,20]. The antinociceptive effects of CoPP were evaluated at 1, 4, 7, and 11 days from its administration. The investigator who made these experiments was blinded to the treatments.

We assessed the effects of CoPP and CORM-2 on the expression of HO-1, NOS1, NOS2, CD11b/c, GFAP, pERK 1/2, ERK 1/2, p-JNK, and JNK in the LC of WT, NOS1-KO, or NOS2-KO mice 14 days after CFA injection by western blot assay. For each genotype, naïve mice treated with vehicle were used as controls ($n = 4$ samples per group).

4.6. Drugs

CoPP was purchased from Frontier scientific (Livchem GmbH & Co., Frankfurt, Germany), and CORM-2 from Sigma-Aldrich (St. Louis, MO, USA). Both compounds were dissolved in dimethyl sulfoxide (DMSO; 1% solution in saline), freshly prepared before use, and injected intraperitoneally at 10 mL/kg of body weight, 3 h before testing, twice a day. Control animals received the same volume of vehicle.

4.7. Statistical Analysis

The SPSS (version 17 for Windows, IBM, Madrid, Spain) was used to perform the statistical analysis. Data were expressed as mean ± standard error of the mean (SEM). For each behavioral test and time evaluated, the antiallodynic and antihyperalgesic effects of CoPP were evaluated by using a three-way analysis of variance (ANOVA) (genotype, treatment, and paw as factors of variation) followed by the pertinent one-way ANOVA and Student–Newman–Keuls test whenever required. For each genotype, alterations in the protein levels of HO-1, NOS1, NOS2, CD11b/c, GFAP, p-ERK 1/2, ERK 1/2, p-JNK, and JNK in the LC of CFA-injected mice treated with CORM-2, CoPP, or vehicle vs. naïve vehicle-treated mice were analyzed by a one-way ANOVA followed by the Student–Newman–Keuls test. A $p < 0.05$ was considered significant.

Author Contributions: Formal analysis, P.M., R.A.C., and J.M.-G.; Funding acquisition, C.R.A.L.-P. and O.P.; Investigation, P.M., R.A.C., J.M.-G., A.F.D., S.P., and S.L.; Supervision, C.R.A.L.-P. and O.P.; Writing, O.P. All authors read and approved the final manuscript.

Funding: This work was supported by Ministerio de Economía y Competitividad, Instituto de Salud Carlos III, Madrid, Spain and Fondo Europeo de Desarrollo Regional (FEDER), Unión Europea [Grants: PS0900968 and PI1400927], and CAPES/PROEX, CNPq, Brasil [401472/2014-0].

Conflicts of Interest: The authors declare no conflict of interest.

Abbreviations

HO-1	Heme oxygenase 1
LC	Locus coeruleus
WT	Wild-type
NOS1	Neuronal nitric oxide synthase
NOS2	Inducible nitric oxide synthase
KO	Knockout
CFA	Complete Freund's adjuvant
CoPP	Cobalt protoporphyrin IX
CORM-2	Tricarbonyldichlororuthenium(II) dimer
ERK 1/2	Extracellular signal-regulated kinase 1/2
JNK	c-Jun N-terminal kinase
GAPDH	Glyceraldehyde-3-phosphate dehydrogenase
GFAP	Glial fibrillary acidic protein
SEM	Standard error of the mean
ANOVA	Analysis of variance

References

1. Carvalho, P.G.; Branco, L.G.; Panissi, C.R. Involvement of the heme oxygenase-carbon monoxide-cGMP pathway in the nociception induced by acute painful stimulus in rats. *Brain Res.* **2011**, *1385*, 107–113. [CrossRef]

2. Hervera, A.; Leánez, S.; Negrete, R.; Motterlini, R.; Pol, O. Carbon monoxide reduces neuropathic pain and spinal microglial activation by inhibiting nitric oxide synthesis in mice. *PLoS ONE* **2012**, *7*, e43693. [CrossRef] [PubMed]

3. Bijjem, K.R.; Padi, S.S.; lal Sharma, P. Pharmacological activation of heme oxygenase (HO)-1/carbon monoxide pathway prevents the development of peripheral neuropathic pain in Wistar rats. *Naunyn Schmiedebergs Arch. Pharmacol.* **2013**, *386*, 79–90. [CrossRef]

4. Castany, S.; Carcolé, M.; Leánez, S.; Pol, O. The role of carbon monoxide on the anti-nociceptive effects and expression of cannabinoid 2 receptors during painful diabetic neuropathy in mice. *Psychopharmacology* **2016**, *233*, 2209–2219. [CrossRef] [PubMed]

5. Takagi, T.; Naito, Y.; Inoue, M.; Akagiri, S.; Mizushima, K.; Handa, O.; Kokura, S.; Ichikawa, H.; Yoshikawa, T. Inhalation of carbon monoxide ameliorates collagen-induced arthritis in mice and regulates the articular expression of IL-1beta and MCP-1. *Inflammation* **2009**, *32*, 83–88. [CrossRef]

6. Bonelli, M.; Savitskaya, A.; Steiner, C.W.; Rath, E.; Bilban, M.; Wagner, O.; Bach, F.H.; Smolen, J.S.; Scheinecker, C. Heme oxygenase-1 end-products carbon monoxide and biliverdin ameliorate murine collagen induced arthritis. *Clin. Exp. Rheumatol.* **2012**, *30*, 73–78.

7. Hervera, A.; Gou, G.; Leánez, S.; Pol, O. Effects of treatment with a carbon monoxide-releasing molecule and a heme oxygenase 1 inducer in the antinociceptive effects of morphine in different models of acute and chronic pain in mice. *Psychopharmacology* **2013**, *228*, 463–477. [CrossRef]

8. Wang, H.; Sun, X. Carbon Monoxide-Releasing Molecule-2 Inhibits Connexin 43-Hemichannel Activity in Spinal Cord Astrocytes to Attenuate Neuropathic Pain. *J. Mol. Neurosci.* **2017**, *63*, 58–69. [CrossRef]

9. Chu, Y.C.; Guan, Y.; Skinner, J.; Raja, S.N.; Johns, R.A.; Tao, Y.X. Effect of genetic knockout or pharmacologic inhibition of neuronal nitric oxide synthase on complete Freund's adjuvant-induced persistent pain. *Pain* **2005**, *119*, 113–123. [CrossRef] [PubMed]

10. De Alba, J.; Clayton, N.M.; Collins, S.D.; Colthup, P.; Chessell, I.; Knowles, R.G. GW274150, a novel and highly selective inhibitor of the inducible isoform of nitric oxide synthase (iNOS), shows analgesic effects in rat models of inflammatory and neuropathic pain. *Pain* **2006**, *120*, 170–181. [CrossRef]

11. Boettger, M.K.; Uceyler, N.; Zelenka, M.; Schmitt, A.; Reif, A.; Chen, Y.; Sommer, C. Differences in inflammatory pain in nNOS-, iNOS- and eNOS-deficient mice. *Eur. J. Pain* **2007**, *11*, 810–818. [CrossRef]

12. Leánez, S.; Hervera, A.; Pol, O. Peripheral antinociceptive effects of mu- and delta-opioid receptor agonists in NOS2 and NOS1 knockout mice during chronic inflammatory pain. *Eur. J. Pharmacol.* **2009**, *602*, 41–49. [CrossRef]

13. Hervera, A.; Negrete, R.; Leánez, S.; Martín-Campos, J.; Pol, O. The role of nitric oxide in the local antiallodynic and antihyperalgesic effects and expression of delta-opioid and cannabinoid-2 receptors during neuropathic pain in mice. *J. Pharmacol. Exp. Ther.* **2010**, *334*, 887–896. [CrossRef]

14. Durante, W.; Schafer, A.I. Carbon monoxide and vascular cell function (review). *Int. J. Mol. Med.* **1998**, *2*, 255–262. [CrossRef]

15. Speranza, L.; Franceschelli, S.; Pesce, M.; Ferrone, A.; Patruno, A.; Riccioni, G.; De Lutiis, M.A.; Felaco, M.; Grilli, A. Negative feedback interaction of HO-1/INOS in PBMC of acute congestive heart failure patients. *J. Biol. Regul. Homeost. Agents* **2013**, *27*, 739–748.

16. Luo, W.; Wang, Y.; Yang, H.; Dai, C.; Hong, H.; Li, J.; Liu, Z.; Guo, Z.; Chen, X.; He, P.; et al. Heme oxygenase-1 ameliorates oxidative stress-induced endothelial senescence via regulating endothelial nitric oxide synthase activation and coupling. *Aging* **2018**, *10*, 1722–1744. [CrossRef]

17. Soriano, R.N.; Kwiatkoski, M.; Batalhao, M.E.; Branco, L.G.; Carnio, E.C. Interaction between the carbon monoxide and nitric oxide pathways in the locus coeruleus during fever. *Neuroscience* **2012**, *206*, 69–80. [CrossRef]

18. Duvigneau, J.C.; Kozlov, A.V. Pathological Impact of the Interaction of NO and CO with Mitochondria in Critical Care Diseases. *Front. Med.* **2017**, *4*, 223. [CrossRef]

19. Steiner, A.A.; Branco, L.G.; Cunha, F.Q.; Ferreira, S.H. Role of the haeme oxygenase/carbon monoxide pathway in mechanical nociceptor hypersensitivity. *Br. J. Pharmacol.* **2001**, *132*, 1673–1682. [CrossRef]
20. Negrete, R.; Hervera, A.; Leánez, S.; Pol, O. Treatment with a carbon monoxide-releasing molecule inhibits chronic inflammatory pain in mice: Nitric oxide contribution. *Psychopharmacology* **2014**, *231*, 853–861. [CrossRef]
21. Liu, X.; Zhang, Z.; Cheng, Z.; Zhang, J.; Xu, S.; Liu, H.; Jia, H.; Jin, Y. Spinal Heme Oxygenase-1 (HO-1) Exerts Antinociceptive Effects Against Neuropathic Pain in a Mouse Model of L5 Spinal Nerve Ligation. *Pain Med.* **2016**, *17*, 220–229. [CrossRef]
22. Riego, G.; Redondo, A.; Leánez, S.; Pol, O. Mechanism implicated in the antiallodynic and antihyperalgesic effects induced by the activation of heme oxygenase 1/carbon monoxide signaling pathway in the central nervous system of mice with neuropathic pain. *Biochem. Pharmacol.* **2018**, *148*, 52–63. [CrossRef]
23. Benarroch, E.E. Locus coeruleus. *Cell. Tissue Res.* **2018**, *373*, 221–232. [CrossRef]
24. Hundahl, C.A.; Kelsen, J.; Dewilde, S.; Hay-Schmidt, A. Neuroglobin in the rat brain (II): Co-localisation with neurotransmitters. *Neuroendocrinology* **2008**, *88*, 183–198. [CrossRef]
25. Cazuza, R.A.; Pol, O.; Leite-Panissi, C.R.A. Enhanced expression of heme oxygenase-1 in the locus coeruleus can be associated with anxiolytic-like effects. *Behav. Brain Res.* **2018**, *336*, 204–210. [CrossRef]
26. Le Maître, E.; Barde, S.S.; Palkovits, M.; Diaz-Heijtz, R.; Hökfelt, T.G. Distinct features of neurotransmitter systems in the human brain with focus on the galanin system in locus coeruleus and dorsal raphe. *Proc. Natl. Acad. Sci. USA* **2013**, *110*, E536–E545.
27. Pablos, P.; Mendiguren, A.; Pineda, J. Contribution of nitric oxide-dependent guanylate cyclase and reactive oxygen species signaling pathways to desensitization of μ-opioid receptors in the rat locus coeruleus. *Neuropharmacology* **2015**, *99*, 422–431. [CrossRef]
28. Ji, R.R.; Nackley, A.; Huh, Y.; Terrando, N.; Maixner, W. Neuroinflammation and Central Sensitization in Chronic and Widespread Pain. *Anesthesiology* **2018**, *129*, 343–366. [CrossRef]
29. Ji, R.R.; Gereau, R.W., 4th; Malcangio, M.; Strichartz, G.R. MAP kinase and pain. *Brain Res. Rev.* **2009**, *60*, 135–148. [CrossRef]
30. Redondo, A.; Chamorro, P.A.F.; Riego, G.; Leánez, S.; Pol, O. Treatment with Sulforaphane Produces Antinociception and Improves Morphine Effects during Inflammatory Pain in Mice. *J. Pharmacol. Exp. Ther.* **2017**, *363*, 293–302. [CrossRef]
31. Imbe, H.; Okamoto, K.; Donishi, T.; Kawai, S.; Enoki, K.; Senba, E.; Kimura, A. Activation of ERK in the locus coeruleus following acute noxious stimulation. *Brain Res.* **2009**, *1263*, 50–57. [CrossRef]
32. Borges, G.; Neto, F.; Mico, J.A.; Berrocoso, E. Reversal of monoarthritis-induced affective disorders by diclofenac in rats. *Anesthesiology* **2014**, *120*, 1476–1490. [CrossRef]
33. Gou, G.; Leánez, S.; Pol, O. The role of gaseous neurotransmitters in the antinociceptive effects of morphine during acute thermal pain. *Eur. J. Pharmacol.* **2014**, *737*, 41–46. [CrossRef]
34. Shen, Y.; Zhang, Z.J.; Zhu, M.D.; Jiang, B.C.; Yang, T.; Gao, Y.J. Exogenous induction of HO-1 alleviates vincristine-induced neuropathic pain by reducing spinal glial activation in mice. *Neurobiol. Dis.* **2015**, *79*, 100–110. [CrossRef]
35. McDonnell, C.; Leánez, S.; Pol, O. The inhibitory effects of cobalt protoporphyrin IX and cannabinoid 2 receptor agonists in type 2 diabetic mice. *Int. J. Mol. Sci.* **2017**, *18*, 2268. [CrossRef]
36. Carcolé, M.; Castany, S.; Leánez, S.; Pol, O. Treatment with a heme oxygenase 1 inducer enhances the antinociceptive effects of μ-opioid, δ-opioid, and cannabinoid 2 receptors during inflammatory pain. *J. Pharmacol. Exp. Ther.* **2014**, *351*, 224–232. [CrossRef]
37. Infante, C.; Díaz, M.; Hernández, A.; Constandil, L.; Pelissier, T. Expression of nitric oxide synthase isoforms in the dorsal horn of monoarthritic rats: Effect of competitive and uncompetitive N-methyl-d-aspartate antagonists. *Arthritis Res. Ther.* **2007**, *9*, R53. [CrossRef]
38. Negrete, R.; Hervera, A.; Leánez, S.; Martín-Campos, J.M.; Pol, O. The antinociceptive effects of JWH-015 in chronic inflammatory pain are produced by nitric oxide-cGMP-PKG-KATP pathway activation mediated by opioids. *PLoS ONE* **2011**, *6*, e26688. [CrossRef]
39. Raghavendra, V.; Tanga, F.Y.; DeLeo, J.A. Complete Freunds adjuvant-induced peripheral inflammation evokes glial activation and proinflammatory cytokine expression in the CNS. *Eur. J. Neurosci.* **2004**, *20*, 467–473. [CrossRef]

40. Mika, J.; Osikowicz, M.; Rojewska, E.; Korostynski, M.; Wawrzczak-Bargiela, A.; Przewlocki, R.; Przewlocka, B. Differential activation of spinal microglial and astroglial cells in a mouse model of peripheral neuropathic pain. *Eur. J. Pharmacol.* **2009**, *623*, 65–72. [CrossRef]

41. Ikeda, H.; Mochizuki, K.; Murase, K. Astrocytes are involved in long-term facilitation of neuronal excitation in the anterior cingulate cortex of mice with inflammatory pain. *Pain* **2013**, *154*, 2836–2843. [CrossRef]

42. Xu, Y.; Cheng, G.; Zhu, Y.; Zhang, X.; Pu, S.; Wu, J.; Lv, Y.; Du, D. Anti-nociceptive roles of the glia-specific metabolic inhibitor fluorocitrate in paclitaxel-evoked neuropathic pain. *Acta Biochim. Biophys. Sin.* **2016**, *48*, 902–908. [CrossRef]

43. Edelmayer, R.M.; Brederson, J.D.; Jarvis, M.F.; Bitner, R.S. Biochemical and pharmacological assessment of MAP-kinase signaling along pain pathways in experimental rodent models: A potential tool for the discovery of novel antinociceptive therapeutics. *Biochem. Pharmacol.* **2014**, *87*, 390–398. [CrossRef]

44. Borges, G.; Miguelez, C.; Neto, F.; Mico, J.A.; Ugedo, L.; Berrocoso, E. Activation of Extracellular Signal-Regulated Kinases (ERK 1/2) in the Locus Coeruleus Contributes to Pain-Related Anxiety in Arthritic Male Rats. *Int. J. Neuropsychopharmacol.* **2017**, *20*, 463. [CrossRef]

45. Jurga, A.M.; Piotrowska, A.; Makuch, W.; Przewlocka, B.; Mika, J. Blockade of P2X4 receptors inhibits neuropathic pain-related behavior by preventing MMP-9 activation and consequently, pronociceptive interleukin release in a rat model. *Front. Pharmacol.* **2017**, *8*, 48. [CrossRef]

46. Chaplan, S.R.; Bach, F.W.; Pogrel, J.W.; Chung, J.M.; Yaksh, T.L. Quantitative assessment of tactile allodynia in the rat paw. *J. Neurosci. Methods* **1994**, *53*, 55–63. [CrossRef]

47. Hargreaves, K.; Dubner, R.; Brown, F.; Flores, C.; Joris, J. A new and sensitive method for measuring thermal nociception in cutaneous hyperalgesia. *Pain* **1988**, *32*, 77–88. [CrossRef]

International Journal of
Molecular Sciences

MDPI

Article

Hyperbilirubinemia in Gunn Rats Is Associated with Decreased Inflammatory Response in LPS-Mediated Systemic Inflammation

Petra Valaskova [1], Ales Dvorak [1], Martin Lenicek [1], Katerina Zizalova [1], Nikolina Kutinova-Canova [2], Jaroslav Zelenka [3], Monika Cahova [4], Libor Vitek [1,5] and Lucie Muchova [1,*]

[1] Institute of Medical Biochemistry and Laboratory Diagnostics, First Faculty of Medicine, Charles University and General University Hospital in Prague, 12108 Prague, Czech Republic; petra.valaskova@lf1.cuni.cz (P.V.); aleshdvorak@gmail.com (A.D.); mleni@centrum.cz (M.L.); katka.ziza@seznam.cz (K.Z.); vitek@cesnet.cz (L.V.)
[2] Institute of Pharmacology, First Faculty of Medicine, Charles University and General University Hospital in Prague, 12800 Prague, Czech Republic; Nikolina.Canova@lf1.cuni.cz
[3] Department of Biochemistry and Microbiology, University of Chemistry and Technology, 16628 Prague, Czech Republic; jar.zelenka@gmail.com
[4] Department of Experimental Diabetology, Institute of Clinical and Experimental Medicine, 14021 Prague, Czech Republic; moca@ikem.cz
[5] 4th Department of Medicine—Department of Gastroenterology and Hepatology, First Faculty of Medicine, Charles University and General University Hospital in Prague, 12808 Prague, Czech Republic
* Correspondence: lucie.muchova@lf1.cuni.cz; +420-224964199

Received: 15 April 2019; Accepted: 4 May 2019; Published: 9 May 2019

Abstract: Decreased inflammatory status has been reported in subjects with mild unconjugated hyperbilirubinemia. However, mechanisms of the anti-inflammatory actions of bilirubin (BR) are not fully understood. The aim of this study is to assess the role of BR in systemic inflammation using hyperbilirubinemic Gunn rats as well as their normobilirubinemic littermates and further in primary hepatocytes. The rats were treated with lipopolysaccharide (LPS, 6 mg/kg intraperitoneally) for 12 h, their blood and liver were collected for analyses of inflammatory and hepatic injury markers. Primary hepatocytes were treated with BR and TNF-α. LPS-treated Gunn rats had a significantly decreased inflammatory response, as evidenced by the anti-inflammatory profile of white blood cell subsets, and lower hepatic and systemic expressions of IL-6, TNF-α, IL-1β, and IL-10. Hepatic mRNA expression of LPS-binding protein was upregulated in Gunn rats before and after LPS treatment. In addition, liver injury markers were lower in Gunn rats as compared to in LPS-treated controls. The exposure of primary hepatocytes to TNF-α with BR led to a milder decrease in phosphorylation of the NF-κB p65 subunit compared to in cells without BR. In conclusion, hyperbilirubinemia in Gunn rats is associated with an attenuated systemic inflammatory response and decreased liver damage upon exposure to LPS.

Keywords: bilirubin; Gunn rats; hyperbilirubinemia; inflammation; LPS; NF-κB

1. Introduction

Bilirubin (BR), the end product of the heme degradation pathway in the intravascular compartment, is an important endogenous antioxidant, and it plays a crucial role in protection against oxidative stress as has been demonstrated in numerous in vitro, in vivo, and clinical studies (for review, see [1]). Recently, it has been shown that BR exerts potent anti-inflammatory and immunomodulatory activities [2]. In fact, mild hyperbilirubinemia has been associated with a reduced risk of diseases linked to increased oxidative stress and chronic inflammation (for review, see [3]).

A wide array of BR anti-inflammatory effects are mediated by multiple mechanisms, and indeed, BR is capable of modulating all stages of both the innate as well as the adaptive immune system [2]. These, predominantly suppressing activities, are aimed against: the complement system [4], damage-associated molecular patterns (DAMPs) signaling [5], Toll-like receptors (TLRs), such as TLR4 (a bacterial lipopolysaccharide (LPS) receptor) [6], macrophage activities [7] as well as B cell-mediated antibody production [8], and differentiation of T cells, including regulatory T cells (Tregs) [9], all with wide-spread potential clinical consequences towards autoimmune diseases [10] and transplant medicine [5,9].

An increasing body of evidence suggests that mildly elevated BR concentrations could suppress production of pro-inflammatory cytokines [5,10,11]. The secretion of cytokines is under the control of nuclear factor kappa B (NF-κB), a master regulator of numerous genes involved in the immune and inflammatory responses [12]. In the canonical pathway, NF-κB is activated by many signals including bacterial LPS, which binds to the LPS-binding protein (LBP), and then interacts with TLR4/CD14 receptors [13]. In resting cells, NF-κB is inactive, located in the cytoplasm bound to its inhibitor IκB. Upon activation, the IκB kinase (IKK) complex activates NF-κB by phosphorylating IκB, resulting in ubiquitination and proteasome degradation of IκB. Active NF-κB then translocates into the nucleus and activates specific genes [14]. Taking into consideration the reported inhibitory effects of BR on protein phosphorylation [15] as well as its general immune system-suppressing activities [10], we hypothesize that BR might also interfere with phosphorylation of NF-κB p65 subunit, and thus prevent translocation of NF-κB into the nucleus.

Therefore, the aim of our study was thus to evaluate the pathophysiological role of BR in LPS-induced inflammation in hyperbilirubinemic Gunn rats and primary hepatocytes isolated from hyper- and normobilirubinemic animals.

2. Results

2.1. Hyperbilirubinemia in Gunn Rats Is Associated with Decreased Systemic Inflammatory Response in LPS-Induced Sepsis

To evaluate the effect of BR on systemic and hepatic inflammation, the complete blood count, as well as serum markers of liver injury, was measured in hyperbilirubinemic Gunn rats as well as in their normobilirubinemic heterozygous littermates. Interestingly, higher white blood cell (WBC) counts were observed after LPS treatment in control rats as compared to in hyperbilirubinemic Gunn animals ($(12.39 \pm 5.26) \times 10^9$/L vs. $(8.70 \pm 1.94) \times 10^9$/L, $p = 0.05$). Following LPS administration, significant increases were detected in the proportions of neutrophils ($396 \pm 301\%$, $p < 0.01$), monocytes ($565 \pm 242\%$, $p < 0.01$), basophils ($338 \pm 271\%$, $p < 0.05$), as well as eosinophils ($448 \pm 419\%$, $p < 0.05$), together with a decrease in the lymphocyte count (up to $23 \pm 13\%$, $p < 0.01$) in control animals. However, these changes were substantially attenuated in hyperbilirubinemic Gunn rats (Figure 1a–f).

Simultaneously, marked changes in the CD4$^+$/CD8$^+$ T cells were observed in both hyperbilirubinemic Gunn rats and control animals upon exposure to LPS. In fact, the CD4$^+$/CD8$^+$ T ratio, a marker of immune activation [16], was 13 times higher in hyperbilirubinemic Gunn rats as compared to in controls ($p < 0.05$) (Figure 1g,h).

To evaluate the effect of hyperbilirubinemia on mediators of systemic inflammation, we first measured mRNA expression of the selected cytokines in the liver tissue as well as in the WBC of control and LPS-treated animals. The lower expressions of liver pro-inflammatory cytokines interleukin-6 (*IL-6*) ($50 \pm 49\%$, $p < 0.05$) and tumor necrosis factor-α (*TNF-α*) ($59 \pm 26\%$, $p < 0.05$) were observed in Gunn rat livers without LPS treatment compared to those in heterozygous littermates. After LPS administration, significantly lower increases in pro-inflammatory *TNF-α* ($34 \pm 21\%$, $p < 0.05$), interleukin-1β (*IL-1β*) ($57 \pm 30\%$, $p < 0.05$), and anti-inflammatory interleukin-10 (*IL-10*) ($40 \pm 22\%$, $p < 0.05$, Figure 2a–d) were detected in Gunn rats as compared to in normobilirubinemic controls 12 h after saline or LPS administration. Similar results in mRNA cytokine expressions were observed also in the WBC. Indeed, the elevation levels of cytokines *IL-6*, *TNF-α*, *IL-1β* and *IL-10* after LPS administration were significantly attenuated in Gunn rats (49 ± 35%, 43 ± 43%, 31 ± 28%, and 24 ± 13%, respectively, $p < 0.05$) compared to that in control animals (Figure 2e–h).

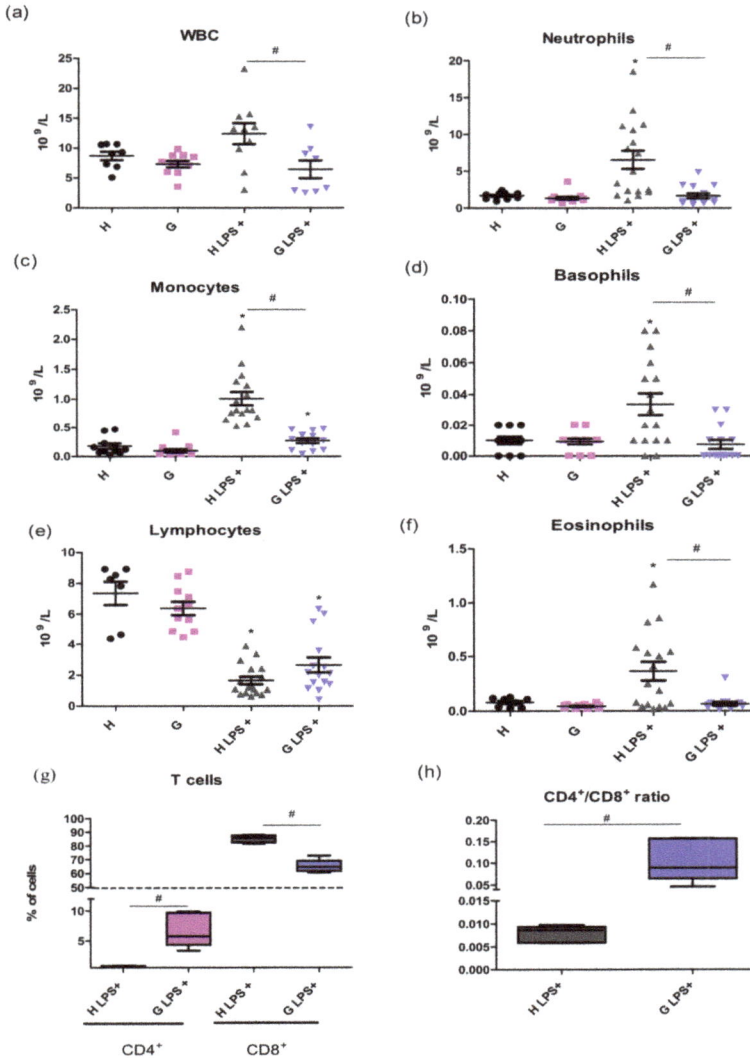

Figure 1. The effect of LPS-induced inflammation on WBC in hyperbilirubinemic Gunn rats. Total WBC cells (**a**) and their subpopulations (**b**–**f**) including T cells count (**g**) and CD4+/CD8+ ratio (**h**) were measured 12 h after LPS administration (6 mg/kg i.p.) in normobilirubinemic heterozygous controls (H or H LPS+) and hyperbilirubinemic Gunn rats (G or G LPS+), respectively. * $p < 0.05$ vs. corresponding control, # $p < 0.05$ vs. LPS-treated group. $n = 8$ animals per group (minimum).

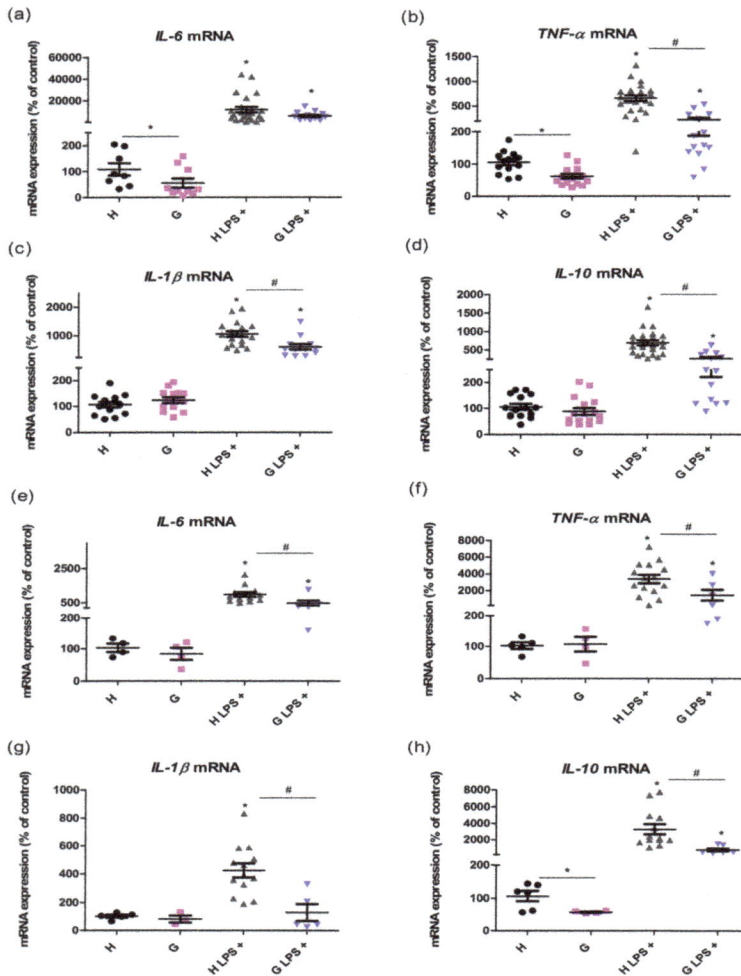

Figure 2. The effects of LPS-induced inflammation on mRNA cytokine expression in the liver and WBC of hyperbilirubinemic Gunn rats. mRNA expressions of pro- and anti-inflammatory cytokines *IL-6*, *TNF-α*, *IL-1β*, and *IL-10* were measured in the liver tissue (**a–d**) and white blood cells (**e–h**) 12 h after saline or LPS administration (6 mg/kg i.p.) in normobilirubinemic heterozygous controls (H or H LPS+) and hyperbilirubinemic Gunn rats (G or G LPS+), respectively. * $p < 0.05$ vs. corresponding control, # $p < 0.05$ vs. LPS-treated group. $n = 5$ animals per group (minimum).

Serum concentrations of selected cytokines were measured to confirm the functional translation of their mRNA expressions. In untreated animals, the concentrations of all tested cytokines were under the limit of detection. However, after LPS treatment, the changes in concentrations of most cytokines followed the pattern of mRNA expressions (although the concentration of IL-1β was under the limit of detection). Compared to that of controls, lower concentrations of IL-6 (35 ± 1%) as well as those of TNF-α (60 ± 56%) and IL-10 (25 ± 23%, $p < 0.05$) were observed in Gunn rats exposed to LPS (Figure 3). This data resulted in a marked difference in the IL-10/TNF-α ratio, a marker of immune homeostasis, between H LPS+ and G LPS+ experimental groups (0.51:0.19, $p < 0.05$).

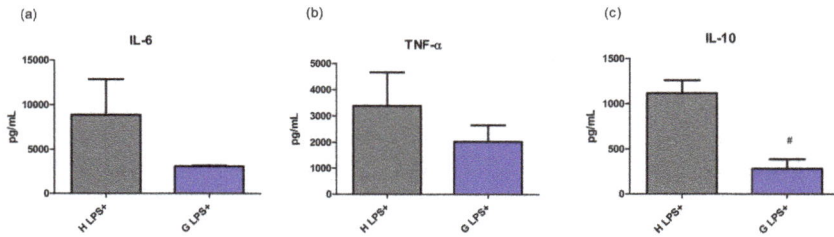

Figure 3. The effect of LPS-induced inflammation on cytokine concentration in serum of hyperbilirubinemic Gunn rats. Concentrations of pro-inflammatory cytokines IL-6, TNF-α, and anti-inflammatory IL-10 were measured 12 h after LPS administration (6 mg/kg i.p.) in normobilirubinemic heterozygous controls (H LPS+) and hyperbilirubinemic Gunn rats (G LPS+), respectively. # $p < 0.05$ vs. LPS-treated group. $n = 5$ animals per group (minimum).

Since the response of an organism to LPS sepsis involves production of LBP, an acute phase protein, by the liver, we tested in whether hyperbilirubinemia might affect production of this mediator. Indeed, *LBP* mRNA expression was upregulated in the liver of Gunn rats compared to in their normobilirubinemic littermates both before ($142 \pm 37\%$, $p < 0.05$) and after LPS treatment ($148 \pm 48\%$, $p < 0.05$, Figure 4a). Based on the results from in vivo experiments, *LBP* expression in primary hepatocytes was assessed upon exposure to LPS. The expression of *LBP* gradually increased starting at 6 h in Gunn primary hepatocytes exposed to 20 and 40 μM BR ($p < 0.05$, Figure 4b). Interestingly, no significant increase in mRNA expression of *LBP* upon incubation with BR was observed in control hepatocytes (Figure S1, Supplemental Materials).

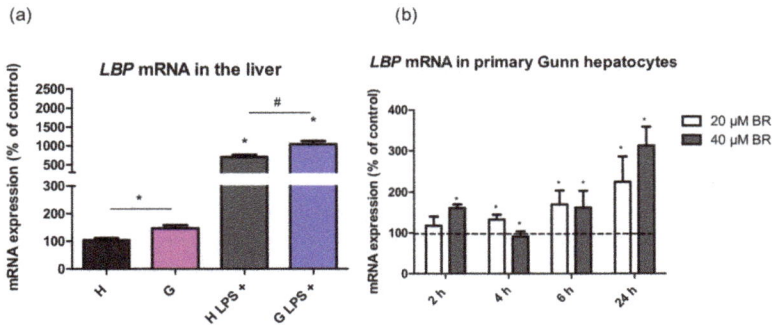

Figure 4. The effects of hyperbilirubinemia on lipopolysaccharide binding protein (*LBP*) mRNA expression in the liver tissues upon exposure to LPS (6 mg/kg i.p.) and in primary hepatocytes. mRNA expression of *LBP* was measured in the liver tissue (**a**) of normobilirubinemic heterozygous controls (H or H LPS+) and hyperbilirubinemic Gunn rats (G or G LPS+), respectively, and in primary hepatocytes (**b**). Primary hepatocytes isolated from Gunn rats were incubated with BR (20 and 40 μM) for 2, 4, 6, and 24 h. (**b**) Values are expressed as % of untreated control cells (100%). * $p < 0.05$ vs. controls, # $p < 0.05$ vs. LPS-treated group. (**a**) $n = 12$ animals per group (minimum); (**b**) $n = 6$ independent cell cultures per group.

Importantly, markers of liver injury such as alanine transaminase (ALT) and aspartate transaminase (AST) activities were lower in the LPS-treated Gunn rats compared to in LPS-treated controls (1.87 ± 1.14 vs. 5.55 ± 3.32 μkat/L, and 4.28 ± 2.26 vs. 6.22 ± 2.88 μkat/L, respectively, $p < 0.05$ for both comparisons, Figure 5a,b).

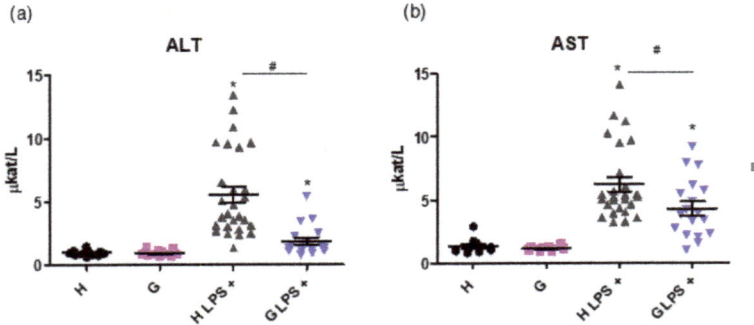

Figure 5. The effect of hyperbilirubinemia and inflammation on markers of the liver injury. ALT (**a**) and AST (**b**) activities, markers of liver injury, were measured in normobilirubinemic heterozygous controls (H or H LPS+) and hyperbilirubinemic Gunn rats (G or G LPS+) 12 h after saline or LPS administration (6 mg/kg i.p.), respectively. * $p < 0.05$ vs. corresponding control, # $p < 0.05$ vs. LPS-treated group. $n = 8$ animals per group (minimum).

2.2. Pretreatment of Primary Hepatocytes with Bilirubin Protects against Inflammation-Induced Cell Death

To assess the underlying mechanisms of anti-inflammatory effect of BR, primary hepatocytes isolated from hyperbilirubinemic Gunn rats and normobilirubinemic heterozygous controls were used for in vitro experiments. Both types of primary liver cells were exposed to BR or/and TNF-α to find out whether constitutive/basal BR could have protective effects on cell viability. No differences in intracellular BR levels were observed between primary hepatocytes isolated from hyperbilirubinemic and normobilirubinemic animals independently of BR treatment (Figure S2a, Supplementary Materials). Nevertheless, primary hepatocytes isolated from hyperbilirubinemic Gunn rats were more resistant to TNF-α-induced cell death as compared to in the control cells (16 ± 10%, $p < 0.05$, Figure S2b), consistent with in vivo data on the effect of hyperbilirubinemia on the liver injury markers described above.

2.3. Effect of Bilirubin on NF-κB Pathway

To examine the role of BR in regulation of NF-κB, a key mediator of inflammatory signaling, we investigated whether BR pre-treatment affects TNF-α-mediated NF-κB activation. Both types of primary hepatocytes were pre-treated with 10–40 μM BR and then exposed to TNF-α. As expected, TNF-α resulted in an increased phosphorylation of the NF-κB p65 subunit. Importantly, pretreatment with BR significantly decreased TNF-α-induced NF-κB p65 subunit phosphorylation (Figure 6a) ($p < 0.05$). No significant changes were detected in total levels of NF-κB p65 protein (Figure 6b), IKKβ protein, and inhibitor IκBα, as well as in phosphorylation of IKKα/β and IκBα after BR and TNF-α treatment (Figure S3). Interestingly, only BR itself increased phosphorylation of IKKα/β (Figure S3).

(a)

(b)

Phosphorylation of p65 NF-κB subunit

Endogenous total level of NF-κB p65 protein

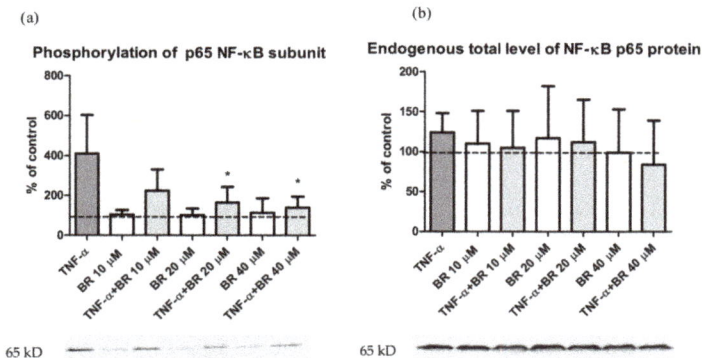

65 kD

65 kD

Figure 6. The effect of bilirubin on NF-κB p65 subunit phosphorylation. Both types of primary hepatocytes were pre-incubated with BR (0–40 μM) for 2 h and then treated with TNF-α (12 ng/mL) for 5 min. Phosphorylated (**a**) and total (**b**) NF-κB p65 subunits were measured by the western blot. Values are expressed as % of untreated control cells (100%). * $p < 0.05$ vs. TNF-α. $n = 6$ independent cell cultures per group.

3. Discussion

BR has been shown to be an important cytoprotective and especially antioxidant molecule at physiological or mildly elevated concentrations [1,3]. Even though in vitro studies as well as clinical observations suggest that BR might also possess considerable anti-inflammatory properties [2], surprisingly scarce data have been published on hyperbilirubinemic animal models of inflammation. In our study, we used a model of LPS-induced sepsis in Gunn rats with plasma-unconjugated bilirubin levels at around 60 μmol/L (compared to heterozygotes with 2 μmol/L).

Interestingly, a marked attenuation of WBC pro-inflammatory response with decreased counts of neutrophils and monocytes was observed after LPS treatment in hyperbilirubinemic Gunn animals, accompanied with substantial changes in the CD4+/CD8+ T cell ratio, an important marker of immune activation [16]. The expansion of CD8+ cells is also driven by the activity of NADPH oxidase (NOX2) [17]. Therefore, the beneficiary CD4+/CD8+ ratio observed in our study could at least partially be due to a previously reported inhibitory effect of BR on NOX2 activity [18].

The major driving force of neutrophil mobilization from bone marrow and other hematopoetic compartments during sepsis are pro-inflammatory cytokines, of which generation was significantly attenuated in our hyperbilirubinemic rats. Since an overabundance of neutrophils during severe inflammation might have serious damaging effects [19], its amelioration seems to contribute to hyperbilirubinemia-induced protection. In addition, overwhelmed cytokine production during sepsis is also considered detrimental. In fact, lower mortality was observed in rats exposed to LPS and treated with a monoclonal antibody against TNF-α [20], and beneficiary effects were also observed for anti-IL-1β as well as anti-IL-6 antibody treatment in other experimental models of sepsis [21,22].

Together with the decreased expression of pro-inflammatory cytokines observed in our septic hyperbilirubinemic animals, there was also reduced production of IL-10. Although IL-10 is generally considered to be an anti-inflammatory cytokine, its overproduction might also be harmful and result in immunosuppression [23]. In fact, serum IL-10 concentrations were demonstrated in a human clinical study to correlate well with the sepsis severity and mortality, as also did the high IL-10/TNF-α ratio [24]. Thus, the balance between IL-10 and TNF-α seems to be important for immune homeostasis maintenance, as demonstrated by a curative effect of blocking of the IL-10 pathway in several models of bacterial infections such as from *Listeria* [25], *Klebsiella* [26], *Pseudomonas* [23], *Streptococcus* [27], or *Mycobacterium* [28]. It has been suggested that this approach seems to be promising as an adjunct therapy for severe septicemias. In our study, the IL-10/TNF-α ratio was markedly lower in hyperbilirubinemic Gunn rats, consistent with the better survival rate in Gunn rats exposed to LPS observed in the previous study by Lanone et al. [29]. In concordance with these observations, significantly lower values of hepatocellular liver injury markers were observed in Gunn rats exposed to LPS, as similarly in previous studies [18,29]. In addition, we have previously shown that BR protects the liver against pro-oxidative effects of elevated bile acids in cholestasis [30], further emphasizing the role of BR in hepatoprotection.

To find the factors contributing to a decreased BR-mediated inflammatory response in Gunn rats to LPS, we investigated the expression of LBP in the liver tissue of our experimental animals. LBP, a plasma protein mainly produced by hepatocytes, plays a crucial role in LPS recognition and signaling and is considered an important mediator of the inflammatory reaction [31]. It binds LPS in plasma and transports it via cluster of differentiation 14 (CD14) to the Toll-like receptor 4 (TLR4)/MD-2 signaling complex triggering a range of pro- and anti-inflammatory responses. Dysregulation of this finely tuned signaling cascade could result in a deleterious effect on organism including sepsis and septic shock [32]. Even though the role of LBP in the activation/inhibition of the inflammatory response is probably a dual one, depending on its serum concentration, it has been described that high LBP levels inhibit LPS-mediated cytokine release and prevent hepatic failure in vivo [33]. In our study, hepatic *LBP* expression was significantly higher in Gunn rats before and after LPS treatment, suggesting a role of hyperbilirubinemia in LBP-mediated LPS signaling. Moreover, the treatment of Gunn primary hepatocytes with BR resulted in an increased *LBP* expression, indicating that BR might affect LBP production, and thus contribute to an attenuated inflammatory response in hyperbilirubinemic subjects.

The production of pro-inflammatory cytokines during sepsis leads to activation of NF-κB [34]. In fact, our experiments on primary rat hepatocytes demonstrated that BR exposure resulted in decreased phosphorylation of the p65 subunit of the NF-κB protein complex, a phenomenon which might be related to both general inhibitory effects of BR on protein phosphorylation [15], as well as inhibition of phosphorylation via suppressed TNF-α signaling [35]. It is thus likely that anti-inflammatory and cytoprotective effects of BR may at least in part be due to attenuation of NF-κB-driven transcription. On the contrary, we did not observe any inhibition of IκB phosphorylation, which has previously been reported, but at much higher BR levels [10]. Moreover, phosphorylation of IKK was increased by BR itself. Our data are consistent with previous reports demonstrating this specific inhibitory effect of both BR [36] and biliverdin [37,38]. It is also interesting to note that the CD4+/CD8+ ratio (similarly to Tregs and myelopoesis), which changes in our hyperbilirubinemic rats, is also regulated by the activity of NF-κB [39].

Furthermore, Gunn rat hepatocytes were more resistant to TNF-α-induced cytotoxicity, although no changes in intracellular BR concentrations were observed compared to control cells. These data suggest that not only BR itself, but also "bilirubin priming", triggering adaptive, para-hormetic mechanisms under hyperbilirubinemic conditions, might significantly contribute to the observed hepatoprotection; however, intensive research is needed to identify these mechanisms.

4. Materials and Methods

4.1. Chemicals and Reagents

BR, the bovine serum albumin (BSA), rat TNF-α, LPS from Escherichia coli 0114:B4, human insulin solution, Williams' E Medium, Collagen type I from rat tail tendon, 2,6-di-tert-butyl-4-methylphenol (BHT), Thiazolyl Blue Tetrazolium Bromide (MTT), RNAlater, and tetrabutyl-ammonium hydroxide (TBA, 40 % in water) were purchased from Sigma-Aldrich (St. Louis, MO, USA); the chloroform (HPLC grade), methanol (HPLC grade), *n*-hexane, ethyl acetate, and acetonitrile were purchased from Merck (Darmstadt, Germany); and 4×-Laemmli sample buffer was from Bio-Rad (Hercules, CA, USA).

As described earlier, the BR was purified before use [40]. For the experiments, BR (2.8 mg) was dissolved in 2 mL of 0.1 M NaOH and immediately mixed with 1 mL of 0.1 M phosphoric acid. The mixture was diluted with a BSA solution (660 µM BSA in 25 mM phosphate buffer, pH: 7.7) to reach a final concentration of 480 µM BR in a phosphate buffer and then serially diluted with a BSA solution to yield solutions with final BR concentrations within the range of 10–40 µM.

4.2. In Vivo Studies

Hyperbilirubinemic adult female Gunn rats and their normomobilirubinemic heterozygous littermates (*n* range: 8–25 per group, weight range: 160–260 g) had access to water and food ad libitum. Gunn rats were kindly provided by Cluster in Biomolecular Medicine (University of Trieste, Italy). All protocols were approved by the Animal Research Committee of the 1st Faculty of Medicine, Charles University, project No. MSMT-25538/2018-2 (29 August 2018) as well as by the Institute of Molecular Genetics of the Academy of the Sciences of the Czech Republic, project No. PP 67/2018 (24 July 2018) and carried out in accordance with the Guide for the Care and Use of Animals of the National Institutes of Health.

The rats were divided into 4 groups: normomobilirubinemic heterozygote (a) and hyperbilirubinemic Gunn (b) experimental groups were treated with LPS (H LPS+/G LPS+, 6 mg/kg intraperitoneally); and normomobilirubinemic heterozygote (c) and hyperbilirubinemic Gunn (d) control groups received vehicle (saline). After 12 h, the animals were anesthetized (xylazin: 16 mg/kg, i.m.) and sacrificed. Blood for further biochemical analyses was obtained from the inferior vena cava and from the aorta for flow cytometry. Relevant organs (liver, heart, lung, kidney, spleen, and brain) were harvested, washed with ice-cold PBS, snap frozen in liquid nitrogen and stored at −80 °C. For RNA analysis, 100 mg of fresh liver was immediately placed in 2-mL microfuge tubes containing RNAlater and stored according to the manufacturer's instructions until analysis.

4.3. Determination of Complete Blood Count with Differential and Serum Biochemical Markers

Complete blood counts were measured from whole blood using an XN-1000™ automatic analyzer (Sysmex, Lincolnshire, IL, USA). Serum biochemical markers (ALT and AST activities) were determined by standard assays using an automatic analyzer (Modular analyzer, Roche Diagnostics GmbH, Mannheim, Germany).

4.4. Determination of Serum Cytokine Concentrations

Determination of serum cytokine concentrations were performed using commercial rat ELISA kits (Duo-Sets kits for IL-1β/IL-1F2, IL-10, TNF-α, and IL-6; Bio-Techne R&D Systems, Minneapolis, MN, USA) according to the manufacturer's instructions.

4.5. Flow Cytometry of Lymphocytes

Blood samples (250 µL of whole blood) were collected in tubes with 3% potassium EDTA. After lysis of the red blood cells (twice) using ACK buffer (0.15 M NH_4Cl, 10 mM $KHCO_3$, and 1 mM EDTA monosodium; pH: 7.3) for 15 and 5 min separately, followed by washing with PBS (twice), the cells were simultaneously stained for effector T cells. Cells were surface-stained using the following anti-rat antibodies: anti-CD45-FITC (OX-1, Thermo Fisher Scientific, Waltham, MA, USA), anti-CD4-BV-786 (OX-35, BD Biosciences, San Jose, CA, USA), anti-CD8α-PerCP-e710 (OX-8, Thermo Fisher Scientific), and anti-CD62L-PE (OX-85, SONY) for CD4/CD8 T cells panel. The cell suspensions were analyzed by flow cytometry (BD LSR II including an FACSFlow Supply of the High Throughput Sampler System, BD Biosciences).

4.6. Gene Expression Analyses

Total RNA from liver tissue and blood was isolated using a GenUP™ Total RNA Kit (Biotechrabbit GmbH, Hennigsdorf, Germany) and a Total RNA Mini Kit (Geneaid Biotech Ltd, New Taipei City, Taiwan), respectively. The quantity and purity of isolated RNA were evaluated spectrophotometrically. cDNA was generated by a High-Capacity cDNA Reverse Transcription Kit (Thermo Fisher Scientific) and stored at −20 °C until analysis. Quantitative real-time PCR was performed using TaqMan® Fast Advanced Master Mix and a TaqMan® Gene Expression Assay Kit for the following genes: *IL-6*

(Rn01410330_m1), *TNF-α* (Rn99999017_m1), *IL-10* (Rn00563409_m1), *IL-1β* (Rn00580432_m1), *LBP* (Rn00567985_m1), and a rat endogenous control β-2 microglobulin (Rn005608865_m1). Results were expressed as the % of controls.

4.7. Primary Rat Hepatocyte Culture

Primary hepatocytes were isolated from anaesthetized Gunn and heterozygote (n = 3 each, weight range: 200–220 g) rats by two-step collagenase perfusion according to a published protocol [41]. Cell viability ranged from 75% to 85% (as trypan blue staining). Hepatocytes were further diluted to 0.8 million cells/mL with William's E medium, supplemented with 1% penicillin/streptomycin, 1% L-glutamine, 0.06% insulin, and 5% fetal bovine serum. Primary hepatocytes were dispensed into a collagen-coated cell culture Petri dishes, 6-well and 96-well plates and allowed to attach for 3 h at 37 °C with 5% CO_2 in the incubator. Unattached cells were removed after 3 h and a new medium was added. The following day, hepatocytes were cultured with complete culture medium containing BR (10–40 μM) and TNF-α (12–100 ng/mL) for 24 h.

4.8. Determination of Cell Viability and Intracellular Bilirubin Levels

All experiments were performed under dim light to minimize BR degradation. Cell viability of primary hepatocytes seeded in 96-well plates was measured using an MTT test after 24 h incubation. Primary hepatocytes harvested from 10 cm Petri dishes were used for determination of intracellular BR as described previously [42].

4.9. Western Blot Analysis

Primary hepatocytes were lysed using a lysis buffer (5 M NaCl, 1 M Tris, pH = 8, 10% Triton-X 100), sonicated for 5 s and centrifuged at a speed of 14,000× g for 10 min (temperature: 4 °C). Supernatants (35–40 μg of protein) were diluted with a loading buffer (4× Laemmli Sample buffer, Bio-Rad, USA), denatured at 95 °C for 10 min, and separated by SDS-PAGE electrophoresis (10%). Proteins were transferred to a nitrocellulose membrane, blocked in 5% BSA in TTBS for 1.5 h and then incubated overnight at 4 °C with primary antibodies anti phospho-NF-κB p65 (Ser536) (dilution, 1:2000 *v/v*), anti NF-κB p65 (dilution, 1:3500 *v/v*), anti IκB-α (dilution, 1:3500 *v/v*), anti phospo-IκB-α (Ser132) (dilution, 1:1500 *v/v*), anti IKKβ (dilution, 1:3500 *v/v*), anti phospho-IKKα/β (Ser176/180) (dilution, 1:1500 *v/v*), as well as anti β-actin (dilution, 1:5000 *v/v*) as a loading control (all antibodies were from Cell Signaling Technology, Danvers, MA, USA). After being washed in TTBS buffer, membranes were incubated with swine anti-rabbit IgG-HRP secondary antibody (Dako, Glostrup, Denmark) and visualized using an ECL kit (LumiGLO®, Cell Signaling Technology). A Fusion Fx7 device and Bio-2D software (Vilber Lourmat, Collegien, France) were used to quantify the signals. Results were normalized to β-actin.

4.10. Statistical Analysis

Student parametric unpaired and paired t-tests were used for comparison of normally distributed data. Non-normally distributed data were analyzed with the Mann–Whitney rank sum test. Group mean differences were analyzed by ANOVA and Kruskal–Wallis tests. Depending on their normality, data are expressed as the mean with SD or the median with interquartile range. Differences were considered statistically significant when $p < 0.05$. Analyses were performed using GraphPad Prism 5.0 statistical software (GraphPad Software, Inc., San Diego, CA, USA).

5. Conclusions

In conclusion, hyperbilirubinemia in Gunn rats is associated with an attenuated systemic inflammatory response and decreased liver damage upon exposure to LPS, an effect associated with a modulation of innate immunity together with decreased production of pro-inflammatory cytokines and NF-κB activation.

Supplementary Materials: Supplementary materials can be found at http://www.mdpi.com/1422-0067/20/9/2306/s1. Figure S1: The effects of BR on LBP mRNA expression in primary heterozygote hepatocytes; Figure S2: The effects of BR and TNF-α on viability of primary hepatocytes; Figure S3: The effect of bilirubin on the NF-κB signaling pathway.

Author Contributions: Conceptualization, L.M. and L.V.; data curation, P.V.; formal analysis, P.V. and M.L.; funding acquisition, L.V.; methodology, P.V., L.M., K.Z., and L.V.; project administration, P.V., A.D., and K.Z.; supervision, L.V.; writing of an original draft, P.V.; writing of review and editing, L.M., M.L., A.D., J.Z., M.C., N.K.C. and L.V. All authors participated in reviewing the manuscript.

Funding: This research was funded by grants GAUK 168216 given by the First Faculty of Medicine, Charles University, Prague, Czech Republic, SVV 260370/2018 given by Charles University, Prague, Czech Republic, and RVO-VFN64165/2018 given by the Czech Ministry of Health.

Acknowledgments: We wish to thank Marie Zadinova, Libuse Slehobrova, Karel Chalupsky, and Peter Neradil for their excellent technical assistance during the animal studies.

Conflicts of Interest: The authors declare no conflicts of interest.

Abbreviations

ALT	Alanine transaminase
AST	Aspartate transaminase
BR	Bilirubin
IKK	IκB kinase
IL-6	Interleukin-6
IL-10	Interleukin-10
IL-1β	Interleukin-1β
LBP	Lipopolysaccharide-binding protein
LPS	Lipopolysaccharide
NF-κB	Nuclear factor kappa B
TLR	Toll-like receptors
TNF-α	Tumor necrosis factor-α
TLR	Toll-like receptor
WBC	White blood cell

References

1. Gazzin, S.; Vitek, L.; Watchko, J.; Shapiro, S.M.; Tiribelli, C. A novel perspective on the biology of bilirubin in health and disease. *Trends Mol. Med.* **2016**, *22*, 758–768. [CrossRef] [PubMed]

2. Jangi, S.; Otterbein, L.; Robson, S. The molecular basis for the immunomodulatory activities of unconjugated bilirubin. *Int. J. Biochem. Cell B* **2013**, *45*, 2843–2851. [CrossRef] [PubMed]

3. Wagner, K.H.; Wallner, M.; Molzer, C.; Gazzin, S.; Bulmer, A.C.; Tiribelli, C.; Vitek, L. Looking to the horizon: The role of bilirubin in the development and prevention of age-related chronic diseases. *Clin. Sci.* **2015**, *129*, 1–25. [CrossRef]

4. Basiglio, C.L.; Arriaga, S.M.; Pelusa, F.; Almara, A.M.; Kapitulnik, J.; Mottino, A.D. Complement activation and disease: Protective effects of hyperbilirubinaemia. *Clin. Sci.* **2010**, *118*, 99–113. [CrossRef] [PubMed]

5. Adin, C.A.; VanGundy, Z.C.; Papenfuss, T.L.; Xu, F.; Ghanem, M.; Lakey, J.; Hadley, G.A. Physiologic doses of bilirubin contribute to tolerance of islet transplants by suppressing the innate immune response. *Cell Transplant.* **2017**, *26*, 11–21. [CrossRef]

6. Idelman, G.; Smith, D.L.H.; Zucker, S.D. Bilirubin inhibits the up-regulation of inducible nitric oxide synthase by scavenging reactive oxygen species generated by the toll-like receptor 4-dependent activation of NADPH oxidase. *Redox Biol.* **2015**, *5*, 398–408. [CrossRef]

7. Vetvicka, V.; Miler, I.; Sima, P.; Taborsky, L.; Fornusek, L. The effect of bilirubin on the Fc receptor expression and phagocytic activity of mouse peritoneal macrophages. *Folia Microbiol.* **1985**, *30*, 373–380. [CrossRef]

8. Nejedla, Z. The development of immunological factors in infants with hyperbilirubinemia. *Pediatrics* **1970**, *45*, 102–104.

9. Rocuts, F.; Zhang, X.Y.; Yan, J.; Yue, Y.A.; Thomas, M.; Bach, F.H.; Czismadia, E.; Wang, H.J. Bilirubin promotes de novo generation of T regulatory cells. *Cell Transplant.* **2010**, *19*, 443–451. [CrossRef]

10. Liu, Y.; Li, P.; Lu, J.; Xiong, W.; Oger, J.; Tetzlaff, W.; Cynader, M. Bilirubin possesses powerful immunomodulatory activity and suppresses experimental autoimmune encephalomyelitis. *J. Immunol.* **2008**, *181*, 1887–1897. [CrossRef]

11. Haga, Y.; Tempero, M.A.; Kay, D.; Zetterman, R.K. Intracellular accumulation of unconjugated bilirubin inhibits phytohemagglutin-induced proliferation and interleukin-2 production of human lymphocytes. *Dig. Dis. Sci.* **1996**, *41*, 1468–1474. [CrossRef]

12. Lawrence, T. The nuclear factor NF-kappaB pathway in inflammation. *Cold Spring Harb. Perspect. Biol.* **2009**, *1*, a001651. [CrossRef]

13. Jerala, R. Structural biology of the LPS recognition. *Int. J. Med. Microbiol.* **2007**, *297*, 353–363. [CrossRef]

14. Siebenlist, U.; Franzoso, G.; Brown, K. Structure, regulation and function of Nf-Kappa-B. *Annu. Rev. Cell Biol.* **1994**, *10*, 405–455. [CrossRef]

15. Hansen, T.W.R.; Mathiesen, S.B.W.; Walaas, S.I. Bilirubin has widespread inhibitory effects on protein phosphorylation. *Pediatr. Res.* **1996**, *39*, 1072–1077. [CrossRef]

16. Bruno, G.; Saracino, A.; Monno, L.; Angarano, G. The Revival of an "Old" Marker: CD4/CD8 Ratio. *Aids Rev.* **2017**, *19*, 81–88.

17. Dhiman, M.; Garg, N.J. P47(phox-/-)mice are compromised in expansion and activation of CD8(+) T cells and susceptible to trypanosoma cruzi infection. *PLoS Pathog.* **2014**, *10*, e1004516. [CrossRef]

18. Wang, W.Z.W.; Smith, D.L.H.; Zucker, S.D. Bilirubin inhibits iNOS expression and NO production in response to endotoxin in rats. *Hepatology* **2004**, *40*, 424–433. [CrossRef] [PubMed]

19. Summers, C.; Rankin, S.M.; Condliffe, A.M.; Singh, N.; Peters, A.M.; Chilvers, E.R. Neutrophil kinetics in health and disease. *Trends Immunol.* **2010**, *31*, 318–324. [CrossRef]

20. Ozer, E.K.; Goktas, M.T.; Kilinc, I.; Toker, A.; Bariskaner, H.; Ugurluoglu, C.; Iskit, A.B. Infliximab alleviates the mortality, mesenteric hypoperfusion, aortic dysfunction, and multiple organ damage in septic rats. *Can. J. Physiol. Pharm.* **2017**, *95*, 866–872. [CrossRef] [PubMed]

21. Ohlsson, K.; Bjork, P.; Bergenfeldt, M.; Hageman, R.; Thompson, R.C. Interleukin-1 receptor antagonist reduces mortality from endotoxin-shock. *Nature* **1990**, *348*, 550–552. [CrossRef]

22. Nullens, S.; Staessens, M.; Peleman, C.; Plaeke, P.; Malhotra-Kumar, S.; Francque, S.; De Man, J.G.; De Winter, B.Y. Beneficial effects of anti-interleukin-6 antibodies on impaired gastrointestinal motility, inflammation and increased colonic permeability in a murine model of sepsis are most pronounced when administered in a preventive setup. *PLoS ONE* **2016**, *11*, e0152914. [CrossRef] [PubMed]

23. Steinhauser, M.E.; Hogaboam, G.M.; Kunkel, S.L.; Lukacs, N.W.; Strieter, R.M.; Standiford, T.J. IL-10 is a major mediator of sepsis-induced impairment in lung antibacterial host defense. *J. Immunol.* **1999**, *162*, 392–399.

24. Gogos, C.A.; Drosou, E.; Bassaris, H.P.; Skoutelis, A. Pro- versus anti-inflammatory cytokine profile in patients with severe sepsis: A marker for prognosis and future therapeutic options. *J. Infect. Dis.* **2000**, *181*, 176–180. [CrossRef]

25. Silva, R.A.; Appelberg, R. Blocking the receptor for interleukin 10 protects mice from lethal listeriosis. *Antimicrob. Agents Chempther.* **2001**, *45*, 1312–1314. [CrossRef]

26. Wang, M.J.; Jeng, K.C.G.; Ping, L.I. Exogenous cytokine modulation or neutralization of interleukin-10 enhance survival in lipopolysaccharide-hyporesponsive C3H/HeJ mice with Klebsiella infection. *Immunology* **1999**, *98*, 90–97. [CrossRef] [PubMed]

27. Van der Poll, T.; Marchant, A.; Keogh, C.B.; Goldman, M.; Lowry, S.F. Interleukin-10 impairs host defense in murine pneumococcal pneumonia. *J. Infect. Dis.* **1996**, *174*, 994–1000. [CrossRef]

28. Jacobs, M.; Brown, N.; Allie, N.; Gulert, R.; Ryffel, B. Increased resistance to mycobacterial infection in the absence of interleukin-10. *Immunology* **2000**, *100*, 494–501. [CrossRef] [PubMed]

29. Lanone, S.; Bloc, S.; Foresti, R.; Almolki, A.; Taille, C.; Callebert, J.; Conti, M.; Goven, D.; Aubier, M.; Dureuil, B.; et al. Bilirubin decreases nos2 expression via inhibition of NAD(P)H oxidase: Implications for protection against endotoxic shock in rats. *FASEB J.* **2005**, *19*, 1890–1892. [CrossRef]

30. Muchova, L.; Vanova, K.; Zelenka, J.; Lenicek, M.; Petr, T.; Vejrazka, M.; Sticova, E.; Vreman, H.J.; Wong, R.J.; Vitek, L. Bile acids decrease intracellular bilirubin levels in the cholestatic liver: Implications for bile acid-mediated oxidative stress. *J. Cell Mol. Med.* **2011**, *15*, 1156–1165. [CrossRef]

31. Su, G.L.; Freeswick, P.D.; Geller, D.A.; Wang, Q.; Shapiro, R.A.; Wan, Y.H.; Billiar, T.R.; Tweardy, D.J.; Simmons, R.L.; Wang, S.C. Molecular-cloning, characterization, and tissue distribution of rat lipopolysaccharide-binding protein - evidence for extrahepatic expression. *J. Immunol.* **1994**, *153*, 743–752.

32. Shimazu, R.; Akashi, S.; Ogata, H.; Nagai, Y.; Fukudome, K.; Miyake, K.; Kimoto, M. MD-2, a molecule that confers lipopolysaccharide responsiveness on Toll-like receptor 4. *J. Exp. Med.* **1999**, *189*, 1777–1789. [CrossRef]

33. Lamping, N.; Dettmer, R.; Schroder, N.W.J.; Pfeil, D.; Hallatschek, W.; Burger, R.; Schumann, R.R. LPS-binding protein protects mice from septic shock caused by LPS or gram-negative bacteria. *J. Clin. Investig.* **1998**, *101*, 2065–2071. [CrossRef] [PubMed]

34. Perkins, N.D. Integrating cell-signalling pathways with NF-kappa B and IKK function. *Nat. Rev. Mol. Cell. Bio.* **2007**, *8*, 49–62. [CrossRef]

35. Mazzone, G.L.; Rigato, I.; Ostrow, J.D.; Tiribelli, C. Bilirubin effect on endothelial adhesion molecules expression is mediated by the NF-kappa B signaling pathway. *Biosci. Trends* **2009**, *3*, 151–157. [PubMed]

36. Soares, M.P.; Seldon, M.P.; Gregoire, I.P.; Vassilevskaia, T.; Berberat, P.O.; Yu, J.; Tsui, T.Y.; Bach, F.H. Heme oxygenase-1 modulates the expression of adhesion molecules associated with endothelial cell activation. *J. Immunol.* **2004**, *172*, 3553–3563. [CrossRef]

37. Gibbs, P.E.M.; Maines, M.D. Biliverdin inhibits activation of NF-kappa B: Reversal of inhibition by human biliverdin reductase. *Int. J. Cancer* **2007**, *121*, 2567–2574. [CrossRef]

38. Nuhn, P.; Mitkus, T.; Ceyhan, G.O.; Kunzli, B.M.; Bergmann, F.; Fischer, L.; Giese, N.; Friess, H.; Berberat, P.O. Heme oxygenase 1-generated carbon monoxide and biliverdin attenuate the course of experimental necrotizing pancreatitis. *Pancreas* **2013**, *42*, 265–271. [CrossRef] [PubMed]

39. Jimi, E.; Strickland, I.; Voll, R.E.; Long, M.X.; Ghosh, S. Differential role of the transcription factor NF-kappa B in selection and survival of CD4(+) and CD8(+) thymocytes. *Immunity* **2008**, *29*, 523–537. [CrossRef]

40. McDonagh, A.F.; Assisi, F. The ready isomerization of bilirubin IX- in aqueous solution. *Biochem. J.* **1972**, *129*, 797–800. [CrossRef]

41. Berry, M.N.; Grivell, A.R.; Grivell, M.B.; Phillips, J.W. Isolated hepatocytes-past, present and future. *Cell Biol. Toxicol.* **1997**, *13*, 223–233. [CrossRef] [PubMed]

42. Zelenka, J.; Lenicek, M.; Muchova, L.; Jirsa, M.; Kudla, M.; Balaz, P.; Zadinova, M.; Ostrow, J.D.; Wong, R.J.; Vitek, L. Highly sensitive method for quantitative determination of bilirubin in biological fluids and tissues. *J. Chromatogr. B* **2008**, *867*, 37–42. [CrossRef] [PubMed]

International Journal of
Molecular Sciences

MDPI

Article

The Role of Heme Oxygenase 1 in the Protective Effect of Caloric Restriction against Diabetic Cardiomyopathy

Maayan Waldman [1,2], Vadim Nudelman [1], Asher Shainberg [3], Romy Zemel [1], Ran Kornwoski [1], Dan Aravot [1], Stephen J. Peterson [4,5,*], Michael Arad [2,†] and Edith Hochhauser [1,*,†]

[1] Cardiac Research Laboratory, Felsenstein Medical Research Institute Petah-Tikva, Sackler Faculty of Medicine, Tel Aviv University, Tel Aviv 49100, Israel; maayanw@gmail.com (M.W.); vadimnud@post.tau.ac.il (V.N.); zemel@post.tau.ac.il (R.Z.); rkornowski@clalit.org.il (R.K.); aravot_dan@clalit.org.il (D.A.)
[2] Levied Heart Center, Sheba Medical Center, Tel Hashomer and Sackler School of Medicine, Tel Aviv University, Ramat Gan 52621, Israel; michael.arad@sheba.health.gov.il
[3] Faculty of Life Sciences, Bar Ilan University, Ramat Gan 5290002, Israel; shaina@mail.biu.ac.il
[4] Department of Medicine, New York Presbyterian Brooklyn Methodist Hospital, Brooklyn, NY 11215, USA
[5] Department of Medicine, Weill Cornell Medicine, New York, NY 14853, USA
* Correspondence: Stp9039@nyp.org (S.J.P.); hochhaus@tauex.tau.ac.il or aditho@clalit.org.il (E.H.)
† These authors equally contributed to this work.

Received: 21 February 2019; Accepted: 10 May 2019; Published: 16 May 2019

Abstract: Type 2 diabetes mellitus (DM2) leads to cardiomyopathy characterized by cardiomyocyte hypertrophy, followed by mitochondrial dysfunction and interstitial fibrosis, all of which are exacerbated by angiotensin II (AT). SIRT1 and its transcriptional coactivator target PGC-1α (peroxisome proliferator-activated receptor-γ coactivator), and heme oxygenase-1 (HO-1) modulates mitochondrial biogenesis and antioxidant protection. We have previously shown the beneficial effect of caloric restriction (CR) on diabetic cardiomyopathy through intracellular signaling pathways involving the SIRT1–PGC-1α axis. In the current study, we examined the role of HO-1 in diabetic cardiomyopathy in mice subjected to CR. Methods: Cardiomyopathy was induced in obese diabetic (*db/db*) mice by AT infusion. Mice were either fed ad libitum or subjected to CR. In an in vitro study, the reactive oxygen species (ROS) level was determined in cardiomyocytes exposed to different glucose levels (7.5–33 mM). We examined the effects of Sn(tin)-mesoporphyrin (SnMP), which is an inhibitor of HO activity, the HO-1 inducer cobalt protoporphyrin (CoPP), and the SIRT1 inhibitor (EX-527) on diabetic cardiomyopathy. Results: Diabetic mice had low levels of HO-1 and elevated levels of the oxidative marker malondialdehyde (MDA). CR attenuated left ventricular hypertrophy (LVH), increased HO-1 levels, and decreased MDA levels. SnMP abolished the protective effects of CR and caused pronounced LVH and cardiac metabolic dysfunction represented by suppressed levels of adiponectin, SIRT1, PPARγ, PGC-1α, and increased MDA. High glucose (33 mM) increased ROS in cultured cardiomyocytes, while SnMP reduced SIRT1, PGC-1α levels, and HO activity. Similarly, SIRT1 inhibition led to a reduction in PGC-1α and HO-1 levels. CoPP increased HO-1 protein levels and activity, SIRT1, and PGC-1α levels, and decreased ROS production, suggesting a positive feedback between SIRT1 and HO-1. Conclusion: These results establish a link between SIRT1, PGC-1α, and HO-1 signaling that leads to the attenuation of ROS production and diabetic cardiomyopathy. CoPP mimicked the beneficial effect of CR, while SnMP increased oxidative stress, aggravating cardiac hypertrophy. The data suggest that increasing HO-1 levels constitutes a novel therapeutic approach to protect the diabetic heart. Brief Summary: CR attenuates cardiomyopathy, and increases HO-1, SIRT activity, and PGC-1α protein levels in diabetic mice. High glucose reduces adiponectin, SIRT1, PGC1-1α, and HO-1 levels in cardiomyocytes, resulting in oxidative stress. The pharmacological activation of HO-1 activity mimics the effect of CR, while SnMP increased oxidative stress and cardiac hypertrophy. These data suggest the critical role of HO-1 in protecting the diabetic heart.

Int. J. Mol. Sci. **2019**, *20*, 2427

Keywords: caloric restriction; Sirtuin 1; Heme Oxygenase-1; PGC-1α; cardiomyopathy; diabetes mellitus

1. Introduction

Diabetes mellitus type 2 (DM2) is associated with excess cardiovascular morbidity and mortality [1,2]. Diastolic dysfunction, reduced myocardial contractility and heart failure are evident as a result of progressive cardiac fibrosis and/or pressure overload [3,4]. Insulin resistance and hyperinsulinemia, hyperglycemia, and elevated free fatty acids are primary factors that lead to cardiomyocyte injury, dysfunction and myocardial lipotoxicity in diabetes [5]. Oxidative stress, mitochondrial dysfunction, abnormal intracellular calcium metabolism [6] and chronic inflammation [7] are mediators of cardiac damage. Angiotensin II (AT) is a potent vasoconstrictor [8]. Endogenous cardiac AT synthesis triggers the development of cardiac hypertrophy [9] irrespective of hypertension [10].

Individuals voluntarily practicing long-term Caloric Restriction (CR) suggest that it favorably affects cardiovascular disease risk factors [11], simultaneously postponing age-related diseases and longevity in animal models [12,13]. Adiponectin, which increases in the plasma after CR [14,15], has been implicated in CR-induced cardioprotection [14]. Sirtuin-1 (SIRT1), a redox-sensitive enzyme, is a member of a large family of class III histone deacetylases (HDAC) [16,17]. It modulates genetic stability, extending the life span of flies, and worms [18]. SIRT1 regulates cellular processes such as apoptosis/cell survival, chromatin remodeling, and gene transcription [19]. SIRT1 activation by CR drives a number of downstream events, including the peroxisome proliferator-activated receptor gamma coactivator 1-alpha (PGC-1α) and the anti-oxidant protein heme oxygenase (HO-1) [20,21]. There are decreased cardiac levels of HO-1 and adiponectin and elevated levels of inflammatory cytokines (Tumor Necrosis Factor α: TNFα) and the oxidative stress marker malondialdehyde (MDA) in the sera of diabetic patients [20]. AT release by the adipocyte and the reduction of HO-1 lead to reactive oxygen species (ROS) and oxidative stress. These factors have a decisive role in obesity-induced injury, mitochondrial dysfunction, and fragmentation [22,23]. Together, these proteins improve metabolic signaling pathways, and blunt pro-inflammatory pathways in mice fed a high-fat, high-calorie diet [24,25]. ROS dependent perturbations associated with metabolic syndrome are influenced by HO activity [26,27]. It is suggested that the genes associated with lipid metabolism, adipocyte differentiation and insulin sensitivity upregulation are influenced by the nuclear transcription factors, peroxisome proliferator-activated receptors (PPARs), i.e., PPARα, γ, and δ [28,29].

Using a murine cardiomyopathy model obtained by stressing the diabetic heart by AT, we reported that CR decreased cardiac hypertrophy and inflammatory markers [30,31]. We also showed that CR affects cardiac remodeling in these mice through molecular mechanisms related to mitochondrial function and an antioxidative signaling pathway mediated by SIRT1 and PGC-1α [31,32]. In the present study, using the same model, we identify HO-1 as a key factor behind the cardioprotective effect of CR. We demonstrate that increased levels of HO-1 improve antioxidant defense and enhance metabolic adaptation through its interaction with SIRT1. Increasing HO-1 mimicked the protective effects of CR on the diabetic heart, while the inhibition of HO activity increased oxidative stress and aggravated pathological hypertrophy.

2. Results

2.1. CR Reduced Oxidative Stress and Increased PGC-1α and HO-1 Levels

We have previously characterized the murine model of diabetic cardiomyopathy by combining diabetes (*db/db* transgenic mice) and AT infusion, and reported the cardioprotective effects of CR [31]. The model of cardiomyopathy in AT-stressed diabetic mice and the protective effect of CR are described in Table 1. Body weight, glucose, Aspartate Aminotransferase (AST), Alanine aminotransferase (ALT), and cholesterol triglycerides were all higher in diabetic mice compared to WT mice. AT induced cardiomyopathy, as demonstrated by both functional and biochemical markers. In order to examine the role of HO-1 in the cardioprotection afforded by CR, SnMP was administrated to the diabetic mice concomitantly with CR. SnMP resulted in increased levels of AST, GOT, and of cholesterol, reversing the beneficial effects of CR. AT with and without diabetes reduced HO-1 levels of cardiac tissue compared to non-treated WT animals, ($p = 0.001$), while CR increased HO-1 ($p = 0.02$, Figure 1). MDA levels were increased in *db/db* + AT mice compared to WT mice ($p = 0.01$), but fell following CR ($p < 0.03$) (Figure 2A).

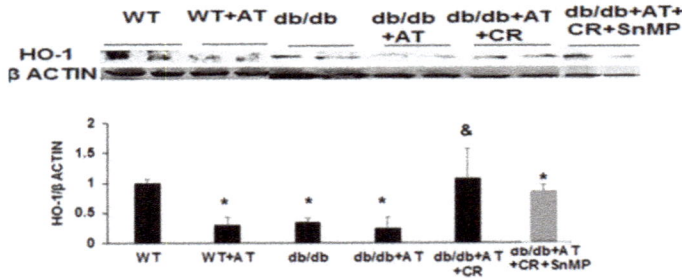

Figure 1. Cardiac heme oxygenase-1 (HO-1) proteins levels: caloric restriction (CR) alleviates oxidative stress through the activation of HO-1. HO-1 was reduced after angiotensin II (AT) treatment in cardiac tissue both in wild-type (WT) and diabetic mice compared to non-treated WT mice ($p = 0.001$), but was elevated after CR. $n = 4$ in each group, * $p < 0.05$ vs. WT, & $p < 0.05$ vs. *db/db* + AT. Values represent mean ± SD.

Table 1. The effect of CR on LV dimension and biochemistry.

	WT $n = 8$	*db/db* $n = 14$	*db/db* + AT $n = 14$	*db/db* + AT + CR $n = 8$	*db/db*+AT+ CR + SnMP $n = 5$
IVS (mm)	0.8 ± 0.1	0.9 ± 0.1	1.1 ± 0.1 #	1 ± 0.1 &	1.3 ± 0.1 $
LVPW (mm)	0.9 ± 0.1	0.9 ± 0.1	1.1 ± 0.2 #	0.9 ± 0.2 &	1.3 ± 0.3 $
LVEDD (mm)	3.6 ± 0.7	3.9 ± 0.2	3.5 ± 0.05 #	4.1 ± 0.4 &	3 ± 1.2 $
LVESD (mm)	2.9 ± 0.2	2.6 ± 0.3	2.4 ± 0.6	2.5 ± 0.5	2.1 ± 0.4
FS (%)	33 ± 14	34 ± 7	34 ± 7	41 ± 10 &	40 ± 4 &
Body Weight (g)	26 ± 3	41 ± 10 *	40 ± 5	33 ± 7 &	36 ± 6
Systolic Blood Pressure (mmHg)	95 ± 21	99 ± 30	148 ± 15 #	114 ± 11 &	138 ± 9 $
Glucose (mg/dL)	137 ± 44	617 ± 93 *	658 ± 107	531 ± 127 &	427 ± 195 &
AST (U/L)	62 ± 25	127 ± 53 *	226 ± 149	99 ± 21 &	138 ± 122
ALT (U/L)	126 ± 42	182 ± 134	281 ± 176	117 ± 32 &	194 ± 198
Cholesterol (mg/dL)	79 ± 24	112 ± 21 *	199 ± 91 #	118 ± 25 &	156 ± 29 $
Triglycerides (mg/dL)	124 ± 57	185 ± 66 *	208 ± 75	127 ± 35 &	188 ± 19 $

Values are mean ± SD. * $p < 0.05$ vs. WT, # $p < 0.05$ vs. *db/db*, & $p < 0.05$ vs. *db/db* + AT, $ $p < 0.05$ vs. *db/db* + AT + CR. IVS, intra ventricular septum; LVPW, left ventricle posterior wall; LVESD, left ventricle end systolic dimension; LVEDD, Left ventricle end diastolic dimension; FS, Fractional shortening.

CR had a beneficial metabolic effect on blood lipids, but that was abolished by SnMP (cholesterol; $p = 0.04$, triglycerides; $p = 0.006$) with no significant effect on both body weight and blood glucose. SnMP resulted in left ventricular hypertrophy (LVH), preventing the protective effect of CR on cardiac hypertrophy ($p = 0.003$). SnMP also increased systolic blood pressure (BP) to the level found in diabetic AT-treated mice without CR ($p = 0.005$) (Table 1), and increased MDA levels (Figure 2A). Adiponectin was reduced in diabetic mice, while AT and CR-treated animals displayed elevated adiponectin levels and SIRT1 activity, which was blocked by SnMP (Figure 2B,C). PGC-1α was reduced in diabetic AT-treated heart tissue ($p < 0.001$). PGC-1α levels were elevated following CR ($p < 0.0001$), but reduced following SnMP treatment (Figure 2D). PPARγ levels were higher in diabetic mice compared to WT. CR reduced PPARγ levels *db/db* +AT hearts. SnMP abolished the beneficial effects of CR, reducing the levels of adiponectin, PGC-1α, and SIRT1 to those of diabetic mice (Figure 2B–D), while increasing PPARγ levels (Figure 2E).

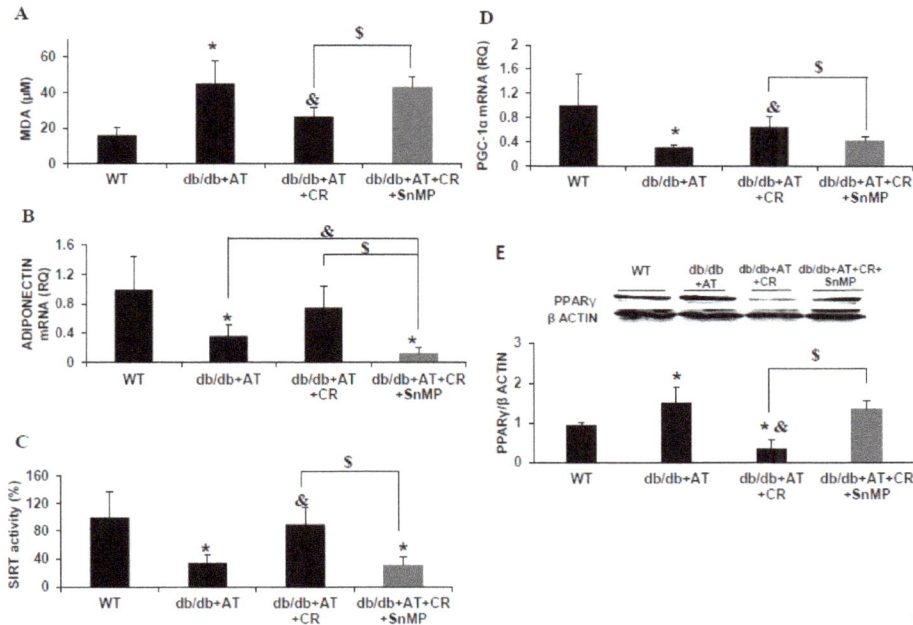

Figure 2. Sn(tin)-mesoporphyrin (SnMP) prevents the beneficial cellular effect of CR. CR diabetic mice were concomitantly treated with SnMP. Malondialdehyde (MDA) levels in the serum were measured using thiobarbituric acid-reactive substances (TBARS) kit, $n = 4$ in each group. $n = 4$ in each group, * $p < 0.007$ vs. WT, & $p = 0.009$ vs. *db/db* + AT, $ $p < 0.005$ vs. *db/db* + AT + CR. Values represent mean ± SD (**A**). Adiponectin (**B**), SIRT1 (**C**), and peroxisome proliferator-activated receptor-γ coactivator (PGC-1α) (**D**) mRNA levels were measured in the cardiac tissue. $n = 4$ in each group, * $p < 0.04$ vs. WT, & $p < 0.05$ vs. *db/db* + AT, $ $p < 0.04$ vs. *db/db* + AT + CR. Values represent mean ± SD. Western blot for peroxisome proliferator-activated receptor y (PPARγ) protein and densitometry analysis of PPARγ normalized to β actin. $n = 4$ in each group, * $p < 0.03$ vs. WT, & $p = 0.004$ vs. *db/db* + AT, $ $p = 0.002$ vs. *db/db* + AT + CR. Values represent mean ± SD (**E**). HO-1 protein levels were reduced in the heart (**F**).

2.2. Cross-Talk between HO-1-SIRT1-and PGC-1α

In order to examine the interaction between HO-1–SIRT1–PGC-1α and their role in glucose metabolism and oxidative stress in the heart, cultured rat neonatal cardiomyocytes exposed to different concentrations of glucose (7.5 mM, 17.5 mM, and 33 mM) were used. Elevated glucose levels led to a concomitant increase in cellular ROS production ($p < 0.03$) (Figure 3(Aa,d,g,B)) and reduction in SIRT1 and PGC-1α proteins levels ($p < 0.05$, Figure 3A,C–F). HO-1 inhibitor SnMP produced a significant reduction in the levels of both SIRT1 ($p < 0.002$) and PGC-1α ($p < 0.009$) (Figure 3C–F), leading to increased ROS production ($p < 0.001$, Figure 4Ab,e,h). The HO-1 inducer, CoPP, increased SIRT1 ($p < 0.008$) and PGC-1α expression ($p < 0.03$, Figure 3D–F) and prevented the glucose-mediated elevation of ROS ($p < 0.002$, Figure 3Ac,f,i). As shown in Figure 3G, the basal levels of HO activity is inhibited by about 70% in bilirubin formation in the presence of SnMP. An increase of glucose levels caused the inhibition of HO activity, and was further potentiated by SnMP, which is clearly observed when glucose reached 33 mM (* $p < 0.01$ vs. control, # $p < 0.001$ vs. SnMP).

SIRT1 inhibition by EX-527 elevated ROS production (Figure 4A,B). PGC-1α and HO-1 protein levels also decreased (Figure 4C–E). Cumulatively, these results indicate a direct bilateral relationship between SIRT1–PGC-1α and HO-1; perturbations in HO activity and HO-1 levels influence upstream molecules e.g., SIRT1. Therefore, SIRT1–PGC-1α and HO-1 form a pathway with a positive feedback loop protecting cardiomyocytes against oxidative stress, which participates in the pathogenesis of diabetic heart disease (Figure 5).

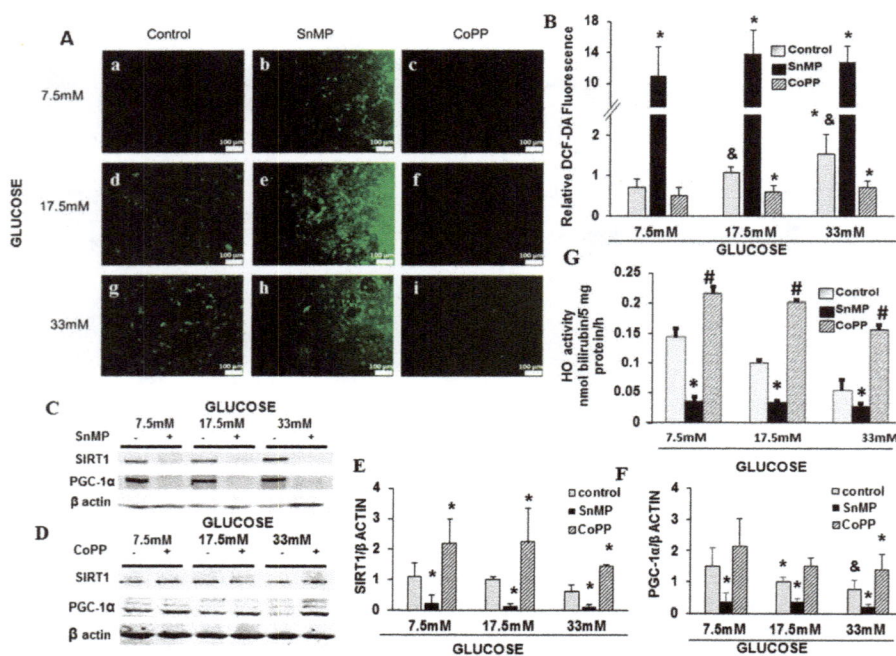

Figure 3. HO-1 is required for the expression of SIRT1 and PGC-1α. Neonatal rat cardiomyocytes were exposed to 7.5 mM, 17.5 mM, and 33 mM of glucose and treated with SnMP or cobalt protoporphyrin (CoPP). Cells were stained with 2′, 7′-dichlorofluorescin diacetate (DCF-DA) (**Aa**–i), and fluorescence was measured using a fluorimeter (**B**). Representative western blots for SIRT1 and PGC-1α for cells treated with SnMP (**C**) and CoPP (**D**), densitometry analysis for SIRT1 (**E**) and PGC-1α (**F**). HO activity in the presence and absence of SnMP and CoPP (**G**). Results were normalized to the group exposed to 17.5 mM of glucose. * $p < 0.05$ vs. control, & $p < 0.05$ vs. 7.5 mM of control, # $p < 0.05$ vs. SnMP. $n = 4$ in each group. Values represent mean ± SD.

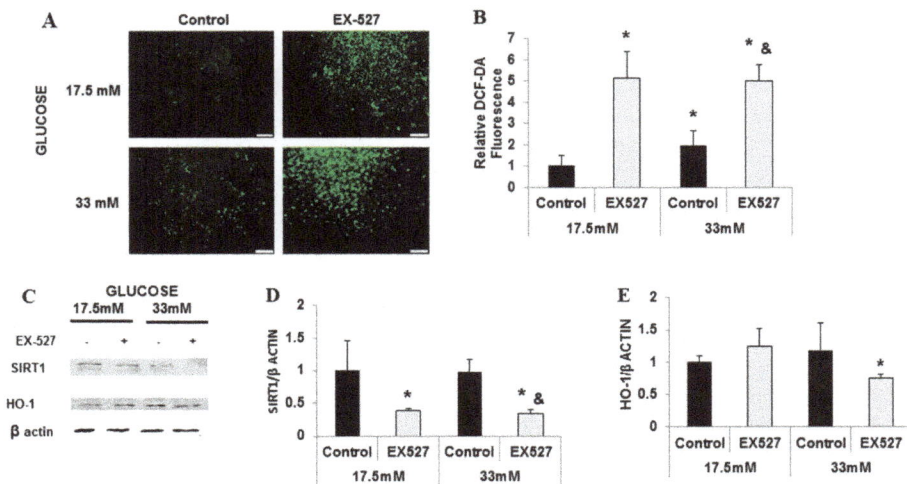

Figure 4. SIRT1 is required for the expression of PGC-1α and HO-1. Neonatal rat cardiomyocytes were exposed to 17.5 mM and 33 mM of glucose and treated with the SIRT1 inhibitor EX-527. Cells were stained with DCF-DA, Scale bar: 100 µm. (**Aa–d**), and fluorescence was measured using fluorimeter (**B**). Representative Western blots for SIRT1, PGC-1α, and HO-1 for cells treated with EX-527 (**C**), densitometry analysis for SIRT1 (**D**), and HO-1 (**E**). Results were normalized to the group exposed to 17.5 mM of glucose. * $p < 0.05$ vs. control, & $p < 0.05$ vs. 33 mM of control, $n = 4$ in each group. Values represent mean ± SD.

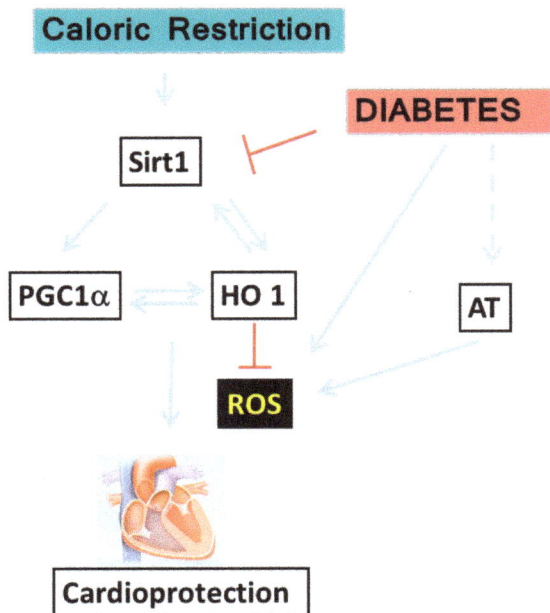

Figure 5. Illustration of CR cellular signaling hereby suggested to participate in the development of Type II diabetic cardiomyopathy: The energetic dysfunction in diabetes increase in the heart together

with elevation in the production of angiotensin, leading to mitochondrial dysfunction, oxidative stress, and inflammation. CR elevates adiponectin and SIRT-1 levels. These lead to the activation of both PGC-1α and HO-1, which together improve mitochondrial function, alleviate the oxidative stress, and reduce inflammation by CR, ameliorating cardiomyopathy. SIRT-1, PGC-1α, and HO-1 form a positive feedback loop elevating each other and thus protecting cardiomyocytes against oxidative stress, which participates in the pathogenesis of diabetic heart disease.

3. Discussion

Obesity affects major segments of the population. We have previously shown that oxidative stress is implicated in the pathogenesis of insulin resistance and its consequent vascular injury. We have emphasized the role of reactive oxygen species in adipocytes that resulted in decreased adiponectin levels, increased inflammation, and decreased adipogenesis. In this study, we identified that the CR-mediated cardio protection effect is dependent on HO-1 expression. Furthermore, we showed a direct link between HO-1-SIRT1 and PGC-1α signaling and the attenuation of diabetic cardiomyopathy.

There are numerous pathways involved in different models of cardiomyopathy [33]. Cardiac hypertrophy is mediated in part by the RAS and TGF-β, which have a central role in cardiac remodeling. Since four-month-old diabetic mice did not develop cardiac hypertrophy or fibrosis, we developed an AT-dependent murine cardiomyopathy model by further stressing the diabetic heart by AT [31]. AT has been reported to induce cardiomyopathy mainly through its profibrotic effects [2,10]. Oxidative stress plays a pivotal role in the development of obesity and the pathogenesis of diabetes [6]. In the current study, SnMP led to a marked hypertrophic remodeling in excess of that present in AT-treated *db/db* mice. Perturbations in adiponectin and PPARγ that are closely related to HO-1 are also involved in the hypertrophic remodeling as well [34,35]. While oxidative stress decreases adiponectin, HO-1 helps increase adiponectin, thereby preventing cardiomyopathy and heart failure development [36]. AT with and without diabetes reduced HO-1 levels of cardiac tissue compared to non-treated WT animals, while CR increased HO-1. MDA levels were increased in *db/db* + AT mice compared to WT mice, but fell following CR, demonstrating the antioxidative role of HO-1 in cardiac tissues.

PGC-1α has been characterized as a master regulator of mitochondrial biogenesis. It acts through several transcription factors, including Nuclear Respiratory Factor (NRF1 and NRF2), which regulate the expression of antioxidant genes, including HO-1 [37]. We have previously shown that the activation of PGC-1α reduced mitochondrial ROS in adipocytes through the induction of HO-1, and that the silencing of PGC-1α prevented the increased levels of HO-1 in these cells [38]. PGC-1α is not activated until it is deacetylated by SIRT1 [39,40], thereby helping antioxidant defenses [41]. The levels of mitochondrial cofactors SIRT1, PGC-1α, and HO-1 were reduced in diabetic AT-treated hearts, while CR elevated these factors. On the contrary, SnMP produced a significant reduction in the levels of both SIRT1 and PGC-1α, leading to increased ROS production. These results establish the link between SIRT1, PGC-1α, and HO-1 signaling that leads to the attenuation of ROS production and diabetic cardiomyopathy.

The downregulation of SIRT1 has been implicated as a contributing factor in metabolic disorders, inducing the metabolic syndrome and DM2 [42]. The SIRT1 protein binds to and represses genes controlled by the fat regulator PPARγ [43]. During CR, fatty acids levels are reduced in diabetic mice, resulting in reduced lipotoxicity [30,44]. CR reduced PPARγ levels, consequently preventing the initiation of the cascade that leads to lipotoxicity that participates in the cardiomyopathy process.

Heme oxygenase exists in two forms, HO-1(inducible) and HO-2 (non-inducible), and is rate limiting in heme degradation to biliverdin, iron, and carbon monoxide. Biliverdin is rapidly converted to bilirubin with positive effects on numerous biological functions [27]. The pleiotropic effects of HO-1 on obesity and cardiovascular disease is well documented [26,45]. HO-1 exhibits a broad spectrum of actions on blood vessel endothelial tissue. This includes increased levels of vasodilation, increased numbers of endothelial progenitor cells, and improved cardiac cell function, while decreasing vasoconstriction and inflammation [26,46]. Increased levels of HO-1 result in increased levels of the antioxidant, bilirubin, and the antiapoptotic, carbon monoxide, which are responsible for neutralizing free radicals, ICAM-1, VCAM-1, TNF, and IL-18 [27]. The role of inflammation and HO-1 in cardiac diabetes using the strepotozocin (STZ) model has been previously described. Myocardial fibrosis and apoptosis, but not inflammation, were found in long-term experimental diabetes STZ [47]. Previously published reports showed that HO-1 induction attenuates glucose-mediated cell growth arrest and apoptosis in human and mice cell line [48]. Additionally, high levels of glucose and hyperglycemia inhibits HO-1 activity and expression, as glucose deprivation increases HO-1 expression [49,50]. CR offers effective protection on later responses such as hypertrophy. Thus, the upregulation of HO-1 in CR offers cytoprotection that is manifested in the amelioration of cardiovascular disease and protection against cardiomyopathy [51,52]. Increased levels of HO-1 through pharmacologic intervention with compounds such as resveratrol and L4F result in cardiac improvement that is akin to that observed with CR. In summary, pharmacologic or genetic interventions to increase HO-1 constitute a novel therapeutic approach to preventing diabetic cardiomyopathy in humans.

In conclusion, in the current study, we show that the cardioprotective effect of CR in diabetic mice involves the increased expression of PGC-1α in association with increased HO-1 and SIRT1 levels. The dependency of the cell survival on HO activity was evident (Figure 3G). Thus, the mechanism of the cell protection in glucose is partially dependent on HO activity. The inhibition of HO activity by SnMP abolished the beneficial effect on cardiac metabolic dysfunction represented by adiponectin, SIRT1, PPARγ, PGC-1α, MDA, and pathological cardiac hypertrophy. The pharmacological inhibition of either HO-1 or SIRT1 of isolated cardiomyocytes was followed by the decreased expression of SIRT1 and PGC-1α and an elevation of PPARγ levels. In contrast, CoPP increased the levels of SIRT1, PGC-1α, and HO-1, and attenuated the myocardial RO, suggesting a mutual symbiotic relationship between these cardioprotective mediators. Prior studies demonstrated that increased levels of HO have been shown to attenuate the expression of inflammatory markers through a number of mechanisms [53–55]. The current data suggest that the increased expression of SIRT1 and PGC-1α is responsible for the increased levels of HO-1. This may be considered as a pivotal axis that is the first line of defense against oxidative stress caused by hyperglycemia, and is essential to protect the diabetic heart from insults. While the field of pharmacological therapies continues to expand, efforts to facilitate weight loss have had limited success. In the present study, we examined the cellular mechanism by which CR protects the diabetic heart. We must understand the underlying cellular mechanisms in order to prevent adverse cardiac remodeling. Our findings are crucial for the development of novel therapeutic approaches such as targeting the HO-1–SIRT1–PGC-1α axis to prevent cardiomyopathy and heart failure, which is a major source of morbidity and mortality in diabetic patients.

4. Materials and Methods

4.1. Animal Model

The animal experiments were approved by the institutional animal care and use committee of Tel Aviv University (M-15-010, 16 February 2015). Homozygous *db/db* mice (C57BLKS/J-*leprdb/leprdb*) and their wild-type (WT) littermates were maintained in a pathogen-free facility on regular rodent chow with free access to water and 12-h light and dark cycles. Homozygous mice were verified by PCR. Male WT or *db/db* mice (12–14 weeks old) were used for the experiments. *db/db* mice develop mild cardiomyopathy at an advanced age [2,56]. To enhance the development of heart disease and obtain a robust phenotype, mice were stressed by ATII as described in other cardiomyopathy models [10]. Mice were divided into the following groups, $n = 5$–14 each in each group: WT, *db/db*, *db/db* + AT, *db/db* + AT + CR, and *db/db* + AT + CR + SnMP.

4.2. Angiotensin

Mice were anesthetized with 2% isoflurane, and an ALZET osmotic pump (Durect Corp., Cupertino, CA, USA) was subcutaneously implanted into each mouse. The osmotic pumps infused angiotensin II (Sigma-Aldrich, St. Louis, MO, USA) at a rate of 1000 ng·kg^{-1}·min^{-1} for 4 weeks.

4.3. Caloric Restriction

Mice were housed in individual cages. Caloric-restricted (CR) mice were fed 90% of their average caloric intake for 2 weeks (10% restriction), followed by 65% of that for an additional 2 weeks (35% restriction). Experiments were conducted after the 4-week period, as we have previously described [57].

4.4. Cell Culture

Rat hearts (Sprague–Dawley 1–2 days old) were removed under sterile conditions and washed three times in phosphate-buffered saline (PBS) to remove excess blood cells. We used rat culture because the rat heart is bigger than the mouse heart; therefore, the yield of cardiomyocytes is higher. The hearts were minced and then gently agitated in a solution of proteolytic enzymes, RDB (Biological Institute, Ness-Ziona, Israel), which was prepared from fig tree extract. RDB was diluted 1:100 in Ca^{2+} and Mg^{2+}-free PBS for a few cycles of 10 min each, as previously described [58]. Dulbecco's modified Eagle's medium (Biological Industries, Kibbutz Beit Haemek, Israel) containing 10% horse serum was added to supernatant suspensions containing dissociated cells. The mixture was centrifuged at 300 g for 5 min. The supernatant was discarded, and the cells were resuspended. The suspension of the cells was diluted to 1.06×10^6 cells/mL, and 1.5 mL of the suspension was placed in 35-mm plastic culture dishes, or 0.5 mL in 24-well plates. The cultures were incubated in a humidified atmosphere of 5% CO$_2$ and 95% air at 37 °C. Confluent monolayers exhibiting spontaneous contractions developed in culture within 2 days [31].

4.5. Experiments with EX-527, CoPP, SnMP, and HO Activity

Cultured cardiomyocytes were incubated with different concentration of glucose (7.5 mM, 17.5 mM, and 33 mM) for 4 days. A glucose concentration of 17.5 mM was considered as control. The SIRT1 inhibitor EX-527 (Cayman Chemical, Ann Arbor MI, USA) was added to the culture for 24 h (10 μM). Cobalt protoporphyrin dichloride (CoPP), 2 μM, which increases HO-1 protein levels, and HO activity and Sn(tin)-mesoporphyrin dichloride (SnMP) 1 μM, inhibits HO activity (Frontier Science, Logan, UT, USA), were dissolved in 0.1 M of sodium citrate buffer, pH 7.8 and added to the cardiomyocyte cultures for 72 h [59]. For the in vivo study, SnMP (2 mg/100 g, intraperitoneal was injected every 4 days concomitantly with AT infusion and CR. HO activity was measured by incubating myocyte in the presence of glucose using the same methods by Da-Silva et al. [60] in which bilirubin, the end product of heme degradation by HO, was extracted with chloroform, and its concentration was determined spectrophotometrically (Dual UV/VIS Beam Spectrophotometer Lambda 25; Perkin-Elmer, Norwalk, CT, USA) using the difference in absorbance at wavelength from λ 460 to λ 530 nm with an absorption coefficient of 40 mmol/L^{-1} and cm^{-1}.

4.6. In Vitro ROS Production Measurement

ROS was detected using a 2′, 7′-dichlorofluorescin diacetate (DCF-DA) reagent (Sigma-Aldrich, St. Louis, MO, USA). This compound is an uncharged cell-permeable molecule. Inside cells, this probe is cleaved by non-specific esterases, forming carboxy dichlorofluoroscein, which is oxidized in the presence of ROS. Cells were loaded with 10 μM of DCF-DA for 30 min at 37 °C and then washed. Fluorescence was monitored with a microplate fluorimeter using wavelengths of 485/538 nm for excitation/emission, respectively.

4.7. Western Blotting

Cardiac tissue was homogenized in lysis buffer and quantified for protein levels using a commercial assay (Bio-Rad, Hercules, CA, USA). Western blotting was performed according to standard procedures, as previously described [31,61]. Protein samples (60 μg) were applied to sodium dodecyl sulfate (SDS) polyacrylamide gel (10–15%), electrophoresed under denaturing conditions and electrotransferred onto nitrocellulose membranes (Bio-Rad). Membranes were blocked with 3% BSA in tris-buffered saline (TBS). Primary antibodies for β actin, GAPDH (Santa Cruz Biotechnology, Dallas, Texas, USA), PGC-1α (ABCAM, Cambridge, UK), HO-1 (Enzo Life Sciences, Farmingdale, NY, USA), and SIRT1 (Merck Millipore Corp, Darmstadt, Germany) were used in TBST with 3% BSA overnight at 4 °C. Dye 680 or 800 secondary antibodies were added at a concentration of 1:10,000 for 1 h at room temperature (LI-COR Biosciences, Lincoln, NE, USA). Detection was carried out with the LI COR Odyssey. Quantification of signals was carried out with the Odyssey program. The ratio between the intensity of the band of the tested protein and the intensity of the corresponding actin or GAPDH band was calculated for the normalization/expression of results.

4.8. RT-PCR

Total RNA was purified from hearts using TRIzol (Ambion, Austin, TX, USA) as per the manufacturer's instructions. The quantity of total RNA was determined by OD260 measurements. cDNA was synthesized from total RNA using the TaqMan High Capacity cDNA Reverse Transcription Kit (Applied Biosystems, Foster City, CA, USA) according to the manufacturer's protocol. Quantitative real-time PCR analysis was performed using the Step One Plus system (Applied Biosystems, Foster City, CA, USA). The primers and TaqMan FAM probes were ordered from Applied Biosystems [31].

Gene	Assay ID
Tbp (TATA BOX)	Mm00446973
Ppargc1 (PGC-1α)	Mm01208835
Adipoq (adiponectin)	Mm00456425

4.9. Serum Thiobarbituric Acid Reactive Substances

Malondialdehyde (MDA) was quantified through a controlled reaction with thiobarbituric acid, generating thiobarbituric acid-reactive substances (TBARS). Thus, lipid peroxidation was determined using the TBARS assay kit (Cayman Chemical, Ann Arbor, MI, USA) according to the manufacturer's instructions.

4.10. SIRT Activity

SIRT activity in the nuclear fraction of cardiac tissue samples was measured using the Universal SIRT activity assay kit (Abcam, Cambridge, UK).

4.11. Statistical Analysis

Animals were assigned to groups randomly. All the values were expressed as mean ± SD. In the in vivo studies, results were normalized to the WT group, and in the in vitro studies, the results were normalized to the 17.5-mM glucose control group. The statistical difference between the two groups was assessed using the two-tailed Student's *t*-test. To compare more than two groups, one-way analysis of variance (ANOVA) with Duncan's multiple comparison option was used. $p < 0.05$ was considered significant.

Author Contributions: All the authors have seen and approved of the data presented. Conceptualization: M.W., M.A., A.S., R.Z., R.K., D.A., S.J.P., E.H. Methodology, M.W., M.A., A.S., R.Z., E.H. Software, M.W., M.A., S.J.P., E.H. validation, M.W., M.A., V.N., E.H.; formal analysis, M.W., M.A., V.N., E.H. Investigation, M.W., M.A., A.S.,

Int. J. Mol. Sci. **2019**, *20*, 2427

R.Z., R.K., D.A., S.J.P., E.H. Resources, M.A., E.H.; Data curation, M.W., M.A., A.S., R.Z., R.K., D.A., S.J.P., E.H. Writing—original draft preparation, M.W., M.A., E.H. Writing—review and editing, M.W., M.A., A.S., R.Z., R.K., D.A., S.J.P., E.H.; Visualization, M.W., M.A., S.J.P., E.H. Supervision, M.A., E.H.; Project administration, M.A., E.H.

Funding: This research received no external funding.

Conflicts of Interest: The authors declare no conflict of interest.

References

1. Sarwar, N.; Gao, P.; Seshasai, S.R.; Gobin, R.; Kaptoge, S.; Di Angelantonio, E.; Ingelsson, E.; Lawlor, D.A.; Selvin, E.; Stampfer, M.; et al. Diabetes mellitus, fasting blood glucose concentration, and risk of vascular disease: A collaborative meta-analysis of 102 prospective studies. *Lancet* **2010**, *375*, 2215–2222.

2. van Bilsen, M.; Daniels, A.; Brouwers, O.; Janssen, B.J.; Derks, W.J.; Brouns, A.E.; Munts, C.; Schalkwijk, C.G.; van der Vusse, G.J.; van Nieuwenhoven, F.A. Hypertension is a conditional factor for the development of cardiac hypertrophy in type 2 diabetic mice. *Plos ONE* **2014**, *9*, e85078. [CrossRef]

3. Asbun, J.; Villarreal, F.J. The pathogenesis of myocardial fibrosis in the setting of diabetic cardiomyopathy. *J. Am. Coll. Cardiol.* **2006**, *47*, 693–700. [CrossRef]

4. Kai, H.; Kuwahara, F.; Tokuda, K.; Imaizumi, T. Diastolic dysfunction in hypertensive hearts: Roles of perivascular inflammation and reactive myocardial fibrosis. *Hypertens. Res.* **2005**, *28*, 483–490. [CrossRef] [PubMed]

5. Zhang, L.; Keung, W.; Samokhvalov, V.; Wang, W.; Lopaschuk, G.D. Role of fatty acid uptake and fatty acid beta-oxidation in mediating insulin resistance in heart and skeletal muscle. *Biochim. Biophys. Acta* **2010**, *1801*, 1–22. [CrossRef]

6. Ansley, D.M.; Wang, B. Oxidative stress and myocardial injury in the diabetic heart. *J. Pathol.* **2012**, *229*, 232–241. [CrossRef]

7. Zhang, C.; Jin, S.; Guo, W.; Li, C.; Li, X.; Rane, M.J.; Wang, G.; Cai, L. Attenuation of diabetes-induced cardiac inflammation and pathological remodeling by low-dose radiation. *Radiat. Res.* **2012**, *175*, 307–321. [CrossRef] [PubMed]

8. Touyz, R.M.; Laurant, P.; Schiffrin, E.L. Effect of magnesium on calcium responses to vasopressin in vascular smooth muscle cells of spontaneously hypertensive rats. *J. Pharmacol. Exp. Ther.* **1998**, *284*, 998–1005.

9. Mazzolai, L.; Nussberger, J.; Aubert, J.F.; Brunner, D.B.; Gabbiani, G.; Brunner, H.R.; Pedrazzini, T. Blood pressure-independent cardiac hypertrophy induced by locally activated renin-angiotensin system. *Hypertension* **1998**, *31*, 1324–1330. [CrossRef]

10. Tokuda, K.; Kai, H.; Kuwahara, F.; Yasukawa, H.; Tahara, N.; Kudo, H.; Takemiya, K.; Koga, M.; Yamamoto, T.; Imaizumi, T. Pressure-independent effects of angiotensin II on hypertensive myocardial fibrosis. *Hypertension* **2004**, *43*, 499–503. [CrossRef] [PubMed]

11. Soare, A.; Weiss, E.P.; Pozzilli, P. Benefits of caloric restriction for cardiometabolic health, including type 2 diabetes mellitus risk. *Diabetes/Metab. Res. Rev.* **2014**, *30* (Suppl. 1), 41–47. [CrossRef]

12. Speakman, J.R.; Mitchell, S.E. Caloric restriction. *Mol. Asp. Med.* **2011**, *32*, 159–221. [CrossRef]

13. Mattison, J.A.; Roth, G.S.; Beasley, T.M.; Tilmont, E.M.; Handy, A.M.; Herbert, R.L.; Longo, D.L.; Allison, D.B.; Young, J.E.; Bryant, M.; et al. Impact of caloric restriction on health and survival in rhesus monkeys from the NIA study. *Nature* **2012**, *489*, 318–321. [CrossRef]

14. Shinmura, K.; Tamaki, K.; Saito, K.; Nakano, Y.; Tobe, T.; Bolli, R. Cardioprotective effects of short-term caloric restriction are mediated by adiponectin via activation of AMP-activated protein kinase. *Circulation* **2007**, *116*, 2809–2817. [CrossRef] [PubMed]

15. Zhu, M.; Miura, J.; Lu, L.X.; Bernier, M.; DeCabo, R.; Lane, M.A.; Roth, G.S.; Ingram, D.K. Circulating adiponectin levels increase in rats on caloric restriction: The potential for insulin sensitization. *Exp. Gerontol.* **2004**, *39*, 1049–1059. [CrossRef]

16. Imai, S.; Armstrong, C.M.; Kaeberlein, M.; Guarente, L. Transcriptional silencing and longevity protein Sir2 is an NAD-dependent histone deacetylase. *Nature* **2000**, *403*, 795–800. [CrossRef]

17. Landry, J.; Slama, J.T.; Sternglanz, R. Role of NAD(+) in the deacetylase activity of the SIR2-like proteins. *Biochem. Biophys. Res. Commun.* **2000**, *278*, 685–690. [CrossRef]

18. Howitz, K.T.; Bitterman, K.J.; Cohen, H.Y.; Lamming, D.W.; Lavu, S.; Wood, J.G.; Zipkin, R.E.; Chung, P.; Kisielewski, A.; Zhang, L.L.; et al. Small molecule activators of sirtuins extend Saccharomyces cerevisiae lifespan. *Nature* **2003**, *425*, 191–196. [CrossRef]

19. Sakamoto, J.; Miura, T.; Shimamoto, K.; Horio, Y. Predominant expression of Sir2alpha, an NAD-dependent histone deacetylase, in the embryonic mouse heart and brain. *FEBS Lett.* **2004**, *556*, 281–286. [CrossRef]

20. Kondo, M.; Shibata, R.; Miura, R.; Shimano, M.; Kondo, K.; Li, P.; Ohashi, T.; Kihara, S.; Maeda, N.; Walsh, K.; et al. Caloric restriction stimulates revascularization in response to ischemia via adiponectin-mediated activation of endothelial nitric-oxide synthase. *J. Biol. Chem.* **2009**, *284*, 1718–1724. [CrossRef] [PubMed]

21. Romashko, M.; Schragenheim, J.; Abraham, N.G.; McClung, J.A. Epoxyeicosatrienoic acid as therapy for diabetic and ischemic cardiomyopathy. *Trends Pharmacol. Sci.* **2016**, *37*, 945–962. [CrossRef]

22. Ayer, A.; Zarjou, A.; Agarwal, A.; Stocker, R. Heme Oxygenases in Cardiovascular Health and Disease. *Physiol. Rev.* **2016**, *96*, 1449–1508. [CrossRef]

23. Lever, J.M.; Boddu, R.; George, J.F.; Agarwal, A. Heme Oxygenase-1 in Kidney Health and Disease. *Antioxid. Redox Signal.* **2016**, *25*, 165–183. [CrossRef] [PubMed]

24. Wood, J.G.; Rogina, B.; Lavu, S.; Howitz, K.; Helfand, S.L.; Tatar, M.; Sinclair, D. Sirtuin activators mimic caloric restriction and delay ageing in metazoans. *Nature* **2004**, *430*, 686–689. [CrossRef]

25. Lagouge, M.; Argmann, C.; Gerhart-Hines, Z.; Meziane, H.; Lerin, C.; Daussin, F.; Messadeq, N.; Milne, J.; Lambert, P.; Elliott, P.; et al. Resveratrol improves mitochondrial function and protects against metabolic disease by activating SIRT1 and PGC-1alpha. *Cell* **2006**, *127*, 1109–1122. [CrossRef]

26. Abraham, N.G.; Junge, J.M.; Drummond, G.S. Translational Significance of Heme Oxygenase in Obesity and Metabolic Syndrome. *Trends Pharmacol. Sci.* **2016**, *37*, 17–36. [CrossRef] [PubMed]

27. Abraham, N.G.; Kappas, A. Pharmacological and clinical aspects of heme oxygenase. *Pharmacol. Rev.* **2008**, *60*, 79–127. [CrossRef] [PubMed]

28. Tyagi, S.; Gupta, P.; Saini, A.S.; Kaushal, C.; Sharma, S. The peroxisome proliferator-activated receptor: A family of nuclear receptors role in various diseases. *J. Adv. Pharm. Technol. Res.* **2011**, *2*, 236–240. [CrossRef] [PubMed]

29. Nawrocki, A.R.; Rajala, M.W.; Tomas, E.; Pajvani, U.B.; Saha, A.K.; Trumbauer, M.E.; Pang, Z.; Chen, A.S.; Ruderman, N.B.; Chen, H.; et al. Mice lacking adiponectin show decreased hepatic insulin sensitivity and reduced responsiveness to peroxisome proliferator-activated receptor gamma agonists. *J. Biol. Chem.* **2006**, *281*, 2654–2660. [CrossRef]

30. Cohen, K.; Waldman, M.; Abraham, N.G.; Laniado-Schwartzman, M.; Gurfield, D.; Aravot, D.; Arad, M.; Hochhauser, E. Caloric restriction ameliorates cardiomyopathy in animal model of diabetes. *Exp. Cell. Res.* **2016**, *350*, 147–153. [CrossRef] [PubMed]

31. Waldman, M.; Cohen, K.; Yadin, D.; Nudelman, V.; Gorfil, D.; Laniado-Schwartzman, M.; Kornwoski, R.; Aravot, D.; Abraham, N.G.; Arad, M.; et al. Regulation of diabetic cardiomyopathy by caloric restriction is mediated by intracellular signaling pathways involving 'SIRT1 and PGC-1alpha'. *Cardiovasc. Diabetol.* **2018**, *17*, 111. [CrossRef] [PubMed]

32. Waldman, M.; Nudelman, V.; Shainberg, A.; Abraham, N.G.; Kornwoski, R.; Aravot, D.; Arad, M.; Hochhauser, E. PARP-1 inhibition protects the diabetic heart through activation of SIRT1-PGC-1alpha axis. *Exp. Cell Res.* **2018**. [CrossRef]

33. Leviner, D.B.; Hochhauser, E.; Arad, M. Inherited cardiomyopathies–Novel therapies. *Pharmacol. Ther.* **2015**, *155*, 36–48. [CrossRef]

34. Zhang, N.; Wei, W.Y.; Liao, H.H.; Yang, Z.; Hu, C.; Wang, S.S.; Deng, W.; Tang, Q.Z. AdipoRon, an adiponectin receptor agonist, attenuates cardiac remodeling induced by pressure overload. *J. Mol. Med. (Berl)* **2018**, *96*, 1345–1357. [CrossRef]

35. Fang, X.; Stroud, M.J.; Ouyang, K.; Fang, L.; Zhang, J.; Dalton, N.D.; Gu, Y.; Wu, T.; Peterson, K.L.; Huang, H.D.; et al. Adipocyte-specific loss of PPARgamma attenuates cardiac hypertrophy. *JCI Insight* **2016**, *1*, e89908. [CrossRef]

36. L'Abbate, A.; Neglia, D.; Vecoli, C.; Novelli, M.; Ottaviano, V.; Baldi, S.; Barsacchi, R.; Paolicchi, A.; Masiello, P.; Drummond, G.S.; et al. Beneficial effect of heme oxygenase-1 expression on myocardial ischemia-reperfusion involves an increase in adiponectin in mildly diabetic rats. *Am. J. Physiol. Heart Circ. Physiol.* **2007**, *293*, H3532–H3541. [CrossRef]

37. Sedgwick, B.; Riches, K.; Bageghni, S.A.; O'Regan, D.J.; Porter, K.E.; Turner, N.A. Investigating inherent functional differences between human cardiac fibroblasts cultured from nondiabetic and Type 2 diabetic donors. *Cardiovasc. Pathol.* **2014**, *23*, 204–210. [CrossRef]

38. Waldman, M.; Bellner, L.; Vanella, L.; Schragenheim, J.; Sodhi, K.; Singh, S.P.; Lin, D.; Lakhkar, A.; Li, J.; Hochhauser, E.; et al. Epoxyeicosatrienoic Acids Regulate Adipocyte Differentiation of Mouse 3T3 Cells, Via PGC-1alpha Activation, Which Is Required for HO-1 Expression and Increased Mitochondrial Function. *Stem Cells Dev.* **2016**, *25*, 1084–1094. [CrossRef]

39. Canto, C.; Gerhart-Hines, Z.; Feige, J.N.; Lagouge, M.; Noriega, L.; Milne, J.C.; Elliott, P.J.; Puigserver, P.; Auwerx, J. AMPK regulates energy expenditure by modulating NAD+ metabolism and SIRT1 activity. *Nature* **2009**, *458*, 1056–1060. [CrossRef] [PubMed]

40. Rodgers, J.T.; Lerin, C.; Haas, W.; Gygi, S.P.; Spiegelman, B.M.; Puigserver, P. Nutrient control of glucose homeostasis through a complex of PGC-1alpha and SIRT1. *Nature* **2005**, *434*, 113–118. [CrossRef] [PubMed]

41. Han, X.; Turdi, S.; Hu, N.; Guo, R.; Zhang, Y.; Ren, J. Influence of long-term caloric restriction on myocardial and cardiomyocyte contractile function and autophagy in mice. *J. Nutr. Biochem.* **2012**, *23*, 1592–1599. [CrossRef]

42. de Kreutzenberg, S.V.; Ceolotto, G.; Papparella, I.; Bortoluzzi, A.; Semplicini, A.; Dalla Man, C.; Cobelli, C.; Fadini, G.P.; Avogaro, A. Downregulation of the longevity-associated protein sirtuin 1 in insulin resistance and metabolic syndrome: Potential biochemical mechanisms. *Diabetes* **2010**, *59*, 1006–1015. [CrossRef]

43. Picard, F.; Kurtev, M.; Chung, N.; Topark-Ngarm, A.; Senawong, T.; Machado De Oliveira, R.; Leid, M.; McBurney, M.W.; Guarente, L. Sirt1 promotes fat mobilization in white adipocytes by repressing PPAR-gamma. *Nature* **2004**, *429*, 771–776. [CrossRef]

44. Takemori, K.; Kimura, T.; Shirasaka, N.; Inoue, T.; Masuno, K.; Ito, H. Food restriction improves glucose and lipid metabolism through Sirt1 expression: A study using a new rat model with obesity and severe hypertension. *Life Sci.* **2011**, *88*, 1088–1094. [CrossRef] [PubMed]

45. Cao, J.; Peterson, S.J.; Sodhi, K.; Vanella, L.; Barbagallo, I.; Rodella, L.F.; Schwartzman, M.L.; Abraham, N.G.; Kappas, A. Heme oxygenase gene targeting to adipocytes attenuates adiposity and vascular dysfunction in mice fed a high-fat diet. *Hypertension* **2012**, *60*, 467–475. [CrossRef]

46. Calay, D.; Mason, J.C. The multifunctional role and therapeutic potential of HO-1 in the vascular endothelium. *Antioxid. Redox Signal.* **2014**, *20*, 1789–1809. [CrossRef]

47. Ares-Carrasco, S.; Picatoste, B.; Benito-Martin, A.; Zubiri, I.; Sanz, A.B.; Sanchez-Nino, M.D.; Ortiz, A.; Egido, J.; Tunon, J.; Lorenzo, O. Myocardial fibrosis and apoptosis, but not inflammation, are present in long-term experimental diabetes. *Am. J. Physiol. Heart Circ. Physiol.* **2009**, *297*, H2109–H2119. [CrossRef] [PubMed]

48. Colombrita, C.; Lombardo, G.; Scapagnini, G.; Abraham, N.G. Heme oxygenase-1 expression levels are cell cycle dependent. *Biochem. Biophys. Res. Commun.* **2003**, *308*, 1001–1008. [CrossRef]

49. Chang, S.H.; Barbosa-Tessmann, I.; Chen, C.; Kilberg, M.S.; Agarwal, A. Glucose deprivation induces heme oxygenase-1 gene expression by a pathway independent of the unfolded protein response. *J. Biol. Chem.* **2002**, *277*, 1933–1940. [CrossRef]

50. Quan, S.; Kaminski, P.M.; Yang, L.; Morita, T.; Inaba, M.; Ikehara, S.; Goodman, A.I.; Wolin, M.S.; Abraham, N.G. Heme oxygenase-1 prevents superoxide anion-associated endothelial cell sloughing in diabetic rats. *Biochem. Biophys. Res. Commun.* **2004**, *315*, 509–516. [CrossRef]

51. Cao, J.; Singh, S.P.; McClung, J.A.; Joseph, G.; Vanella, L.; Barbagallo, I.; Jiang, H.; Falck, J.R.; Arad, M.; Shapiro, J.I.; et al. EET intervention on Wnt1, NOV, and HO-1 signaling prevents obesity-induced cardiomyopathy in obese mice. *Am. J. Physiol. Heart Circ. Physiol.* **2017**, *313*, H368–H380. [CrossRef]

52. Singh, S.P.; McClung, J.A.; Bellner, L.; Cao, J.; Waldman, M.; Schragenheim, J.; Arad, M.; Hochhauser, E.; Falck, J.R.; Weingarten, J.A.; et al. CYP-450 Epoxygenase Derived Epoxyeicosatrienoic Acid Contribute To Reversal of Heart Failure in Obesity-Induced Diabetic Cardiomyopathy via PGC-1 alpha Activation. *Cardiovasc. Pharm. Open Access* **2018**, *7*, 233.

53. Raffaele, M.; Li Volti, G.; Barbagallo, I.A.; Vanella, L. Therapeutic Efficacy of Stem Cells Transplantation in Diabetes: Role of Heme Oxygenase. *Front. Cell Dev. Biol.* **2016**, *4*, 80. [CrossRef]

54. Rodella, L.F.; Vanella, L.; Peterson, S.J.; Drummond, G.; Rezzani, R.; Falck, J.R.; Abraham, N.G. Heme oxygenase-derived carbon monoxide restores vascular function in type 1 diabetes. *Drug Metab. Lett.* **2008**, *2*, 290–300. [CrossRef]

55. Cao, J.; Sodhi, K.; Inoue, K.; Quilley, J.; Rezzani, R.; Rodella, L.; Vanella, L.; Germinario, L.; Stec, D.E.; Abraham, N.G.; et al. Lentiviral-human heme oxygenase targeting endothelium improved vascular function in angiotensin II animal model of hypertension. *Hum. Gene* **2011**, *22*, 271–282. [CrossRef]

56. Daniels, A.; van Bilsen, M.; Janssen, B.J.; Brouns, A.E.; Cleutjens, J.P.; Roemen, T.H.; Schaart, G.; van der Velden, J.; van der Vusse, G.J.; van Nieuwenhoven, F.A. Impaired cardiac functional reserve in type 2 diabetic db/db mice is associated with metabolic, but not structural, remodelling. *Acta Physiol. (Oxf.)* **2010**, *200*, 11–22. [CrossRef]

57. Levy, E.; Kornowski, R.; Gavrieli, R.; Fratty, I.; Greenberg, G.; Waldman, M.; Birk, E.; Shainberg, A.; Akirov, A.; Miskin, R.; et al. Long-Lived alphaMUPA Mice Show Attenuation of Cardiac Aging and Leptin-Dependent Cardioprotection. *PloS ONE* **2015**, *10*, e0144593. [CrossRef]

58. Avlas, O.; Srara, S.; Shainberg, A.; Aravot, D.; Hochhauser, E. Silencing cardiomyocyte TLR4 reduces injury following hypoxia. *Exp. Cell Res.* **2016**, *348*, 115–122. [CrossRef]

59. Issan, Y.; Kornowski, R.; Aravot, D.; Shainberg, A.; Laniado-Schwartzman, M.; Sodhi, K.; Abraham, N.G.; Hochhauser, E. Heme oxygenase-1 induction improves cardiac function following myocardial ischemia by reducing oxidative stress. *PloS ONE* **2014**, *9*, e92246. [CrossRef]

60. Da Silva, J.L.; Tiefenthaler, M.; Park, E.; Escalante, B.; Schwartzman, M.L.; Levere, R.D.; Abraham, N.G. Tin-mediated heme oxygenase gene activation and cytochrome P450 arachidonate hydroxylase inhibition in spontaneously hypertensive rats. *Am. J. Med. Sci* **1994**, *307*, 173–181. [CrossRef]

61. Waldman, M.; Hochhauser, E.; Fishbein, M.; Aravot, D.; Shainberg, A.; Sarne, Y. An ultra-low dose of tetrahydrocannabinol provides cardioprotection. *Biochem. Pharmacol.* **2013**, *85*, 1626–1633. [CrossRef] [PubMed]

International Journal of
Molecular Sciences

MDPI

Article

Protective Effects of Caffeic Acid Phenethyl Ester (CAPE) and Novel Cape Analogue as Inducers of Heme Oxygenase-1 in Streptozotocin-Induced Type 1 Diabetic Rats

Valeria Sorrenti [1,*], Marco Raffaele [1], Luca Vanella [1], Rosaria Acquaviva [1], Loredana Salerno [2], Valeria Pittalà [2], Sebastiano Intagliata [3] and Claudia Di Giacomo [1]

[1] Department of Drug Science, Biochemistry Section, University of Catania, 95125 Catania, Italy;
    marco.raffaele@hotmail.com (M.R.); lvanella@unict.it (L.V.); racquavi@unict.it (R.A.);
    cdigiaco@unict.it (C.D.G.)
[2] Department of Drug Science, Pharmaceutical Chemistry Section, University of Catania, 95125 Catania, Italy;
    lsalerno@unict.it (L.S.); vpittala@unict.it (V.P.)
[3] Department of Medicinal Chemistry, College of Pharmacy, University of Florida, Gainesville, FL 32610, USA;
    s.intagliata@cop.ufl.edu
* Correspondence: sorrenti@unict.it; Tel.: +39-095-738-4115

Received: 24 April 2019; Accepted: 15 May 2019; Published: 17 May 2019

Abstract: Type 1 diabetes mellitus (T1D) is a chronic autoimmune disease resulting in the destruction of insulin producing β-cells of the pancreas, with consequent insulin deficiency and excessive glucose production. Hyperglycemia results in increased levels of reactive oxygen species (ROS) and nitrogen species (RNS) with consequent oxidative/nitrosative stress and tissue damage. Oxidative damage of the pancreatic tissue may contribute to endothelial dysfunction associated with diabetes. The aim of the present study was to investigate if the potentially protective effects of phenethyl ester of caffeic acid (CAPE), a natural phenolic compound occurring in a variety of plants and derived from honeybee hive propolis, and of a novel CAPE analogue, as heme oxygenase-1 (HO-1) inducers, could reduce pancreatic oxidative damage induced by excessive amount of glucose, affecting the nitric oxide synthase/dimethylarginine dimethylaminohydrolase (NOS/DDAH) pathway in streptozotocin-induced type 1 diabetic rats. Our data demonstrated that inducible nitric oxide synthase/gamma-Glutamyl-cysteine ligase (iNOS/GGCL) and DDAH dysregulation may play a key role in high glucose mediated oxidative stress, whereas HO-1 inducers such as CAPE or its more potent derivatives may be useful in diabetes and other stress-induced pathological conditions.

Keywords: Type 1 diabetes mellitus (T1D); Pancreatic oxidative damage; Heme oxygenase-1 (HO-1) inducers; Caffeic acid phenethyl ester (CAPE); Reactive oxygen species (ROS); Dimethylarginine dimethylaminohydrolase-1 (DDAH-1); Inducible nitric oxide synthase (iNOS); Gamma-Glutamyl-Cysteine Ligase (GGCL)

---

1. Introduction

Diabetes mellitus (DM) is a chronic syndrome of impaired carbohydrate, protein, and fat metabolism caused by insufficient secretion of insulin and/or defects in insulin action in tissues due to insulin resistance. Type 1 diabetes mellitus (T1D) is a chronic autoimmune disease resulting in the destruction of insulin producing β-cells of the pancreas, with consequent insulin deficiency and excessive glucose production [1,2]. Although insulin resistance is traditionally linked to type 2 diabetes mellitus (T2D), intense inflammatory activities characterized by the presence of cytokines, apoptotic cells, immune cell infiltration, amyloid deposits, and fibrosis may result also in T2D due to loss of β-cells and reduced insulin production [3]. Moreover, irrespective of the type, DM is a complex metabolic disease, often associated with long-term complications, including vascular complications, and affecting many tissues [4–9]. In the diabetic status, exposure to high levels of glucose cause a marked reduction in endothelial cell (EC)-released NO [10], with consequent vascular dysfunction [11]. Previous studies have shown that endogenous arginine analogs may play a regulatory role in the arginine/NO pathway [12]. Asymmetric NG, NG-dimethyl-ʟ-arginine (ADMA) is an endogenous inhibitor of all isoforms of nitric oxide synthase (NOS). Elevated ADMA levels have been identified as a biomarker of endothelial dysfunction [13], suggesting that plasma ADMA is significantly associated with cardiovascular risk. ADMA metabolism is related to its generation from protein breakdown and to its cleavage by dimethylarginine dimethylaminohydrolase (DDAH) into citrulline and dimethylamine [14]. Two distinct isoforms of DDAH have been described so far, DDAH-1 and DDAH-2, with distinct tissue distribution [15,16]. It has been reported that overproduction of reactive oxygen species (ROS) leads to downregulation of DDAH-1 and -2, as well as ADMA accumulation by inhibiting DDAH enzyme, which can be prevented by antioxidants [17,18]. In most tissues, hyperglycemia results in increased levels of ROS and nitrogen species (RNS). Without adequate compensatory response by endogenous antioxidant systems, a redox imbalance occurs, leading to the activation of specific pathways that can amplify the damage. It has been reported that in diabetic patients the increase in oxidative stress is associated with a decline in cellular antioxidant defenses [7]. The transcription factor called Nrf2 (nuclear factor erythroid-derived 2) is referred to as the "master regulator" of the antioxidant response; it modulates the expression of hundreds of genes, including those with a promoter region containing an antioxidant response element (ARE) [19], such as heme oxygenase-1 (HO-1), DDAH-1, DDAH-2, gamma-Glutamyl-cysteine ligase (GGCL) [20–23] and other antioxidant/detoxifying enzymes. The pharmacological manipulation of Nrf2 may represent a target in treating metabolic disorders such as diabetes. This research aims to elucidate some biochemical and metabolic aspects of diabetes, identifying any changes in the capacity of antioxidant defense, in an experimental in vivo model of diabetes. In addition, although some experimental data showed unwanted effects of HO-1 induction in diabetic models [24–26], as it is also evident that all molecules capable of inducing the biosynthesis of HO-1 may represent potential protective agents, natural compounds and synthetic derivatives of natural molecules could be a valid approach for use as adjuvants in antidiabetic therapy [27]. The phenethyl ester of caffeic acid (CAPE), a natural phenolic compound occurring in a variety of plants and derived from honeybee hive propolis has many beneficial properties (anti-carcinogenic, anti-viral, anti-inflammatory, anti-oxidant) [28,29], however, the mechanisms of pleiotropism of CAPE are not fully understood and are partially attributed to the ability to induce HO-1 expression [30]. Our previous, in vitro, study showed that CAPE and small focused series of CAPE analogues were HO-1 inducers. Some of tested compounds were more potent HO-1 inducers than CAPE.

Particularly, 3-(3,4-dihydroxyphenyl)-(2E)-2-propenoic acid 2-(3,4-dimethoxyphenyl) ethyl ester (VP961) was the most potent (Figure 1). Moreover, VP961 is the first known compound able to directly activate the HO-1 enzyme and to induce its protein expression at the same time [31].

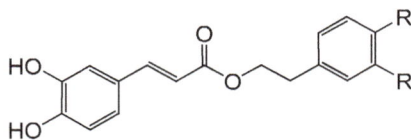

CAPE: R = H
VP961: R = OCH$_3$

Figure 1. Chemical structure of caffeic acid phenethyl ester (CAPE) and VP961-CAPE derivative.

The aim of the present study was to investigate if the potentially protective effect of CAPE as an HO-1 inducer could reduce pancreatic oxidative damage induced by excessive amount of glucose, affecting the NOS/DDAH pathway in streptozotocin-induced type 1 diabetic (STZ) rats. Moreover, because to date only limited strctural CAPE analogues have been examined in vivo [32,33], the protective effect of CAPE derivative VP961, more potent in vitro than the parent compound CAPE as an HO-1 inducer, was investigated in the same animal model mentioned above.

2. Results

2.1. Body Weight, Blood Glucose Content, Food Intake, Water Intake, and Volume of Urine Excreted

2.1.1. The Effects of CAPE and VP961 on Animal Body Weight

Table 1 shows the time course of the changes in body weight during the experimental period. The diabetic control rats (STZ) displayed a marked decrease in body weight with respect to the normal control rats. Body weight was significantly increased in CAPE- or VP961-treated STZ rats with respect to the diabetic control.

Table 1. Effects of CAPE and VP961 on body weight during the experimental period.

Groups	T0	8 Days	15 Days	21 Days
	Body Weight (g)	Body Weight (g)	Body Weight (g)	Body Weight (g)
Control	231 ± 3	265 ± 5	300 ± 7	335 ± 11
STZ	228 ± 5	238 ± 7*	262 ± 5 *	280 ± 3 *
STZ/CAPE	220 ± 3	256 ± 9	291 ± 3	329 ± 4
STZ/VP961	226 ± 5	253 ± 8	291 ± 4	318 ± 6

Values are mean ± standard deviation (S.D.) of three independent experiments performed in triplicate. * $p < 0.05$ vs. normal control rats.

2.1.2. The Effects of CAPE and VP961 on Blood Glucose Content

Figure 2 shows the time course of the changes in blood glucose content during the experimental period. After two days, a significant increase was observed in both diabetic control rats (STZ) and in CAPE- or VP961-treated STZ rats with respect to normal control rats. A significant reduction in blood glucose content was observed in CAPE- or VP961-treated STZ rats with respect to diabetic control rats after 8, 15, and 21 days of treatment.

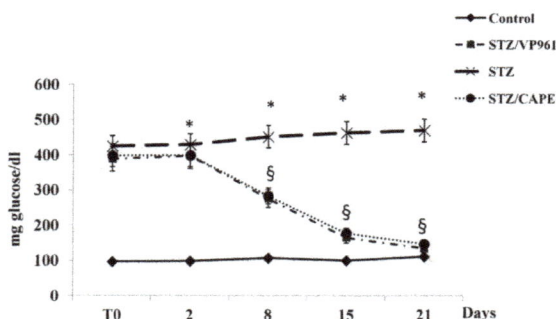

Figure 2. Effects of CAPE and VP961 on blood glucose content during the experimental period. Values are mean ± standard deviation (S.D.) of three independent experiments performed in triplicate. * $p < 0.05$ vs. normal control rats; § $p < 0.05$ vs. diabetic control rats (STZ).

2.1.3. The Effects of CAPE and VP961 on Water Intake, Volume of Urine Excreted, and Food Intake

The time course of the changes in water intake and volume of urine excreted during the experimental period shows that after two days, a significant increase was observed in both diabetic control rats and in CAPE- or VP961-treated STZ rats with respect to normal control rats. A significant reduction in water intake and volume of urine excreted was observed in CAPE- or VP961-treated STZ rats with respect to diabetic control rats (STZ) after 8, 15, and 21 days of treatment (Figures 3 and 4).

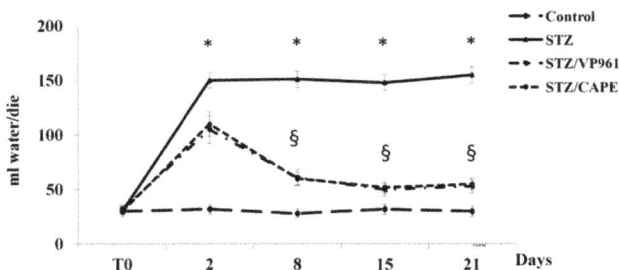

Figure 3. Effects of CAPE and VP961 on water intake during the experimental period. Values are mean ± standard deviation (S.D.) of three independent experiments performed in triplicate. * $p < 0.05$ vs. normal control rats; § $p < 0.05$ vs. diabetic control rats (STZ).

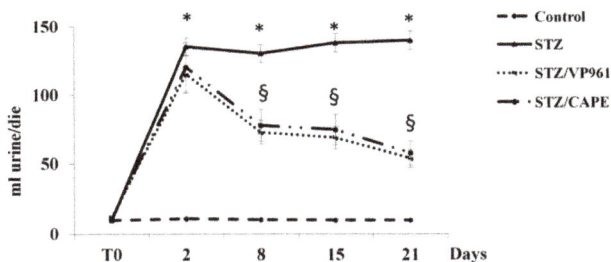

Figure 4. Effects of CAPE and VP961 on volume of urine excreted during the experimental period. Values are mean ± standard deviation (S.D.) of three independent experiments performed in triplicate. * $p < 0.05$ vs. normal control rats; § $p < 0.05$ vs. diabetic control rats (STZ).

The food intake of normal rats was higher with respect to STZ rats (normal control = 25 ± 2 g/day; diabetic control rats (STZ) = 35 ± 3 g/day). The food intake of STZ rats treated with CAPE or VP961 was similar to that of normal rats.

2.2. *Plasma Insulin, RSH, LOOH, ADMA, and Nitrite/Nitrate Levels*

As shown in Table 2, the plasmatic insulin and non-proteic thiol groups (RSH) levels were significantly lower in the diabetic control rats (STZ) than that in the non-STZ rats. Treatment with CAPE or VP961 significantly increased these levels. The levels of lipid hydroperoxide (LOOH), an oxidative stress biomarker, in the plasma of diabetic control rats (STZ) were significantly elevated compared with non-STZ rats; however, these levels were significantly decreased upon receiving CAPE or VP961. STZ rats had increased plasmatic ADMA and NO_2^-/NO_3^- levels compared to the normal control group. CAPE or VP961 treatment in STZ rats significantly reduced ADMA and NO_2^-/NO_3^- levels with respect to control STZ rats.

Table 2. Plasmatic insulin, non-proteic thiol groups, lipid hydroperoxide, Asymmetric NG, NG-dimethyl-L-arginine (RSH, LOOH, ADMA), and NO_2^-/NO_3^- levels.

PLASMA	Insulin (ng/mL)	RSH (nmoles/mL)	LOOH (nmoles/mL)	ADMA (nmoles/mL)	NO_2^-/NO_3^- (nmoles/mL)
Control	1.0 ± 0.03	140 ± 10	15 ± 2	0.1 ± 0.02	0.75 ± 0.03
STZ	0.45 ± 0.05 *	80 ± 7 *	30 ± 5 *	0.9 ± 0.03 *	1.5 ± 0.05 *
STZ/CAPE	0.82 ± 0.03 **	130 ± 9 **	18 ± 3 **	0.6 ± 0.02 **	0.8 ± 0.03 **
STZ/VP961	0.78 ± 0.07 **	120 ± 8 **	17 ± 4 **	0.3 ± 0.02 **±	0.7 ± 0.04 **

Values are mean ± standard deviation (S.D.) of three independent experiments performed in triplicate. * $p < 0.05$ vs. normal control rats; ** $p < 0.05$ vs. diabetic control rats (STZ).

2.3. Pancreatic RSH, LOOH, ADMA, and Nitrite/Nitrate Levels

Concerning the pancreatic RSH content, the diabetic control rats (STZ) showed a marked decrease compared with non-diabetic control rats. This content was significantly increased by CAPE or VP961 treatment, as shown in Table 3. The levels of pancreatic LOOH, an oxidative stress biomarker, in diabetic control rats (STZ) were significantly elevated compared with non-STZ rats; however, these levels were significantly decreased upon receiving CAPE or VP961. STZ rats had increased pancreatic ADMA and NO_2^-/NO_3^- levels compared to the normal control group. CAPE or VP961 treatment in STZ rats significantly reduced ADMA and NO_2^-/NO_3^- levels respect to control STZ rats (Table 3).

Table 3. Pancreatic RSH, LOOH, ADMA, and NO_2^-/NO_3^- levels.

PANCREAS	RSH (nmoles/mg prot.)	LOOH (nmoles/mg prot.)	ADMA (nmoles/mg prot.)	NO_2^-/NO_3^- (nmoles/mg prot.)
Control	28 ± 2	0.2 ± 0.03	20 ± 0.8	4 ± 0.9
STZ	12 ± 1 *	1 ± 0.04 *	200 ± 5 *	12 ± 2 *
STZ/CAPE	27 ± 2 **	0.4 ± 0.02 **	50 ± 4 **	6 ± 0.8 **
STZ/VP961	26 ± 3 **	0.3 ± 0.03 **	53 ± 3 **	5 ± 0.9 **

Values are mean ± standard deviation (S.D.) of three independent experiments performed in triplicate. * $p < 0.05$ vs. normal control rats; ** $p < 0.05$ vs. diabetic control rats (STZ).

2.4. Pancreatic HO-1, DDAH-1, GGCL, iNOS Protein Expressions

The expression levels of antioxidant enzyme-related proteins, such as HO-1 and GGCL, in diabetic control rats (STZ) were very low (Figure 5, Panels B-C). In more detail, HO-1 protein was weakly expressed both in STZ rats and in non-STZ rats, however, CAPE or VP961 administration in STZ rats resulted in a significant upregulation (Figure 5, Panel B).

The expression levels of GGCL in STZ rats were significantly lower than those of non-STZ rats, as shown in Figure 5 (Panel C). The decreased protein expression of GGCL in STZ rats was increased by CAPE or VP961 administration.

The expression levels of iNOS protein in diabetic control rats were significantly higher than those of non-STZ rats, as shown in Figure 6 (Panel B). The increased protein expression of iNOS in STZ rats was decreased by CAPE or VP961 administration.

The expression levels of DDAH-1 protein in diabetic control rats were significantly lower than those of non-STZ rats, as shown in Figure 6 (Panel C). The decreased protein expression of DDAH-1 in STZ rats was increased by CAPE or VP961 administration.

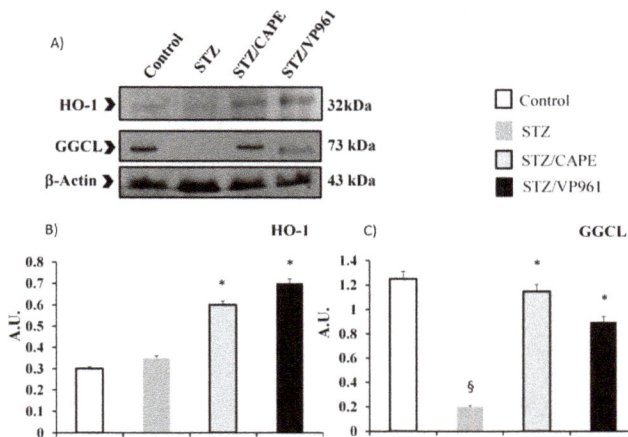

Figure 5. Representative Western blotting of HO-1 and GGCL protein expressions (Panel **A**). Densitometric quantification of HO-1 and GGCL protein expressions in the pancreas of non-STZ rats (control), STZ rats, and CAPE- or VP961-treated STZ rats (CAPE/STZ; VP961/STZ) (Panel **B**,**C**). Values are mean ± standard deviation (S.D.) of three independent experiments performed in triplicate. * $p < 0.05$ vs. diabetic control rats (STZ); § $p < 0.05$ vs. normal control rats.

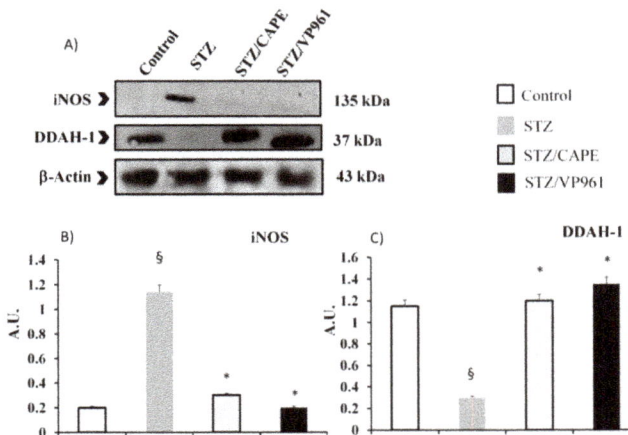

Figure 6. Representative Western blotting of iNOS and DDAH-1 protein expressions (Panel **A**). Densitometric quantification of iNOS and DDAH-1 protein expressions in the pancreas of non-STZ rats (control), STZ rats, and CAPE- or VP961-treated STZ rats (CAPE/STZ; VP961/STZ) (Panel **B**,**C**). Values are mean ± standard deviation (S.D.) of three independent experiments performed in triplicate. * $p < 0.05$ vs. diabetic control rats (STZ); § $p < 0.05$ vs. normal control rats.

3. Discussion

ROS and RNS are well recognized for playing a dual role in human pathology as both deleterious and beneficial species [34]. In addition, it is often difficult to distinguish whether oxidative reactions occurring during a disease process are the cause, by participating in the initial pathogenetic mechanisms of tissue damage, or if they appear only as one of the final effects of the process [35]. Attempts have been made to reduce oxidative damage related to diabetes complications, but the results of administration of antioxidants were disappointing [36,37]; for these reasons, currently research is aiming at the identification of so-called "indirect antioxidants," and the stimulation and strengthening of endogenous antioxidant defenses [38,39]. In recent years, much attention has been focused on phyto-constituents present in fruits, vegetables, and medicinal herbs, and, in particular, on some plant secondary metabolites such as phenolic and terpene compounds [40]. There are now many published studies on their antioxidant activities or their ability to enhance endogenous antioxidant defenses by modulating the cellular redox state of plant-derived substances [41–48], but their potential beneficial effects on human health are not confined to their antioxidant action; in fact, numerous interesting biological activities could reveal new roles of these compounds in the prevention and treatment of certain diseases, such as metabolic syndrome and/or diabetes complications [37,49,50]. However, it is important to note that most studies were conducted using cell models, while few results were obtained using in vivo models [36,37]. Type 1 diabetes leads to high blood glucose levels (hyperglycemia) that can cause serious health complications [51]. Although hyperglycemic damage is a multifactorial process, data in the literature suggest that oxidative/nitrosative stress and stress-activated signaling pathways might represent a unifying hypothesis [7]. In diabetic patients, the overproduction of ROS and RNS is associated with iNOS overexpression, which might contribute to stress-induced pancreatic cell death [52,53].

In our experimental conditions, the significant increase of plasmatic and pancreatic LOOH and nitrite/nitrate levels, markers of oxidative/nitrosative stress, induced by low insulin content and consequent hyperglycemia, may be related to upregulation of pancreatic iNOS protein.

Moreover, oxidative stress is also related to depletion of antioxidant defenses, which also contributes to many of the complications of diabetes, including vascular complications. It has been reported that the overproduction of free radicals could cause damage and apoptosis of pancreatic islet β-cells and reduction of insulin secretion [54]. Bruce et al. reported that HO-1 mRNA expression is significantly reduced in T2D patients [55], whereas upregulation of the HO system increases pancreatic β-cell insulin release and reduces hyperglycemia in different diabetic models [56]. In vitro and in vivo studies have demonstrated that CAPE has many beneficial properties, including anti-hyperglycemic and antioxidant properties [57–61]. In our experimental conditions, body weight of CAPE- or VP961-treated STZ rats was significantly increased compared to the diabetic control rats. Moreover, treatment of STZ rats with CAPE or VP961 significantly reduced blood glucose levels, increased plasmatic insulin levels, and decreased plasmatic and pancreatic LOOH and nitrite/nitrate levels with respect to control STZ rats. The reduction of plasmatic and pancreatic nitrite/nitrate levels may be related to iNOS downregulation in CAPE- or VP961-treated STZ rats. These results suggest that the effects of CAPE or VP961 may be due to their protecting the pancreatic tissue from damage. Moreover, the significant increase of plasmatic and pancreatic LOOH induced by low insulin content and consequent hyperglycemia may be related to downregulation of pancreatic antioxidant defenses, both enzymatic and nonenzymatic, such as HO-1, GGCL, and RSH. Under physiological conditions, Nrf2 locates in the cytoplasm and binds to its inhibitor, kelch-like ECH associated protein 1 (KEAP1). Upon exposure of cells to natural phenolic compounds, Nrf2 is freed from KEAP1 and translocates into the nucleus to bind to antioxidant-responsive elements (ARE) in the genes encoding antioxidant enzymes such as heme oxygenase-1 (HO-1) DDAH-1, DDAH-2, gamma-Glutamyl-cysteine ligase (GGCL) [20–22], and other antioxidant/detoxifying enzymes. Our experimental data showed that HO-1 protein at basal levels was weakly expressed in the pancreatic tissue of control rats. These data are in agreement with the data of Li et al. [62]. Since HO-1 protein was also weakly expressed in the pancreatic tissue of STZ rats, any downregulation could not be detectable. However, the levels of HO-1 protein were significantly increased by CAPE or VP961 administration. According to Ye et al. [63], HO-1 induction may be protective of pancreatic β-cells because of the scavenging of free heme, the antioxidant effects of the end product bilirubin, or the generation of carbon monoxide, which might have insulin-secretion-promoting effects and inhibitory effects on nitric oxide synthase. Moreover, VP961 in vivo was, slightly but significantly, more potent than CAPE as an HO-1 inducer. In our experimental conditions in pancreas, GGCL protein, which catalyzes the first and also limiting step in the synthesis of the antioxidant glutathione (GSH), was downregulated in STZ rats, but the levels of this protein were significantly increased by CAPE or VP961 administration. The increased expressions of GGCL induced in STZ rats treated with CAPE or VP961 are related to increased levels of plasmatic and pancreatic RSH and to decreased levels of plasmatic and pancreatic LOOH. Our results demonstrate that in vivo CAPE is more potent than VP961 as a GGCL inducer.

It has been reported that overproduction of ROS leads to downregulation of DDAH-1 and -2, as well as ADMA accumulation, by inhibiting the DDAH enzyme, which can be prevented by antioxidants [17,18]. Numerous experimental data have shown that DDAH activities are crucial in the regulation of ADMA metabolism [64–66] and in the prevention of endothelial dysfunction. Newsholme et al. reported that oxidative stress and ADMA accumulation could lead to pancreatic β-cell dysfunction and decreased insulin secretion, thus compounding the problematic metabolic status of diabetes [67]. In our experimental conditions, DDAH-1 protein, the main isoform of DDAH expressed in pancreas, was also downregulated in STZ rats, but the levels of this protein were significantly increased by CAPE or VP961 administration. The increased expression of DDAH-1 induced in STZ rats treated with CAPE or VP961 may be due to Nrf2 translocation into the nucleus and to its binding to antioxidant-responsive elements (ARE) in the genes encoding DDAH-1. DDAH-1 upregulation is related to decreased levels of plasmatic and pancreatic ADMA. Our results demonstrate that VP961 was more potent in vivo than CAPE as a DDAH-1 inducer. Overall, our data demonstrated that in an animal model of T1D, CAPE or VP961 treatment may reverse the diabetes-induced oxidative stress in rat pancreas.

4. Materials and Methods

4.1. Animal Model

All animal procedures were performed in accordance with the Guidelines for Care and Use of Laboratory Animals of "Catania University," and experiments were approved by the Animal Ethics Committee (project code N.170, Italy; 1 October 2016) of "MINISTRY OF HEALTH (Directorate General for Animal Health and Veterinary Medicines) (Italy)". Thirty day old Wistar rats were purchased from Charles River Labs (Lecco, Italy). The rats were maintained under a 12 h light/dark cycle, and housed in a controlled temperature (24 ± 2 °C) and humidity (50 ± 5%) environment. After several days of adaptation, the rats were divided into normal and diabetic groups. The experimental diabetes was induced by intraperitoneal (i.p.) injection of streptozotocin (50 mg/kg body weight in a 10 mM citrate buffer, pH 4.5). One week after the injection, we verified the occurrence of hyperglycemia; animals with blood glucose >140 mg/dL were placed in individual metabolic cages; body weight, amount of water and food taken, and volume of urine excreted were recorded daily. Non-fasting blood samples were collected twice per week by tail bleeding into heparinized tubes. In the plasma samples, the glucose concentrations were determined. Rats were distributed in four groups: group I included six untreated animals that were considered the normal control group; group II included six diabetic animals considered the diabetic control group; group III included six diabetic animals orally treated with a non-toxic dose (30 mg/Kg) of the alcoholic extract of CAPE; and group IV included six diabetic animals orally treated with a non-toxic dose (30 mg/Kg) of the alcoholic extract of CAPE derivative VP961. Control groups (diabetic and non-diabetic rats) received the same volume of ethanol as vehicle.

After 21 days, animals were sacrificed by an overdose of anesthetic, and blood and pancreas tissues were immediately removed and frozen for biochemical assays.

4.2. Measurement of Glucose and Insulin in the Plasma

Plasmatic glucose and insulin levels were measured using, respectively, a commercial glucose ELISA kit (CrystalChem, Zaandam, the Netherlands) and commercial insulin ELISA kit (ALPCO, Salem, NH, USA) in accordance with the manufacturer's instructions. Results are reported respectively as mg glucose/dl of plasma and ng insulin/mL of plasma.

4.3. Plasmatic and Pancreatic Nitrite/Nitrate Determination

Quantification of nitrite, the stable metabolite of nitric oxide, was measured colorimetrically via Griess reaction. Aliquots of plasma or pancreas homogenates were preincubated for 30 min at room temperature with 50 µM nicotinamide adenine dinucleotide phosphate (Sigma-Aldrich, St. Louis, MO, USA) and 24 mU nitrate reductase (Roche Diagnostics Gmbh, Mannheim, Germany), and then the samples were treated with 0.2 U lactate dehydrogenase (Roche) and 0.5 mol sodium pyruvate for 10 min. The coloration was developed by adding Griess reagent (Merck KGaA, Darmstadt, Germany; 1:1, vol/vol). Finally, after 10 min at room temperature, absorbance was recorded by 96 well plate microtiter at λ 540 nm. Nitrite levels were determined using a standard curve and expressed as nanomoles of NO_2^-/NO_3^- per ml of plasma or NO_2^-/NO_3^- per milligram of protein. Protein concentration was measured using TAKE 3 nanodrop.

4.4. Plasmatic and Pancreatic ADMA Determination

Plasma and tissue ADMA concentration was determined in plasma or pancreas homogenates using a commercially available enzyme-linked immunosorbent assay kit (DLD Diagnostika GmbH, Hamburg, Germany) according to the manufacturer's instructions. Results are reported as nmoles ADMA/mL of plasma or nmoles ADMA/mg prot.

4.5. Determination of Plasmatic and Pancreatic Lipid Hydroperoxide Levels

Plasma and pancreatic levels of lipid hydroperoxide were evaluated following the oxidation of Fe^{+2} to Fe^{+3} in the presence of xylenol orange at λ 560 nm, as previously described [68]. Results are reported as nmoles LOOH/mL of plasma or nmoles LOOH/mg prot.

4.6. Non-Proteic Thiol Groups Determination

Plasma and pancreatic levels of non-proteic thiol groups were measured, in 200 µL of plasma or pancreatic homogenate, using a spectrophotometric assay, as previously described [68]. Results are reported as nmoles RSH/mL of plasma or nmoles RSH/mg prot.

4.7. Western Blotting

Western blotting analysis was performed as previously described [69,70]. Briefly, tissues were homogenized in lysis buffer (50 mM Tris-HCl, 10 mM EDTA, 1% *v/v* Triton X-100, 1% phenylmethylsulfonyl fluoride (PMSF), 0.05 mM pepstatin A, and 0.2 mM leupeptin) and tissue homogenates (30 µg proteins) were loaded onto 12% SDS-polyacrylamide (SDS-PAGE) gels and subjected to electrophoresis (120 V, 90 min). The separated proteins were transferred to nitrocellulose membranes (Bio-Rad, Hercules, CA, USA). After transfer, the blots were incubated with Li-COR blocking buffer for 1 h, followed by overnight incubation with primary antibodies directed against HO-1 (1:1000) [Enzo Life Sciences, Plymouth Meeting, PA)], GGCL (1:1000) [Abcam, Cambridge, United Kingdom], DDAH-1 (1:5000) [Calbiochem EMD Biosciences (Darmstadt, Germany)], iNOS (1:1000) (SantaCruz Biotechnology, Santa Cruz, CA, USA) and β-actin (Cell Signaling Technology, Inc., Danvers, MA, USA). After washing with TBS, the blots were incubated for 1 h with the secondary antibody (1:1000). Protein detection was carried out using a secondary infrared fluorescent dye-conjugated antibody absorbing at λ 800 and λ 700 nm. The blots were visualized using an Odyssey Infrared imaging scanner (LI-COR Biosciences), and quantified by densitometric analysis performed after normalization with β-actin. Results are expressed as arbitrary units (A.U.).

4.8. Statistical Analysis

Data are reported as mean ± standard deviation (S.D.) values of at least three independent experiments. The results were analyzed for statistical significance using ANOVA, followed by Bonferroni's post hoc test. A p-value < 0.05 was considered as significant.

Int. J. Mol. Sci. **2019**, *20*, 2441

5. Conclusions

Overproduction and/or insufficient removal of free radicals results in different pathological conditions, including diabetes [71]. Diabetes mellitus increases oxidative stress in pancreatic tissue. Oxidative damage of the pancreatic tissue may contribute to endothelial dysfunction associated with diabetes. It can be concluded that CAPE and VP961 inhibit lipid peroxidation and regulate antioxidant enzyme-related proteins in STZ rats. Moreover, iNOS/GGCL and DDAH dysregulation may play a key role in high glucose mediated oxidative stress, whereas HO-1 inducers such as CAPE or its derivatives may be useful in treating diabetes and other stress-induced pathological conditions.

Author Contributions: Conceptualization, V.S., L.V., R.A. and C.D.G.; Formal analysis, M.R.; Funding acquisition, V.S., R.A., L.S., V.P. and C.D.G.; Investigation, M.R.; Methodology, L.S., V.P. and S.I.; Project administration, V.S., C.D.G.; Supervision, V.S.; Writing—original draft, V.S., L.V., R.A. and C.D.G.; Writing—review & editing, V.S.

Funding: This work was supported by 1) Research Funding for University, Italian FIR, project code 75DEDE, Italy) and 2) Project authorized by the Ministry of Health (Directorate General for Animal Health and Veterinary Medicines) (project code n. 170, Italy)

Conflicts of Interest: The authors declare no conflict of interest.

References

1. Pinto, A.; Tuttolomondo, A.; Di Raimondo, D.; Fernández, P.; La Placa, S.; Di Gati, M.; Licata, G. Cardiovascular risk profile and morbidity in subjects affected by type 2 diabetes mellitus with and without diabetic foot. *Metab. Clin. Exp.* **2008**, *57*, 676–682. [CrossRef] [PubMed]
2. Hermans, M.P. Diabetes and the Endothelium. *Acta Clin. Belg.* **2007**, *62*, 97–101. [CrossRef]
3. Butler, A.E.; Janson, J.; Bonner-Weir, S.; Ritzel, R.; Rizza, R.A.; Butler, P.C. Beta-cell deficit and increased beta-cell apoptosis in humans with type 2 diabetes. *Diabetes* **2003**, *52*, 102–110. [CrossRef] [PubMed]
4. Sobrevia, L.; Mann, G. Dysfunction of the endothelial nitric oxide signalling pathway in diabetes and hyperglycaemia. *Exp. Physiol.* **1997**, *82*, 423–452. [CrossRef]
5. Lorenzi, M.; Cagliero, E.; Toledo, S. Glucose Toxicity for Human Endothelial Cells in Culture: Delayed Replication, Disturbed Cell Cycle, and Accelerated Death. *Diabetes* **1985**, *34*, 621–627. [CrossRef]
6. Ceriello, A.; dello Russo, P.; Amstad, P.; Cerutti, P. High Glucose Induces Antioxidant Enzymes in Human Endothelial Cells in Culture. Evidence Linking Hyperglycemia and Oxidative Stress. *Diabetes* **1996**, *45*, 471–477. [CrossRef]
7. Evans, J.; Goldfine, I.D.; Maddux, B.A.; Grodsky, G.M. Oxidative Stress and Stress-Activated Signaling Pathways: A Unifying Hypothesis of Type 2 Diabetes. *Endocr. Rev.* **2002**, *23*, 599–622. [CrossRef]
8. Lash, J.M.; Nase, G.P.; Bohlen, H.G. Acute hyperglycemia depresses arteriolar NO formation in skeletal muscle. *Am. J. Physiol. Circ. Physiol.* **1999**, *277*, 1513–1520. [CrossRef] [PubMed]
9. Cosenza, J.; Sorrenti, V.; Acquaviva, R.; Di Giacomo, C. Dietary Compounds, Epigenetic Modifications and Metabolic Diseases. *Chem. Boil.* **2017**, *11*, 17.
10. Taubert, D.; Rosenkranz, A.; Berkels, R.; Roesen, R. Acute effects of glucose and insulin on vascular endothelium. *Diabetologia* **2004**, *47*, 2059–2071. [CrossRef] [PubMed]
11. Elahi, M.M.; Kong, Y.X.; Matata, B.M. Oxidative stress as a mediator of cardiovascular disease. *Oxidative Med. Cell. Longev.* **2009**, *2*, 259–269. [CrossRef]
12. Jin, J.-S.; D'Alecy, L.G. Central and Peripheral Effects of Asymmetric Dimethylarginine, an Endogenous Nitric Oxide Synthetase Inhibitor. *J. Cardiovasc. Pharmacol.* **1996**, *28*, 439–446. [CrossRef]
13. Ito, A.; Tsao, P.S.; Adimoolam, S.; Kimoto, M.; Ogawa, T.; Cooke, J.P. Novel mechanism for endothelial dysfunction: Dysregulation of dimethylarginine dimethylaminohydrolase. *Circulation* **1999**, *99*, 3092–3095. [CrossRef] [PubMed]
14. Tran, C.T.; Leiper, J.M.; Vallance, P. The Ddah/Adma/Nos Pathway. *Atheroscler* **2003**, *4* (Suppl. 4), 33–40. [CrossRef]
15. Leiper, J.M.; Maria, J.S.; Chubb, A.; MacAllister, R.J.; Charles, I.G.; Whitley, G.S.J.; Vallance, P. Identification of two human dimethylarginine dimethylaminohydrolases with distinct tissue distributions and homology with microbial arginine deiminases. *Biochem. J.* **1999**, *343*, 209–214. [CrossRef]

16. Sorrenti, V.; Mazza, F.; Campisi, A.; Vanella, L.; Volti, G.; Giacomo, C.; Di Giacomo, C. High Glucose-Mediated Imbalance of Nitric Oxide Synthase and Dimethylarginine Dimethylaminohydrolase Expression in Endothelial Cells. *Curr. Neurovascular Res.* **2006**, *3*, 49–54. [CrossRef]

17. Tain, Y.-L.; Kao, Y.-H.; Hsieh, C.-S.; Chen, C.-C.; Sheen, J.-M.; Lin, I.-C.; Huang, L.-T. Melatonin blocks oxidative stress-induced increased asymmetric dimethylarginine. *Radic. Boil. Med.* **2010**, *49*, 1088–1098. [CrossRef]

18. Palm, F.; Onozato, M.L.; Luo, Z.; Wilcox, C.S. Dimethylarginine dimethylaminohydrolase (DDAH): Expression, regulation, and function in the cardiovascular and renal systems. *Am. J. Physiol. Circ. Physiol.* **2007**, *293*, H3227–H3245. [CrossRef]

19. Magesh, S.; Chen, Y.; Hu, L. Small molecule modulators of Keap1-Nrf2-ARE pathway as potential preventive and therapeutic agents. *Med. Rev.* **2012**, *32*, 687–726. [CrossRef] [PubMed]

20. Maines, M.D. Heme oxygenase: Function, multiplicity, regulatory mechanisms, and clinical applications. *FASEB J.* **1988**, *2*, 2557–2568. [CrossRef]

21. Luo, Z.; Aslam, V.; Welch, W.J.; Wilcox, C.S. Activation of Nuclear Factor Erythroid 2-Related Factor 2 Coordinates Dimethylarginine Dimethylaminohydrolase/Ppar-Gamma/Endothelial Nitric Oxide Synthase Pathways That Enhance Nitric Oxide Generation in Human Glomerular Endothelial Cells. *Hypertension* **2015**, *65*, 896–902. [CrossRef]

22. Wild, A.C.; Moinova, H.R.; Mulcahy, R.T. Regulation of gamma-glutamylcysteine synthetase subunit gene expression by the transcription factor Nrf2. *J. Boil. Chem.* **1999**, *274*, 33627–33636. [CrossRef]

23. Raffaele, M.; Volti, G.L.; Barbagallo, I.A.; Vanella, L. Therapeutic Efficacy of Stem Cells Transplantation in Diabetes: Role of Heme Oxygenase. *Front. Cell Dev. Boil.* **2016**, *4*, 17. [CrossRef] [PubMed]

24. Csepanyi, E.; Czompa, A.; Szabados-Furjesi, P.; Lekli, I.; Balla, J.; Balla, G.; Tosaki, A.; Bak, I. The Effects of Long-Term, Low- and High-Dose Beta-Carotene Treatment in Zucker Diabetic Fatty Rats: The Role of HO-1. *Int. J. Mol. Sci.* **2018**, *19*, 1132. [CrossRef] [PubMed]

25. Farhangkhoee, H.; Khan, Z.A.; Mukherjee, S.; Cukiernik, M.; Barbin, Y.P.; Karmazyn, M.; Chakrabarti, S. Heme oxygenase in diabetes-induced oxidative stress in the heart. *J. Mol. Cell. Cardiol.* **2003**, *35*, 1439–1448. [CrossRef] [PubMed]

26. Chen, S.; Khan, Z.A.; Barbin, Y.; Chakrabarti, S. Pro-oxidant Role of Heme Oxygenase in Mediating Glucose-induced Endothelial Cell Damage. *Free. Radic. Res.* **2004**, *38*, 1301–1310. [CrossRef]

27. Liu, L.; Puri, N.; Raffaele, M.; Schragenheim, J.; Singh, S.P.; Bradbury, J.A.; Bellner, L.; Vanella, L.; Zeldin, D.C.; Cao, J.; et al. Ablation of soluble epoxide hydrolase reprogram white fat to beige-like fat through an increase in mitochondrial integrity, HO-1-adiponectin in vitro and in vivo. *Prostaglandins Lipid Mediat.* **2018**, *138*, 1–8. [CrossRef]

28. Kurek-Górecka, A.; Rzepecka-Stojko, A.; Górecki, M.; Stojko, J.; Sosada, M.; Świerczek-Zięba, G. Structure and Antioxidant Activity of Polyphenols Derived from Propolis. *Molecules* **2013**, *19*, 78–101. [CrossRef]

29. Barbagallo, I.; Volti, G.; Sorrenti, V.; Giacomo, C.; Acquaviva, R.; Raffaele, M.; Galvano, F.; Vanella, L. Caffeic Acid Phenethyl Ester Restores Adipocyte Gene Profile Expression Following Lipopolysaccharide Treatment. *Lett. Drug Discov.* **2017**, *14*, 481–487. [CrossRef]

30. Kamiya, T.; Izumi, M.; Hara, H.; Adachi, T. Propolis Suppresses CdCl2-Induced Cytotoxicity of COS7 Cells through the Prevention of Intracellular Reactive Oxygen Species Accumulation. *Boil. Pharm.* **2012**, *35*, 1126–1131. [CrossRef]

31. Pittala, V.; Vanella, L.; Salerno, L.; Di Giacomo, C.; Acquaviva, R.; Raffaele, M.; Romeo, G.; Modica, M.N.; Prezzavento, O.; Sorrenti, V. Novel Caffeic Acid Phenethyl Ester (Cape) Analogues as Inducers of Heme Oxygenase-1. *Curr. Pharm. Des.* **2017**, *23*, 1. [CrossRef] [PubMed]

32. Weng, Y.-C.; Chuang, S.-T.; Lin, Y.-C.; Chuang, C.-F.; Chi, T.-C.; Chiu, H.-L.; Kuo, Y.-H.; Su, M.-J. Caffeic Acid Phenylethyl Amide Protects against the Metabolic Consequences in Diabetes Mellitus Induced by Diet and Streptozocin. *Evidence-Based Complement. Altern. Med.* **2012**, *2012*, 1–12. [CrossRef] [PubMed]

33. Guo, X.; Shen, L.; Tong, Y.; Zhang, J.; Wu, G.; He, Q.; Yu, S.; Ye, X.; Zou, L.; Zhang, Z.; et al. Antitumor activity of caffeic acid 3,4-dihydroxyphenethyl ester and its pharmacokinetic and metabolic properties. *Phytomedicine* **2013**, *20*, 904–912. [CrossRef]

34. Di Meo, S.; Reed, T.T.; Venditti, P.; Victor, V.M. Role of ROS and RNS Sources in Physiological and Pathological Conditions. *Oxidative Med. Cell. Longev.* **2016**, *2016*, 1–44. [CrossRef]

35. Srivastava, K.K.; Kuma, R. Stress, Oxidative Injury and Disease. *Indian J. Clin. Biochem.* **2015**, *30*, 3–10. [CrossRef]

36. Barbagallo, I.; Galvano, F.; Frigiola, A.; Cappello, F.; Riccioni, G.; Murabito, P.; D'Orazio, N.; Torella, M.; Gazzolo, D.; Volti, G.L. Potential Therapeutic Effects of Natural Heme Oxygenase-1 Inducers in Cardiovascular Diseases. *Antioxidants Redox Signal.* **2013**, *18*, 507–521. [CrossRef] [PubMed]

37. Pittala, V.; Salerno, L.; Romeo, G.; Acquaviva, R.; Di Giacomo, C.; Sorrenti, V. Therapeutic Potential of Caffeic Acid Phenethyl Ester (Cape) in Diabetes. *Curr. Med. Chem.* **2018**, *25*, 4827–4836. [CrossRef] [PubMed]

38. Turrens, J.F. The potential of antioxidant enzymes as pharmacological agents in vivo. *Xenobiotica* **1991**, *21*, 1033–1040. [CrossRef] [PubMed]

39. Talalay, P.; Dinkova-Kostova, A.T.; Dinkova-Kostova, A.T. Direct and indirect antioxidant properties of inducers of cytoprotective proteins. *Mol. Nutr. Food Res.* **2008**, *52*, 128–138.

40. Pisoschi, A.M.; Pop, A.; Cimpeanu, C.; Predoi, G. Antioxidant Capacity Determination in Plants and Plant-Derived Products: A Review. *Oxidative Med. Cell. Longev.* **2016**, *2016*, 1–36. [CrossRef] [PubMed]

41. Russo, A.; Acquaviva, R.; Campisi, A.; Sorrenti, V.; Di Giacomo, C.; Virgata, G.; Barcellona, M.; Vanella, A. Bioflavonoids as antiradicals, antioxidants and DNA cleavage protectors. *Cell Boil. Toxicol.* **2000**, *16*, 91–98. [CrossRef]

42. Di Giacomo, C.; Acquaviva, R.; Piva, A.; Sorrenti, V.; Vanella, L.; Piva, G.; Casadei, G.; La Fauci, L.; Ritieni, A.; Bognanno, M.; et al. Protective Effect of Cyanidin 3-*O*-Beta-D-Glucoside on Ochratoxin a-Mediated Damage in the Rat. *Br. J. Nutr.* **2007**, *98*, 937–943. [CrossRef]

43. Di Giacomo, C.; Acquaviva, R.; Santangelo, R.; Sorrenti, V.; Vanella, L.; Li Volti, G.; D'Orazio, N.; Vanella, A.; Galvano, F. Effect of Treatment with Cyanidin-3-*O*-Beta-D-Glucoside on Rat Ischemic/Reperfusion Brain Damage. Evid Based Complement. *Alternat. Med.* **2012**, *2012*, 285750.

44. Salerno, L.; Modica, M.N.; Pittalà, V.; Romeo, G.; Siracusa, M.A.; Di Giacomo, C.; Sorrenti, V.; Acquaviva, R. Antioxidant Activity and Phenolic Content of Microwave-Assisted Solanum melongena Extracts. *Sci. World. J.* **2014**, *2014*, 719486. [CrossRef]

45. Acquaviva, R.; Di Giacomo, C.; Vanella, L.; Santangelo, R.; Sorrenti, V.; Barbagallo, I.; Genovese, C.; Mastrojeni, S.; Ragusa, S.; Iauk, L.; et al. Antioxidant Activity of Extracts of Momordica Foetida Schumach. et Thonn. *Molecules* **2013**, *18*, 3241–3249. [CrossRef]

46. Sorrenti, V.; Di Giacomo, C.; Acquaviva, R.; Bognanno, M.; Grilli, E.; D'Orazio, N.; Galvano, F. Dimethylarginine Dimethylaminohydrolase/Nitric Oxide Synthase Pathway in Liver and Kidney: Protective Effect of Cyanidin 3-*O*-Beta-D-Glucoside on Ochratoxin-a Toxicity. *Toxins* **2012**, *4*, 353–363. [CrossRef]

47. Di Giacomo, C.; Acquaviva, R.; Sorrenti, V.; Vanella, A.; Grasso, S.; Barcellona, M.L.; Galvano, F.; Vanella, L.; Renis, M. Oxidative and Antioxidant Status in Plasma of Runners: Effect of Oral Supplementation with Natural Antioxidants. *J. Med. Food* **2009**, *12*, 145–150. [CrossRef]

48. Pittala, V.; Vanella, L.; Salerno, L.; Romeo, G.; Marrazzo, A.; Sorrenti, V.; Di Giacomo, C. Effects of Polyphenolic Derivatives on Heme Oxygenase-System in Metabolic Dysfunctions. *Med. Chem.* **2018**, *25*, 1577–1595. [CrossRef] [PubMed]

49. Sorrenti, V.; Randazzo, C.L.; Caggia, C.; Ballistreri, G.; Romeo, F.V.; Fabroni, S.; Timpanaro, N.; Raffaele, M.; Vanella, L. Beneficial Effects of Pomegranate Peel Extract and Probiotics on Pre-adipocyte Differentiation. *Front. Microbiol.* **2019**, *10*, 660. [CrossRef] [PubMed]

50. Volti, G.L.; Salomone, S.; Sorrenti, V.; Mangiameli, A.; Urso, V.; Siarkos, I.; Galvano, F.; Salamone, F. Effect of silibinin on endothelial dysfunction and ADMA levels in obese diabetic mice. *Cardiovasc. Diabetol.* **2011**, *10*, 62. [CrossRef] [PubMed]

51. Kawasaki, E. Type 1 Diabetes and Autoimmunity. *Clin. Pediatr. Endocrinol.* **2014**, *23*, 99–105. [CrossRef] [PubMed]

52. Sayed, L.H.; Badr, G.; Omar, H.M.; El-Rahim, A.M.A.; Mahmoud, M.H. Camel whey protein improves oxidative stress and histopathological alterations in lymphoid organs through Bcl-XL/Bax expression in a streptozotocin-induced type 1 diabetic mouse model. *Biomed. Pharmacother.* **2017**, *88*, 542–552. [CrossRef] [PubMed]

53. Al Dubayee, M.S.; AlAyed, H.; Almansour, R.; Alqaoud, N.; Alnamlah, R.; Obeid, D.; Alshahrani, A.; Zahra, M.M.; Nasr, A.; Al-Bawab, A.; et al. Differential Expression of Human Peripheral Mononuclear Cells Phenotype Markers in Type 2 Diabetic Patients and Type 2 Diabetic Patients on Metformin. *Front. Endocrinol.* **2018**, *9*, 537. [CrossRef]

54. Roh, S.S.; Kwon, O.J.; Yang, J.H.; Kim, Y.S.; Lee, S.H.; Jin, J.S.; Jeon, Y.D.; Yokozawa, T.; Kim, H.J. Allium Hookeri Root Protects Oxidative Stress-Induced Inflammatory Responses and Beta-Cell Damage in Pancreas of Streptozotocin-Induced Diabetic Rats. *BMC Complement. Altern Med.* **2016**, *16*, 63. [CrossRef]

55. Bruce, C.R.; Carey, A.L.; Hawley, J.A.; Febbraio, M.A. Intramuscular Heat Shock Protein 72 and Heme Oxygenase-1 mRNA Are Reduced in Patients with Type 2 Diabetes: Evidence That Insulin Resistance Is Associated with a Disturbed Antioxidant Defense Mechanism. *Diabetes* **2003**, *52*, 2338–2345. [CrossRef]

56. Tiwari, S.; Ndisang, J. The Heme Oxygenase System and Type-1 Diabetes. *Curr. Pharm. Des.* **2014**, *20*, 1328–1337. [CrossRef] [PubMed]

57. Abduljawad, S.H.; El-Refaei, M.F.; El-Nashar, N.N. Protective and anti-angiopathy effects of caffeic acid phenethyl ester against induced type 1 diabetes in vivo. *Int. Immunopharmacol.* **2013**, *17*, 408–414. [CrossRef]

58. Okutan, H.; Özçelik, N.; Yilmaz, H.R.; Uz, E. Effects of caffeic acid phenethyl ester on lipid peroxidation and antioxidant enzymes in diabetic rat heart. *Clin. Biochem.* **2005**, *38*, 191–196. [CrossRef]

59. Celik, S.; Erdogan, S. Caffeic acid phenethyl ester (CAPE) protects brain against oxidative stress and inflammation induced by diabetes in rats. *Mol. Cell. Biochem.* **2008**, *312*, 39–46. [CrossRef]

60. Yilmaz, H.R.; Uz, E.; Yucel, N.; Altuntas, I.; Özçelik, N.; Yilmaz, H.R. Protective effect of caffeic acid phenethyl ester (CAPE) on lipid peroxidation and antioxidant enzymes in diabetic rat liver. *J. Biochem. Mol. Toxicol.* **2004**, *18*, 234–238. [CrossRef]

61. Park, S.H.; Min, T.S. Caffeic acid phenethyl ester ameliorates changes in IGFs secretion and gene expression in streptozotocin-induced diabetic rats. *Life Sci.* **2006**, *78*, 1741–1747. [CrossRef]

62. Li, Y.; Pan, Y.; Gao, L.; Zhang, J.; Xie, X.; Tong, Z.; Li, B.; Li, G.; Lu, G.; Li, W. Naringenin Protects against Acute Pancreatitis in Two Experimental Models in Mice by NLRP3 and Nrf2/HO-1 Pathways. *Mediat. Inflamm.* **2018**, *2018*, 1–13.

63. Ye, J.; Laychock, S.G. A Protective Role for Heme Oxygenase Expression in Pancreatic Islets Exposed to Interleukin-1beta. *Endocrinology* **1998**, *139*, 4155–4163. [CrossRef]

64. Ferrigno, A.; Di Pasqua, L.G.; Berardo, C.; Richelmi, P.; Vairetti, M. Liver plays a central role in asymmetric dimethylarginine-mediated organ injury. *World J. Gastroenterol.* **2015**, *21*, 5131–5137. [CrossRef]

65. Hu, T.; Chouinard, M.; Cox, A.L.; Sipes, P.; Marcelo, M.; Ficorilli, J.; Li, S.; Gao, H.; Ryan, T.P.; Michael, M.D.; et al. Farnesoid X Receptor Agonist Reduces Serum Asymmetric Dimethylarginine Levels through Hepatic Dimethylarginine Dimethylaminohydrolase-1 Gene Regulation. *J. Boil. Chem.* **2006**, *281*, 39831–39838. [CrossRef] [PubMed]

66. Lanteri, R.; Acquaviva, R.; Di Giacomo, C.; Sorrenti, V.; Destri, G.L.; Santangelo, M.; Vanella, L.; Di Cataldo, A. Rutin in rat liver ischemia/reperfusion injury: Effect on DDAH/NOS pathway. *Microsurgery* **2007**, *27*, 245–251. [CrossRef] [PubMed]

67. Newsholme, P.; Homem De Bittencourt, P.I.; O' Hagan, C.; De Vito, G.; Murphy, C.; Krause, M.S. Exercise and Possible Molecular Mechanisms of Protection from Vascular Disease and Diabetes: The Central Role of Ros and Nitric Oxide. *Clin. Sci.* **2009**, *118*, 341–349. [CrossRef]

68. Acquaviva, R.; Lanteri, R.; Destri, G.L.; Caltabiano, R.; Vanella, L.; Lanzafame, S.; Di Cataldo, A.; Volti, G.L.; Di Giacomo, C. Beneficial effects of rutin and L-arginine coadministration in a rat model of liver ischemia-reperfusion injury. *Am. J. Physiol. Liver Physiol.* **2009**, *296*, 664–670. [CrossRef] [PubMed]

69. Vanella, L.; Russo, G.I.; Cimino, S.; Fragalà, E.; Favilla, V.; Volti, G.L.; Barbagallo, I.; Sorrenti, V.; Morgia, G. Correlation Between Lipid Profile and Heme Oxygenase System in Patients with Benign Prostatic Hyperplasia. *Urology* **2014**, *83*, 1444.e7–1444.e13. [CrossRef] [PubMed]

70. Barbagallo, I.; Giallongo, C.; Volti, G.L.; Distefano, A.; Camiolo, G.; Raffaele, M.; Salerno, L.; Pittala, V.; Sorrenti, V.; Avola, R.; et al. Heme Oxygenase Inhibition Sensitizes Neuroblastoma Cells to Carfilzomib. *Mol. Neurobiol.* **2019**, *56*, 1451–1460. [CrossRef]

71. Phaniendra, A.; Jestadi, D.B.; Periyasamy, L. Free Radicals: Properties, Sources, Targets, and Their Implication in Various Diseases. *Indian J. Clin. Biochem.* **2015**, *30*, 11–26. [CrossRef] [PubMed]

International Journal of
Molecular Sciences

MDPI

Communication

Heme Oxygenase-1 Inhibition Sensitizes Human Prostate Cancer Cells towards Glucose Deprivation and Metformin-Mediated Cell Death

Marco Raffaele [1], Valeria Pittalà [2], Veronica Zingales [1], Ignazio Barbagallo [1], Loredana Salerno [2], Giovanni Li Volti [3], Giuseppe Romeo [2], Giuseppe Carota [1], Valeria Sorrenti [1] and Luca Vanella [1,*]

[1] Department of Drug Science, Biochemistry Section, University of Catania, 95125 Catania, Italy; marco.raffaele@hotmail.com (M.R.); veronica_zingales@libero.it (V.Z.); ignazio.barbagallo@unict.it (I.B.); giuseppe-carota@outlook.it (G.C.); sorrenti@unict.it (V.S.)
[2] Department of Drug Science, Pharmaceutical Chemistry Section, University of Catania, 95125 Catania, Italy; vpittala@unict.it (V.P.); l.salerno@unict.it (L.S.); gromeo@unict.it (G.R.)
[3] Department of Biomedical and Biotechnological Sciences, University of Catania, 95125 Catania, Italy; livolti@unict.it
* Correspondence: lvanella@unict.it; Tel.: +39-095-738-4077

Received: 18 April 2019; Accepted: 23 May 2019; Published: 27 May 2019

Abstract: High levels of heme oxygenase (HO)-1 have been frequently reported in different human cancers, playing a major role in drug resistance and regulation of cancer cell redox homeostasis. Metformin (MET), a drug widely used for type 2 diabetes, has recently gained interest for treating several cancers. Recent studies indicated that the anti-proliferative effects of metformin in cancer cells are highly dependent on glucose concentration. The present work was directed to determine whether use of a specific inhibitor of HO-1 activity, alone or in combination with metformin, affected metastatic prostate cancer cell viability under different concentrations of glucose. MTT assay and the xCELLigence system were used to evaluate cell viability and cell proliferation in DU145 human prostate cancer cells. Cell apoptosis and reactive oxygen species were analyzed by flow cytometry. The activity of HO-1 was inhibited using a selective imidazole-based inhibitor; genes associated with antioxidant systems and cell death were evaluated by qRT-PCR. Our study demonstrates that metformin suppressed prostate cancer growth in vitro and increased oxidative stress. Disrupting the antioxidant HO-1 activity, especially under low glucose concentration, could be an attractive approach to potentiate metformin antineoplastic effects and could provide a biochemical basis for developing HO-1-targeting drugs against solid tumors.

Keywords: prostate cancer; heme oxygenase; metformin; apoptosis; ER stress; HO-1 activity inhibitor

1. Introduction

Heme oxygenase-1 (HO-1) is the inducible isoform of heme oxygenase, the first rate-limiting enzyme in the degradation of heme to free iron, carbon monoxide (CO), and biliverdin [1].

HO-1 is present at low levels in many tissues and is highly upregulated by numerous stimuli, such as heme, heavy metals, UV irradiation, reactive oxygen species (ROS), polyphenols, and inflammatory cytokines [2,3]. HO-1 is mainly localized in microsomes, but it has also been demonstrated to be differently localized in caveolae, mitochondria, and nuclei [4–6].

Endogenous induction of HO-1 is widely acknowledged as an adaptive cellular response, able to counteract oxidative stress. Moreover, HO-1-derived metabolites have several protective effects on cells and tissues against injuries related to pathological conditions like diabetes, obesity, and cardiovascular diseases [7–13].

High levels of HO-1 have been frequently reported in different human cancers [14–16], playing a major role in drug resistance and regulation of cancer cell redox homeostasis [17–22].

Elevated HO-1 levels have been shown in many cancers, as reported by Jozkowicz et al. [23]. A previous study by Florczyk et al. [24] revealed that enhanced activity of biliverdin reductase may protect cells in stressful conditions arising from anti-cancer drugs, cisplatin, and doxorubicin. Additionally, data from Banerjee et al. [25] demonstrate that HO-1 is up-regulated in renal cancer cells as a survival strategy against chemotherapeutic drugs and promotes growth of tumor cells by inhibiting both apoptosis and autophagy. Thus, application of chemotherapeutic drugs along with HO-1 inhibitor may elevate therapeutic efficiency by reducing the cytoprotective effects of HO-1 and by simultaneous induction of both apoptosis and autophagy.

Metformin (MET), a drug widely used for type 2 diabetes, has recently gained interest for treating several cancers [26]. The anti-proliferative effects of metformin, reported in several cancers including breast, colon, glioma, ovarian, pancreatic, and prostate cancer [27–29], have been mainly associated with the capacity of MET to inhibit mitochondrial respiration and consequently increasing glycolysis rates [30] and to arrest cell cycle and inducing caspase-dependent apoptosis [31,32].

An inverse relationship has been found between the progress of prostate cancer and type 2 diabetes in patients who use metformin.

Additionally, MET use had a trend of improving survival for prostate cancer patients [33,34]. Metformin was reported to reduce prostate cancer growth prominently under a high fat diet, acting through the modulation of several tumoral-associated processes in a xenograft model of human cell lines, using immunodeficient mice [35].

MET treatment and caloric restriction increase the AMP/ATP ratio and activate AMP-activated protein kinase (AMPK), switching cells from an anabolic to a catabolic state. Treatment of breast cancer cells with MET significantly decreased cholesterol content with concomitant inhibition of various cholesterol regulatory genes [36], suggesting that drugs affecting cholesterol synthesis, such as simvastatin/atorvastatin, could be used in the treatment of cancer [37].

Recent studies showed, in vitro, that MET, in combination with simvastatin, induced G1-phase cell cycle arrest [38], while, in vivo, treatment of mice with a combination of metformin and atorvastatin caused stronger inhibition than either drug used individually on the growth of PC-3 tumors [39]. Furthermore, it has been reported that combination treatment of MET with valproic acid was more effective at slowing prostate tumor growth in vivo compared to either drug alone, in mouse xenografts [40].

MET has been proposed to operate as an agonist of Sirtuin-1, a nicotinamide adenine dinucleotide (NAD+)-dependent deacetylase that mimics most of the metabolic responses to calorie restriction [41].

Most primary and metastatic human cancers show significantly increased glucose uptake because of their enhanced glucose consumption. Cancer cells are more dependent on glucose for energy production than normal cells [42,43]. Recent studies have indicated that the anti-proliferative effects of metformin in cancer cells are highly dependent on the glucose concentration [44,45].

Furthermore, it has been shown that metformin significantly decreases the intracellular glutathione levels and enhanced sensitivity of esophageal squamous cell carcinoma to cisplatin [46], suggesting that regulation of the antioxidant defenses represents a key target for cancer therapy. In this regard, new evidence has shown the involvement of TIGAR (TP53-induced glycolysis and apoptosis regulator) in glutathione restoration [47].

A growing body of independent evidence supports the association between metabolic alterations and the development and progression of prostate cancer, as well as the promising role of MET in controlling prostate cancer outcomes.

The present work was directed to determine whether use of a specific inhibitor of HO-1 activity, alone or in combination with MET, would affect metastatic prostate cancer cell viability under different concentrations of glucose.

2. Results

2.1. Effect of Metformin on Cell Viability

It has been shown that MET is selectively toxic to p53-deficient cells and provides a potential mechanism for the reduced incidence of tumors observed in patients being treated with metformin [48]. Results obtained by Gonnissen et al. showed that p53-mutant cells were more resistant to MET than PC3 and 22Rv1 [49].

In order to determine whether MET affects proliferation of human p53-mutant prostate cancer cells, we analyzed the effect of the drug on DU145, a p53-mutant cell line [50]. DU145 was treated with different concentrations of MET (3–50 mM) for 24 h, then the cell viability was assessed by MTT assay (Figure 1). In the presence of the highest concentration of MET (50 mM), the cell viability was reduced by about 50% compared to treatment with the other concentrations, and about 80% compared to untreated cells. Consistent with previous studies, cell treatment with metformin showed significant cytotoxicity. The concentration of 10 mM was used in the following experiments.

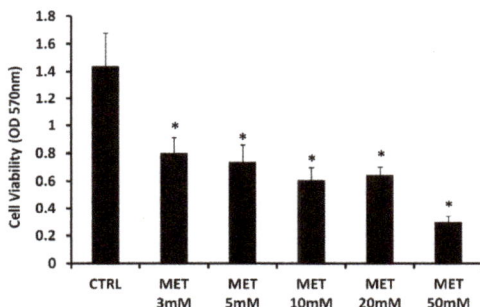

Figure 1. Metformin (MET) caused a decrease in cell viability. Cell viability, determined using MTT assay, of androgen-independent human prostate cancer cells (DU145), control untreated (CTRL) and treated with Metformin at different concentrations (3, 5, 10, 20, and 50 mM) for 24 h. Data are the means ± SD of three experiments performed in triplicate * $p < 0.05$ versus DU145 untreated cells.

2.2. Real-Time Analysis of Cell Proliferation in Presence of Metformin and Different Glucose Concentrations

In order to study the effect of MET on DU145 cells proliferation in conditions of glucose deprivation, dynamic changes in cell index were monitored using the xCELLigence system upon exposure to 1 mM glucose (G1 control (CTRL)) or 25 mM glucose (G25 CTRL) for 48 h. The DU145 cells were untreated and treated with 10 mM MET. The cell index of both control groups of untreated cells followed the same trend across the 48 h. After 24 h, the cell index of cell groups treated with MET followed different trends, showing a noticeable gap at the end of 48 h.

As shown in Figure 2, low glucose concentration enhanced MET cytotoxicity in DU145 cells at 24 h and 48 h.

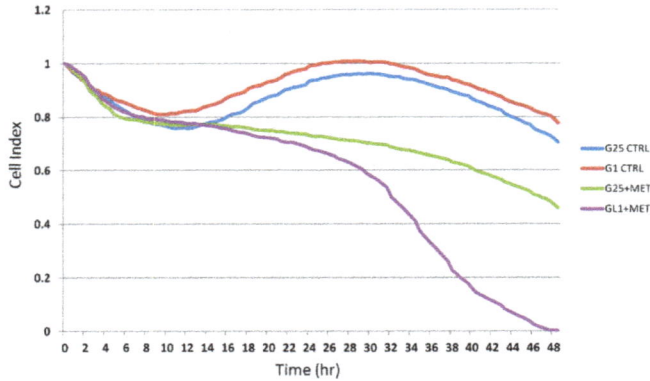

Figure 2. Metformin decrease cell proliferation in presence of different glucose concentrations. DU145 proliferation in the different groups recorded in real time, using the xCELLigence system. The cells showed growth with 1 mM glucose (G1) or 25 mM glucose (G25) in presence and absence of 10 mM metformin (MET).

2.3. Effect of Metformin on HO-1, CHOP, BAX, and Sirtuins mRNA Expression

The effect of MET on HO-1, CHOP and BAX, genes related to the endoplasmic reticulum stress and apoptosis activation, was assessed by measuring mRNA levels. Their gene expression followed the same trend, with an increased level after MET administration, compared to the control group (Figure 3A–C). In order to analyze the effect of metformin on the pathway related to apoptosis regulation, mRNA levels of different sirtuins were assessed. Treatment with 10 mM MET led to an increase of Sirt1 levels, more pronounced in the highest concentration of glucose. Conversely, Sirt3 and Sirt5 levels were reduced after treatment, in both concentrations of glucose. The low concentration of glucose caused a significant decrease of Sirt3 and Sirt5 levels, even in the absence of MET (Figure 3D–F).

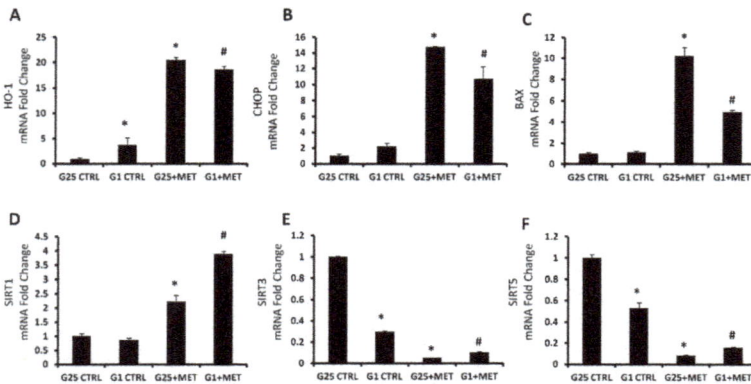

Figure 3. MRNA expression of HO-1 (**A**), CHOP (**B**), BAX (**C**), SIRT1 (**D**), SIRT3 (**E**), and SIRT5 (**F**), of control cells with 25 mM glucose (G25 CTRL), control cells with 1 mM glucose (G1 CTRL), G25 treated with 10 mM metformin (G25 + MET), and G1 treated with 10 mM metformin (G1 + MET). Results are mean ± SD, * $p < 0.05$ vs. G25 CTRL, # $p < 0.05$ vs. G1 CTRL.

2.4. Metformin Enhances the Apoptosis Rate of DU145 in the Presence of a Selective HO-1 Activity Inhibitor

The apoptosis of DU145 cells was measured using Annexin V staining and flow cytometry analysis after 12 h of treatment. As shown in Figure 4, in both concentrations of glucose, the rate of apoptotic

DU145 cells was significantly increased after treatment with MET. The co-treatment with a selective inhibitor of HO-1 activity (VP1347) caused a strong enhancement of apoptosis levels. The treatment with VP1347 alone did not demonstrate significant differences to the untreated control. The decreased level of live cells was more evident in the co-treatment group, indicating a synergistic effect (G1 CDI = 0.90; G25 CDI = 0.95) between MET and VP1347.

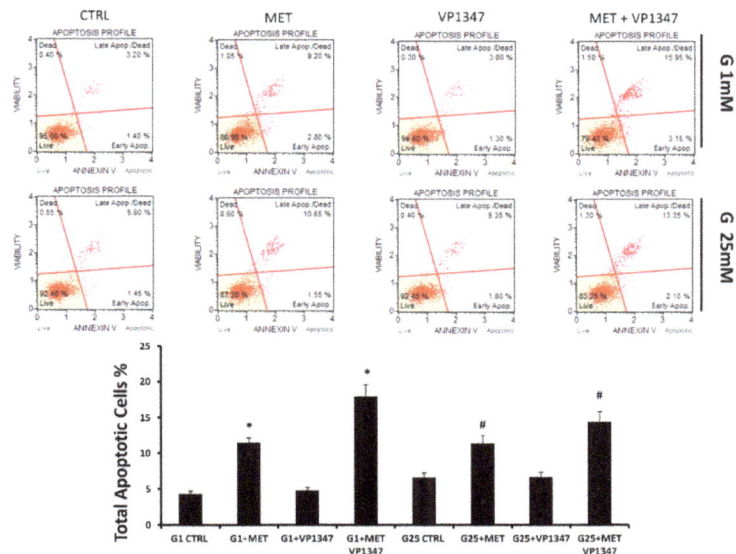

Figure 4. Effect of metformin and HO-1 activity inhibitor VP1347 on DU145 cells apoptosis. Cells were incubated with 1 mM glucose (G1) or 25 mM glucose (G25) in the presence and absence of 10 mM metformin (MET) and 10 μM VP1347 for 12 h. Apoptosis was evaluated by cytometry, using the Muse Annexin V and Dead Cell Assay Kit. The graph showed the total apoptotic cells percentage in the different groups. * $p < 0.05$ vs. G1 CTRL, # $p < 0.05$ vs. G25 CTRL.

2.5. Effect of Metformin on Oxidative Stress Regulation Pathway

At the 1 mM glucose concentration, MET treatment caused a significant decrease of glutathione (GSH) levels compared to the control, especially when combined with VP1347. The treatment with VP1347 alone was not able to reduce the GSH levels. In the presence of the highest concentration of glucose, all treatments showed a reduction of GSH levels without difference within the groups (Figure 5A). TIGAR and Gamma-Glutamylcysteine Synthetase (GCLC) mRNA expression was measured after 6 h of treatment with MET and VP1347 in both concentrations of glucose (Figure 5B,C). At the 1 mM glucose concentration, TIGAR levels were significantly reduced when treated with MET and VP1347, used alone or in combination, compared to the control. GCLC mRNA expression showed a marked increase only in the group treated with metformin. At high glucose concentration, TIGAR levels were strongly reduced in the presence of MET and VP1347, particularly when used as co-treatment. GCLC levels were decreased in all treated groups compared to the control.

2.6. Effect of Metformin and HO-1 Activity Inhibitor on ROS Production

The quantitative measurement of cells undergoing oxidative stress was evaluated by cytometry, using the Muse Oxidative Stress Kit after a 6 h treatment with MET and VP1347. As shown in Figure 6, in both concentrations of glucose, a positive ROS level was notably increased after MET and VP1347 administration compared to the control. The decrease of ROS M1 values was more evident in the co-treatment group, indicating a synergistic effect (G1 CDI = 0.92; G25 CDI = 0.86) between MET and VP1347.

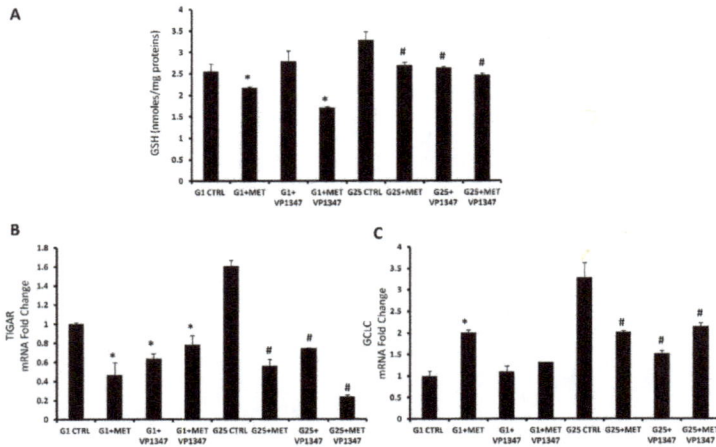

Figure 5. Metformin regulates oxidative stress pathway. (**A**) Thiol groups in DU145 cells treated for 24 h with 10 mM metformin and 10 uM VP1347. Thiol groups are expressed as nmol/mg protein. Values represent the means ± SD of three experiments performed in triplicate. * $p < 0.05$, significant result vs. untreated DU145 cells. (**B,C**) MRNA expression of TIGAR and GCLC of control cells with 25 mM glucose (G25), control cells with 1 mM glucose (G1), in the presence and absence of 10 mM metformin (MET) and 10 μM VP1347 for 6 h. Results are mean ± SD, * $p < 0.05$ vs. G25 CTRL, # $p < 0.05$ vs. G1 CTRL.

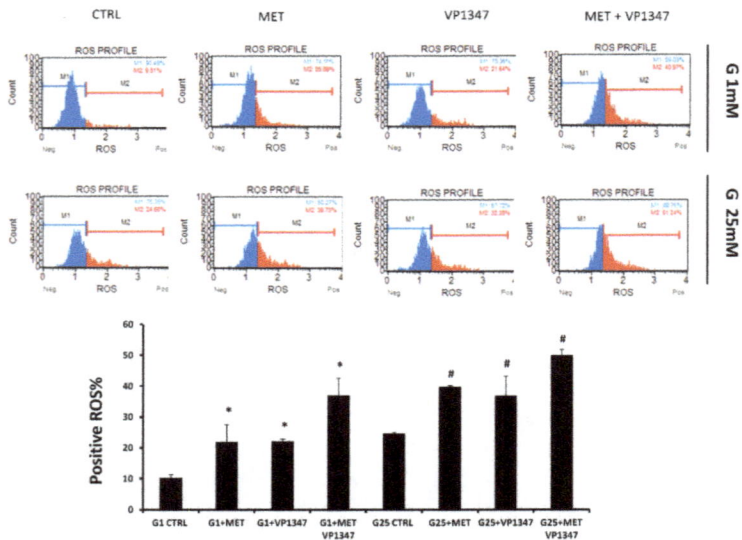

Figure 6. Effect of Metformin and HO-1 activity inhibitor VP1347 on DU145 cells ROS production. Cells were incubated with 1 mM glucose (G1) or 25 mM glucose (G25) in the presence and absence of 10 mM metformin (MET) and 10 μM VP1347 for 6 h. The quantitative measurement of cells undergoing oxidative stress was evaluated by cytometry, using the Muse Oxidative Stress Kit. The graph showed the positive ROS percentage in the different groups. * $p < 0.05$ vs. G1 CTRL, # $p < 0.05$ vs. G25 CTRL.

3. Discussion

In diabetic patients, metformin decreases plasma glucose concentration mainly by lowering hepatic gluconeogenesis and glucose output. This effect is followed by an increase in glucose uptake and the amelioration of insulin resistance [51]. MET works by targeting the enzyme AMPK (AMP activated protein kinase), which induces muscles to take up glucose from the blood. A recent breakthrough has found the upstream regulator of AMPK to be a protein kinase known as LKB1, a well-recognized tumor suppressor [52].

AMP-activated protein kinase (AMPK) activators have been in use for many years to treat type 2 diabetes, but recent data demonstrate that these compounds can inhibit AKT-derived pro-survival effects and induce apoptosis in cancer cells [53].

Several data suggest that MET, an AMPK inducer, could protect from cancer and inhibit breast and glial tumor cell proliferation [52,54,55] through inhibition of the mitochondrial complex I activity and cellular respiration. An essential role of the electron transport chain in cell proliferation has been reported, due to its ability to enable the biosynthesis of aspartate, a proteinogenic amino acid and a precursor in purine and pyrimidine synthesis [56].

In this study, we show that MET not only is a very potent inhibitor of human prostate cancer cell growth, but its effect, in vitro, is potentiated by low glucose treatment and HO-1 activity inhibition.

Although many studies have shown the antineoplastic properties of MET, the mechanisms of action have not been clearly defined yet. As an anticancer agent, MET weakly induces cancer cell apoptosis. However, as shown in Figure 2, under cell culture conditions with low glucose, MET decreases cell proliferation. Starvation from glutamine or glucose for short periods resulted in cell cycle arrest and apoptosis induction by means of reactive oxygen species generation and mitochondrial dysfunction [57,58].

Glucose deprivation has been previously shown to increase MET-induced cell death in cancer cells [59,60].

Glucose deprivation, a cell condition that occurs in solid tumors, activates the unfolded protein response (UPR), which allows the cell to survive under stress conditions [61].

Cancer cells in solid tumors are not supplied sufficient glucose because they are often distant from blood vessels. Adaptive response mechanisms are required for cancer cells to survive in the tumor microenvironment. The UPR in cancer cells plays an important role in their survival and results in tumor malignancies and antitumor drug resistance [62]. If cancer cells have no adaptive response, such as the UPR, activated by endoplasmic reticulum (ER) stress, they would not be able to escape death under glucose deprivation conditions [63].

The acute increase of *CHOP* expression leads to activation of the mitochondria-mediated apoptosis pathway [64]. Therefore, ER stress-dependent apoptosis has been recently reported as a promising therapeutic pathway to target for inducing cancer cell death [65].

Previous studies have shown that metformin induces ER stress and UPR-related genes and inhibited cancer cell proliferation in a dose-dependent manner [66–68]. Aside from IRE1a, PERK, ATF6a, and CHOP regulation, ER stress has been associated with HO-1 induction, which represents a cytoprotective adaptive response to survive stringent conditions [69].

Previously, it has also been reported that the high expression of HO-1 is associated with tumor invasiveness and poor clinical outcome in non-small cell lung cancer patients [16].

Our previous studies showed that the chemotherapeutic drugs, carfilzomib and bortezomib, both increased HO-1 protein and gene expression via the activation of the UPR response, triggered by ER stress [17,70]. As shown in Figure 3, although glucose deprivation already increased HO-1 levels, MET significantly further induced HO-1, CHOP, and BAX mRNA levels, suggesting metformin can induce ER stress and cell apoptosis.

Sirtuins are homologs of the yeast *SIR2* gene, and their function as regulators in a wide range of biological processes is mostly associated with a nicotinamide adenine dinucleotide (NAD+)-dependent deacetylation [71]. Recent studies have shown that seven sirtuins (SIRT 1–7) are linked to tumor growth regulation. Tumor suppression is associated with upregulation of SIRT1 levels, whereas SIRT3 and SIRT5 act differently as tumor promoters [72–74]. DNA damage, exerted by anti-proliferative drugs and the consequent increase of oxidative stress, could induce the expression of transcription factor E2F1, which increases the transcription of SIRT1 and other several apoptotic proteins [75]. Our results showed that MET treatment, both in the presence of 25 mM and 1 mM glucose concentration, increased levels of SIRT1 mRNA, as a demonstration of its anti-proliferative effect [76–78]. SIRT3 and SIRT5 are localized to mitochondria and, as previously published, metformin primarily acts through the mitochondrial respiratory chain inhibition [79,80]. Dysfunctional mitochondria probably represent the main cause of the observed decrease of sirtuins 3 and 5 mRNA in prostate cancer cells [71,81–83].

In this study, we evaluated MET-induced apoptosis in prostate cancer cells, using Annexin cytometry analysis. As shown in Figure 4, MET induced apoptosis in prostate cancer cells and, consistently with previous published results [17,84], HO-1 activity inhibition was able to further increase the cytotoxic effect induced by metformin in DU145 cells, indicating a synergistic effect between the two drugs.

HO-1 represents the inducible isoform of one of the main cytoprotective systems against oxidative stress and inflammation [23,85,86]. When normal cells are exposed to stress conditions, HO-1 induction is the physiological response in order to guarantee regulation of redox homeostasis. For this reason, an adequate expression of this protein is necessary to confer a basal protection against oxidative stress [87,88]. Cancer cells are often characterized by an overexpression of HO-1 and an enhancement of the cytoprotection systems [16,89–91]. It has been reported that metformin alters cellular responses to oxidative stress [92], and the direct activity on mitochondria leads to ROS promotion [30,93–95].

The combined treatment with MET and a selective HO-1 activity inhibitor, under cell culture conditions with low and high glucose, demonstrated a significant decrease of GSH levels, TIGAR, and GCLC mRNA levels. Several reports showed the key role of these factors in the oxidative stress management [96,97]. TIGAR is a protein involved in the switch from glycolysis to the pentose phosphate pathway that promotes the production of cellular nicotinamide adenine dinucleotide phosphate (NADPH) [47,98,99]. NADPH levels enhancement, caused by TIGAR upregulation, leads to an increased restoration of GSH [100]. On the contrary, it has been shown that TIGAR knockdown decreased GSH and NADPH production and increased the levels of ROS [101]. In order to confirm the loss of the physiological protection from oxidative stress, ensured by GSH, we measured the ROS level in the presence of MET and the selective HO-1 activity inhibitor. We observed a considerable increase of ROS levels after treatment with the combination of MET and the HO-1 activity inhibitor. These results suggest that HO-1 inhibition may enhance metformin cytotoxicity through triggering ER stress-associated apoptosis and that ROS is also involved in the activation of ER stress, as schematized in the Figure 7.

In conclusion, our study demonstrates that metformin suppressed prostate cancer growth in vitro and increased oxidative stress under low and high glucose conditions.

Our findings show that disrupting the antioxidant HO-1 activity, especially under low glucose concentration, could be an attractive approach to potentiate MET antineoplastic effects and could provide a biochemical basis for developing HO-1-targeting drugs against solid tumors.

Further investigation of metformin's molecular mechanism and targets will reveal its potential application as a monotherapy or part of a polytherapy associated with a clinically approved HO-1 inhibitor in cancer treatment.

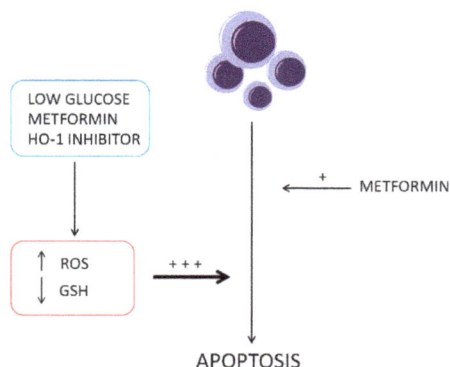

Figure 7. Schematic description of metformin and HO-1 activity inhibitor (VP1347) synergistic effect to induce apoptosis in DU145 cells.

4. Materials and Methods

4.1. Cell Culture

The human androgen-independent prostate cancer cell line DU145 was cultured in Dulbecco's modified Eagle's medium (DMEM) 4.5 g/L glucose, supplemented with 10% FBS, 1% penicillin, and streptomycin at 37 °C and 5% CO_2. The cells were purchased from American Type Culture Collection (Rockville, MD, USA).

4.2. Cell Viability Assay

DU145 cells were seeded at a concentration of 2×10^5 cells per well of a 96-well, flat-bottomed 200 µL microplate. Cells were incubated at 37 °C in a 5% CO_2 humidified atmosphere and maintained in the presence and absence of different concentrations (3, 5, 10, and 50 mM) of metformin (Sigma Chemical Co., St Louis, MO, USA) for 24 h. Three hours before the end of the treatment time, 20 µL of 0.5% 3-(4,5-dimethylthiazol-2-yl)-2,5-diphenyltetrazolium bromide (MTT) in phosphate-buffered saline (PBS) was added to each microwell. After incubation with the reagent, the supernatant was removed and replaced with 100 µL DMSO to dissolve the formazan crystals produced. The amount of formazan is proportional to the number of viable cells present. The optical density was measured using a microplate spectrophotometer reader (Synergy HT, BioTek) at $\lambda = 570$ nm.

4.3. Real-Time Monitoring of Proliferation

Real-time monitoring of cell proliferation was performed using the xCELLigence RTCA DP system [102]. E-plate 16, used with the xCELLigence system, is a single-use 16-well cell culture plate with bottom surfaces covered with microelectrode sensors (0.2 cm^2 well surface area; 243 ± 5 µL maximum volume). Real-time changes in electrical impedance were measured using the gold microelectrodes and expressed as cell index", defined as (Rn-Rb)/15, where Rb is the background impedance and Rn is the impedance of the well with cells. Negative control groups (wells containing 200 µL culture medium without cells with cell index values around 0) were tested in every experiment; however, they were not shown in figures in order to simplify the representations. Before seeding cells into E-plate 16, the background impedance was measured after the addition of 100 µL medium and a 30 min incubation period at room temperature. Following the seeding of the appropriate number of cells into the wells, the plate incubated at room temperature for 30 min in order to allow cell settling. Cell proliferation was monitored every 30 min for 48 h.

4.4. Oxidative Stress Assay

The quantitative measurement of cellular populations undergoing oxidative stress was performed using the Muse Oxidative Stress Kit (Merck Millipore, Billerica, MA, USA), according to the manufacturer's instructions. This assay utilizes dihydroethidium (DHE), which is cell membrane-permeable and, upon reaction with superoxide anions, undergoes oxidation to form DNA-binding fluorophore. The kit determines the percentage of cells that are negative [ROS(−)] and positive [ROS(+)] for reactive oxygen species. Briefly, after 6 h of treatment, 1×10^6 cells/mL were harvested, washed with PBS, and then incubated in the dark at 37 °C for 30 min with the Muse Oxidative Stress Reagent working solution, which contained DHE. The count and percentage of cells undergoing oxidative stress were quantified using the Muse Cell Analyzer and Muse analysis software (Merck Millipore, USA).

4.5. Annexin V Assay

Apoptosis determination by Annexin V staining was carried out using Muse Cell Analyzer with the kit provided by the manufacturer Merck Millipore. In brief, 1×10^6 cells were seeded in six-well plates and, after overnight adherence, they were treated and incubated with 10 mM MET and VP1347, 1-{4-[(4-bromobenzyl)oxy]phenyl}-2-(1H-imidazol-1-yl)ethanol, a HO-1 activity non-competitive inhibitor, that binds to the complex enzyme/substrate. The compound was synthesized by Salerno et al. from the University of Catania as previously described [103].

It presented an $IC_{50} < 1$ μM on HO-1 and showed a >100 selectivity toward HO-2. After 12 h, the cells were detached by trypsinization, centrifuged, and resuspended in PBS. Cell suspension (100 μL) was added with 100 μL of Annexin V reagent and incubated for 20 min at room temperature, after which the cells were analyzed for apoptosis.

4.6. RSH Evaluation

For total thiol groups (RSH) determination, DU145 cells were cultured for 24 h in the presence or absence of 10 mM MET and VP1347. Determination of RSH was performed as previously described [104]. In short, this spectrophotometric assay is based on the reaction of thiol groups with 2,2-dithio-bis-nitrobenzoic acid (DTNB) in absolute ethanol to give a colored compound absorbing at $\lambda = 412$ nm. We then carried out the removal of proteins with an excess of absolute ethanol, followed by centrifugation at $3000 \times g$ for 10 min at room temperature. Each value represents the mean ± SD of three experimental determinations for each sample. Results were expressed as nanomoles per milligram of protein.

4.7. RNA Extraction and qRT-PCR

RNA was extracted by Trizol reagent (Invitrogen, Carlsbad, CA, USA). First strand cDNA was then synthesized with Applied Biosystem (Foster City, CA, USA) reverse transcription reagent. Quantitative real-time PCR was performed in Step One Fast Real-Time PCR System Applied Biosystems, using the SYBR Green PCR MasterMix (Life Technologies, Monza, Italy). The specific PCR products were detected by the fluorescence of SYBR Green, the double stranded DNA binding dye. The relative mRNA expression level was calculated by the threshold cycle (Ct) value of each PCR product and normalized with that of actin by using a comparative $2^{-\Delta\Delta Ct}$ method.

4.8. Statistical Analysis

Statistical significance ($p < 0.05$) of differences between experimental groups was determined by the Fisher method for analysis of multiple comparisons. For comparison between treatment groups, the null hypothesis was tested by either single-factor analysis of variance (ANOVA) for multiple groups, or the unpaired t-test for two groups, and the data are presented as mean ± SD.

Int. J. Mol. Sci. **2019**, *20*, 2593

The coefficient of drug interaction (CDI) has been calculated as follows: CDI = AB/(A × B). According to the values of each group, AB is the ratio of the combination groups to control group; A or B is the ratio of the single agent group to control group. Thus, CDI value <1, =1, or >1 indicates that the drugs are synergistic, additive, or antagonistic, respectively.

Author Contributions: M.R., V.Z., L.V., I.B., and V.S., conceptualization, writing original draft preparation; M.R., V.Z., and G.C., investigation and formal analysis; V.S., V.P., L.S., and G.R., funding acquisition and synthesis of the inhibitor; G.L.V., L.V., resources and writing—review and editing; L.V., supervision and project administration.

Funding: This work was supported by Research Funding for University of Catania, Italy (Piano per la Ricerca, project code 57722172111).

Conflicts of Interest: The authors declare no conflict of interest.

References

1. Abraham, N.G.; Kappas, A. Pharmacological and Clinical Aspects of Heme Oxygenase. *Pharmacol. Rev.* **2008**, *60*, 79–127. [CrossRef]
2. Foresti, R.; Green, C.J.; Motterlini, R. Generation of Bile Pigments by Haem Oxygenase: A Refined Cellular Strategy in Response to Stressful Insults. *Biochem. Soc. Symp.* **2004**, 177–192. [CrossRef]
3. Pittala, V.; Vanella, L.; Salerno, L.; Romeo, G.; Marrazzo, A.; di Giacomo, C.; Sorrenti, V. Effects of Polyphenolic Derivatives on Heme Oxygenase-System in Metabolic Dysfunctions. *Curr. Med. Chem.* **2018**, *25*, 1577–1595. [CrossRef]
4. Vanella, L.; Barbagallo, I.; Tibullo, D.; Forte, S.; Zappala, A.; Volti, G.L. The Non-Canonical Functions of the Heme Oxygenases. *Oncotarget* **2016**, *7*, 69075–69086. [CrossRef] [PubMed]
5. Converso, D.P.; Taille, C.; Carreras, M.C.; Jaitovich, A.; Poderoso, J.J.; Boczkowski, J. Ho-1 Is Located in Liver Mitochondria and Modulates Mitochondrial Heme Content and Metabolism. *FASEB J.* **2006**, *20*, 1236–1238. [CrossRef] [PubMed]
6. Kim, H.P.; Wang, X.; Galbiati, F.; Ryter, S.W.; Choi, A.M. Caveolae Compartmentalization of Heme Oxygenase-1 in Endothelial Cells. *FASEB J.* **2004**, *18*, 1080–1089. [CrossRef]
7. Rodella, L.F.; Vanella, L.; Peterson, S.J.; Drummond, G.; Rezzani, R.; Falck, J.R.; Abraham, N.G. Heme Oxygenase-Derived Carbon Monoxide Restores Vascular Function in Type 1 Diabetes. *Drug Metab. Lett.* **2008**, *2*, 290–300. [CrossRef]
8. Vanella, L.; Sanford, C., Jr.; Kim, D.H.; Abraham, N.G.; Ebraheim, N. Oxidative Stress and Heme Oxygenase-1 Regulated Human Mesenchymal Stem Cells Differentiation. *Int. J. Hypertens.* **2012**, *2012*, 890671. [CrossRef] [PubMed]
9. Liu, L.; Puri, N.; Raffaele, M.; Schragenheim, J.; Singh, S.P.; Bradbury, J.A.; Bellner, L.; Vanella, L.; Zeldin, D.C.; Cao, J.; et al. Ablation of Soluble Epoxide Hydrolase Reprogram White Fat to Beige-Like Fat through an Increase in Mitochondrial Integrity, Ho-1-Adiponectin in Vitro and in Vivo. *Prostaglandins Other Lipid Mediat.* **2018**, *138*, 1–8. [CrossRef]
10. Pittala, V.; Vanella, L.; Salerno, L.; di Giacomo, C.; Acquaviva, R.; Raffaele, M.; Romeo, G.; Modica, M.N.; Prezzavento, O.; Sorrenti, V. Novel Caffeic Acid Phenethyl Ester (Cape) Analogues as Inducers of Heme Oxygenase-1. *Curr. Pharm. Des.* **2017**, *23*, 2657–2664. [CrossRef] [PubMed]
11. Raffaele, M.; Volti, G.L.; Barbagallo, I.A.; Vanella, L. Therapeutic Efficacy of Stem Cells Transplantation in Diabetes: Role of Heme Oxygenase. *Front. Cell Dev. Biol.* **2016**, *4*, 80. [CrossRef]
12. Stec, D.E.; John, K.; Trabbic, C.J.; Luniwal, A.; Hankins, M.W.; Baum, J.; Hinds, T.D., Jr. Bilirubin Binding to Pparalpha Inhibits Lipid Accumulation. *PLoS ONE* **2016**, *11*, e0153427. [CrossRef]
13. Pittala, V.; Salerno, L.; Romeo, G.; Acquaviva, R.; di Giacomo, C.; Sorrenti, V. Therapeutic Potential of Caffeic Acid Phenethyl Ester (Cape) in Diabetes. *Curr. Med. Chem.* **2018**, *25*, 4827–4836. [CrossRef]
14. Dey, S.; Sayers, C.M.; Verginadis, I.I.; Lehman, S.L.; Cheng, Y.; Cerniglia, G.J.; Tuttle, S.W.; Feldman, M.D.; Zhang, P.J.; Fuchs, S.Y.; et al. Atf4-Dependent Induction of Heme Oxygenase 1 Prevents Anoikis and Promotes Metastasis. *J. Clin. Investig.* **2015**, *125*, 2592–2608. [CrossRef] [PubMed]
15. Song, J.; Zhang, X.; Liao, Z.; Liang, H.; Chu, L.; Dong, W.; Zhang, X.; Ge, Q.; Liu, Q.; Fan, P.; et al. 14-3-3zeta Inhibits Heme Oxygenase-1 (Ho-1) Degradation and Promotes Hepatocellular Carcinoma Proliferation: Involvement of Stat3 Signaling. *J. Exp. Clin. Cancer Res.* **2019**, *38*, 3. [CrossRef]

16. Nitti, M.; Piras, S.; Marinari, U.M.; Moretta, L.; Pronzato, M.A.; Furfaro, A.L. Ho-1 Induction in Cancer Progression: A Matter of Cell Adaptation. *Antioxidants* **2017**, *6*, 29. [CrossRef]

17. Barbagallo, I.; Giallongo, C.; Volti, G.L.; Distefano, A.; Camiolo, G.; Raffaele, M.; Salerno, L.; Pittala, V.; Sorrenti, V.; Avola, R.; et al. Heme Oxygenase Inhibition Sensitizes Neuroblastoma Cells to Carfilzomib. *Mol. Neurobiol.* **2019**, *56*, 1451–1460. [CrossRef]

18. Tibullo, D.; Barbagallo, I.; Giallongo, C.; la Cava, P.; Parrinello, N.; Vanella, L.; Stagno, F.; Palumbo, G.A.; Volti, G.L.; di Raimondo, F. Nuclear Translocation of Heme Oxygenase-1 Confers Resistance to Imatinib in Chronic Myeloid Leukemia Cells. *Curr. Pharm. Des.* **2013**, *19*, 2765–2770. [CrossRef]

19. Salerno, L.; Romeo, G.; Modica, M.N.; Amata, E.; Sorrenti, V.; Barbagallo, I.; Pittala, V. Heme Oxygenase-1: A New Druggable Target in the Management of Chronic and Acute Myeloid Leukemia. *Eur. J. Med. Chem.* **2017**, *142*, 163–178. [CrossRef]

20. Schaefer, B.; Behrends, S. Translocation of Heme Oxygenase-1 Contributes to Imatinib Resistance in Chronic Myelogenous Leukemia. *Oncotarget* **2017**, *8*, 67406–67421. [CrossRef] [PubMed]

21. Greish, K.F.; Salerno, L.; Al Zahrani, R.; Amata, E.; Modica, M.N.; Romeo, G.; Marrazzo, A.; Prezzavento, O.; Sorrenti, V.; Rescifina, A.; et al. Novel Structural Insight into Inhibitors of Heme Oxygenase-1 (Ho-1) by New Imidazole-Based Compounds: Biochemical and in Vitro Anticancer Activity Evaluation. *Molecules* **2018**, *23*, 1209. [CrossRef]

22. Sorrenti, V.; Pittala, V.; Romeo, G.; Amata, E.; Dichiara, M.; Marrazzo, A.; Turnaturi, R.; Prezzavento, O.; Barbagallo, I.; Vanella, L.; et al. Targeting Heme Oxygenase-1 with Hybrid Compounds to Overcome Imatinib Resistance in Chronic Myeloid Leukemia Cell Lines. *Eur. J. Med. Chem.* **2018**, *158*, 937–950. [CrossRef]

23. Jozkowicz, A.; Was, H.; Dulak, J. Heme Oxygenase-1 in Tumors: Is It a False Friend? *Antioxid. Redox Signal.* **2007**, *9*, 2099–2117. [CrossRef]

24. Florczyk, U.; Golda, S.; Zieba, A.; Cisowski, J.; Jozkowicz, A.; Dulak, J. Overexpression of Biliverdin Reductase Enhances Resistance to Chemotherapeutics. *Cancer Lett.* **2011**, *300*, 40–47. [CrossRef]

25. Banerjee, P.; Basu, A.; Wegiel, B.; Otterbein, L.E.; Mizumura, K.; Gasser, M.; Waaga-Gasser, A.M.; Choi, A.M.; Pal, S. Heme Oxygenase-1 Promotes Survival of Renal Cancer Cells through Modulation of Apoptosis- and Autophagy-Regulating Molecules. *J. Biol. Chem.* **2012**, *287*, 32113–32123. [CrossRef]

26. Zingales, V.; Distefano, A.; Raffaele, M.; Zanghi, A.; Barbagallo, I.; Vanella, L. Metformin: A Bridge between Diabetes and Prostate Cancer. *Front. Oncol.* **2017**, *7*, 243. [CrossRef]

27. Hankinson, S.J.; Fam, M.; Patel, N.N. A Review for Clinicians: Prostate Cancer and the Antineoplastic Properties of Metformin. *Urol. Oncol.* **2017**, *35*, 21–29. [CrossRef]

28. Wright, J.L.; Stanford, J.L. Metformin Use and Prostate Cancer in Caucasian Men: Results from a Population-Based Case-Control Study. *Cancer Causes Control* **2009**, *20*, 1617–1622. [CrossRef]

29. Lu, C.C.; Chiang, J.H.; Tsai, F.J.; Hsu, Y.M.; Juan, Y.N.; Yang, J.S.; Chiu, H.Y. Metformin Triggers the Intrinsic Apoptotic Response in Human Ags Gastric Adenocarcinoma Cells by Activating Ampk and Suppressing Mtor/Akt Signaling. *Int. J. Oncol.* **2019**, *54*, 1271–1281. [CrossRef]

30. Andrzejewski, S.; Gravel, S.P.; Pollak, M.; St-Pierre, J. Metformin Directly Acts on Mitochondria to Alter Cellular Bioenergetics. *Cancer Metab.* **2014**, *2*, 12. [CrossRef]

31. Cantrell, L.A.; Zhou, C.; Mendivil, A.; Malloy, K.M.; Gehrig, P.A.; Bae-Jump, V.L. Metformin Is a Potent Inhibitor of Endometrial Cancer Cell Proliferation–Implications for a Novel Treatment Strategy. *Gynecol. Oncol.* **2010**, *116*, 92–98. [CrossRef]

32. Ben Sahra, I.; Laurent, K.; Loubat, A.; Giorgetti-Peraldi, S.; Colosetti, P.; Auberger, P.; Tanti, J.F.; le Marchand-Brustel, Y.; Bost, F. The Antidiabetic Drug Metformin Exerts an Antitumoral Effect in Vitro and in Vivo through a Decrease of Cyclin D1 Level. *Oncogene* **2008**, *27*, 3576–3586. [CrossRef]

33. Clements, A.; Gao, B.; Yeap, S.H.; Wong, M.K.; Ali, S.S.; Gurney, H. Metformin in Prostate Cancer: Two for the Price of One. *Ann. Oncol.* **2011**, *22*, 2556–2560. [CrossRef] [PubMed]

34. Richards, K.A.; Liou, J.I.; Cryns, V.L.; Downs, T.M.; Abel, E.J.; Jarrard, D.F. Metformin Use Is Associated with Improved Survival for Patients with Advanced Prostate Cancer on Androgen Deprivation Therapy. *J. Urol.* **2018**, *200*, 1256–1263. [CrossRef]

35. Sarmento-Cabral, A.; Fernando, L.; Gahete, M.D.; Castano, J.P.; Luque, R.M. Metformin Reduces Prostate Tumor Growth, in a Diet-Dependent Manner, by Modulating Multiple Signaling Pathways. *Mol. Cancer Res.* **2017**, *15*, 862–874. [CrossRef]

36. Sharma, A.; Bandyopadhayaya, S.; Chowdhury, K.; Sharma, T.; Maheshwari, R.; Das, A.; Chakrabarti, G.; Kumar, V.; Mandal, C.C. Metformin Exhibited Anticancer Activity by Lowering Cellular Cholesterol Content in Breast Cancer Cells. *PLoS ONE* **2019**, *14*, e0209435. [CrossRef]

37. Hindler, K.; Cleeland, C.S.; Rivera, E.; Collard, C.D. The Role of Statins in Cancer Therapy. *Oncologist* **2006**, *11*, 306–315. [CrossRef]

38. Babcook, M.A.; Sramkoski, R.M.; Fujioka, H.; Daneshgari, F.; Almasan, A.; Shukla, S.; Nanavaty, R.R.; Gupta, S. Combination Simvastatin and Metformin Induces G1-Phase Cell Cycle Arrest and Ripk1- and Ripk3-Dependent Necrosis in C4-2b Osseous Metastatic Castration-Resistant Prostate Cancer Cells. *Cell Death Dis.* **2014**, *5*, e1536. [CrossRef]

39. Wang, Z.S.; Huang, H.R.; Zhang, L.Y.; Kim, S.; He, Y.; Li, D.L.; Farischon, C.; Zhang, K.; Zheng, X.; Du, Z.Y.; et al. Mechanistic Study of Inhibitory Effects of Metformin and Atorvastatin in Combination on Prostate Cancer Cells in Vitro and in Vivo. *Biol. Pharm. Bull.* **2017**, *40*, 1247–1254. [CrossRef]

40. Tran, L.N.K.; Kichenadasse, G.; Morel, K.L.; Lavranos, T.C.; Klebe, S.; Lower, K.M.; Ormsby, R.J.; Elliot, D.J.; Sykes, P.J. The Combination of Metformin and Valproic Acid Has a Greater Anti-Tumoral Effect on Prostate Cancer Growth in Vivo Than Either Drug Alone. *In Vivo* **2019**, *33*, 99–108. [CrossRef]

41. Cuyas, E.; Verdura, S.; Llorach-Pares, L.; Fernandez-Arroyo, S.; Joven, J.; Martin-Castillo, B.; Bosch-Barrera, J.; Brunet, J.; Nonell-Canals, A.; Sanchez-Martinez, M.; et al. Metformin Is a Direct Sirt1-Activating Compound: Computational Modeling and Experimental Validation. *Front. Endocrinol. (Lausanne)* **2018**, *9*, 657. [CrossRef] [PubMed]

42. Pusapati, R.V.; Daemen, A.; Wilson, C.; Sandoval, W.; Gao, M.; Haley, B.; Baudy, A.R.; Hatzivassiliou, G.; Evangelista, M.; Settleman, J. Mtorc1-Dependent Metabolic Reprogramming Underlies Escape from Glycolysis Addiction in Cancer Cells. *Cancer Cell* **2016**, *29*, 548–562. [CrossRef] [PubMed]

43. Granja, S.; Pinheiro, C.; Reis, R.M.; Martinho, O.; Baltazar, F. Glucose Addiction in Cancer Therapy: Advances and Drawbacks. *Curr. Drug Metab.* **2015**, *16*, 221–242. [CrossRef] [PubMed]

44. Bikas, A.; Jensen, K.; Patel, A.; Costello, J., Jr.; McDaniel, D.; Klubo-Gwiezdzinska, J.; Larin, O.; Hoperia, V.; Burman, K.D.; Boyle, L.; et al. Glucose-Deprivation Increases Thyroid Cancer Cells Sensitivity to Metformin. *Endocr. Relat. Cancer* **2015**, *22*, 919–932. [CrossRef]

45. Zordoky, B.N.; Bark, D.; Soltys, C.L.; Sung, M.M.; Dyck, J.R. The Anti-Proliferative Effect of Metformin in Triple-Negative Mda-Mb-231 Breast Cancer Cells Is Highly Dependent on Glucose Concentration: Implications for Cancer Therapy and Prevention. *Biochim. Biophys. Acta* **1840**, *2014*, 1943–1957. [CrossRef]

46. Li, P.D.; Liu, Z.; Cheng, T.T.; Luo, W.G.; Yao, J.; Chen, J.; Zou, Z.W.; Chen, L.L.; Ma, C.; Dai, X.F. Redox-Dependent Modulation of Metformin Contributes to Enhanced Sensitivity of Esophageal Squamous Cell Carcinoma to Cisplatin. *Oncotarget* **2017**, *8*, 62057–62068. [CrossRef] [PubMed]

47. Dodson, M.; Darley-Usmar, V.; Zhang, J. Cellular Metabolic and Autophagic Pathways: Traffic Control by Redox Signaling. *Free Radic. Biol. Med.* **2013**, *63*, 207–221. [CrossRef]

48. Buzzai, M.; Jones, R.G.; Amaravadi, R.K.; Lum, J.J.; DeBerardinis, R.J.; Zhao, F.; Viollet, B.; Thompson, C.B. Systemic Treatment with the Antidiabetic Drug Metformin Selectively Impairs P53-Deficient Tumor Cell Growth. *Cancer Res.* **2007**, *67*, 6745–6752. [CrossRef]

49. Gonnissen, A.; Isebaert, S.; McKee, C.M.; Muschel, R.J.; Haustermans, K. The Effect of Metformin and Gant61 Combinations on the Radiosensitivity of Prostate Cancer Cells. *Int. J. Mol. Sci.* **2017**, *18*, 399. [CrossRef]

50. Knowell, A.E.; Patel, D.; Morton, D.J.; Sharma, P.; Glymph, S.; Chaudhary, J. Id4 Dependent Acetylation Restores Mutant-P53 Transcriptional Activity. *Mol. Cancer* **2013**, *12*, 161. [CrossRef]

51. Salani, B.; del Rio, A.; Marini, C.; Sambuceti, G.; Cordera, R.; Maggi, D. Metformin, Cancer and Glucose Metabolism. *Endocr. Relat. Cancer* **2014**, *21*, R461–R471. [CrossRef]

52. Evans, J.M.; Donnelly, L.A.; Emslie-Smith, A.M.; Alessi, D.R.; Morris, A.D. Metformin and Reduced Risk of Cancer in Diabetic Patients. *BMJ* **2005**, *330*, 1304–1305. [CrossRef] [PubMed]

53. Mukhopadhyay, S.; Chatterjee, A.; Kogan, D.; Patel, D.; Foster, D.A. 5-Aminoimidazole-4-Carboxamide-1-Beta-4-Ribofuranoside (Aicar) Enhances the Efficacy of Rapamycin in Human Cancer Cells. *Cell Cycle* **2015**, *14*, 3331–3339. [CrossRef]

54. Schneider, M.B.; Matsuzaki, H.; Haorah, J.; Ulrich, A.; Standop, J.; Ding, X.Z.; Adrian, T.E.; Pour, P.M. Prevention of Pancreatic Cancer Induction in Hamsters by Metformin. *Gastroenterology* **2001**, *120*, 1263–1270. [CrossRef]

55. Zakikhani, M.; Dowling, R.; Fantus, I.G.; Sonenberg, N.; Pollak, M. Metformin Is an Amp Kinase-Dependent Growth Inhibitor for Breast Cancer Cells. *Cancer Res.* **2006**, *66*, 10269–10273. [CrossRef] [PubMed]

56. Birsoy, K.; Wang, T.; Chen, W.W.; Freinkman, E.; Abu-Remaileh, M.; Sabatini, D.M. An Essential Role of the Mitochondrial Electron Transport Chain in Cell Proliferation Is to Enable Aspartate Synthesis. *Cell* **2015**, *162*, 540–551. [CrossRef] [PubMed]

57. Visagie, M.H.; Mqoco, T.V.; Liebenberg, L.; Mathews, E.H.; Mathews, G.E.; Joubert, A.M. Influence of Partial and Complete Glutamine-and Glucose Deprivation of Breast-and Cervical Tumorigenic Cell Lines. *Cell Biosci.* **2015**, *5*, 37. [CrossRef] [PubMed]

58. Mukhopadhyay, S.; Saqcena, M.; Foster, D.A. Synthetic Lethality in Kras-Driven Cancer Cells Created by Glutamine Deprivation. *Oncoscience* **2015**, *2*, 807–808.

59. Menendez, J.A.; Oliveras-Ferraros, C.; Cufi, S.; Corominas-Faja, B.; Joven, J.; Martin-Castillo, B.; Vazquez-Martin, A. Metformin Is Synthetically Lethal with Glucose Withdrawal in Cancer Cells. *Cell Cycle* **2012**, *11*, 2782–2792. [CrossRef]

60. Saito, S.; Furuno, A.; Sakurai, J.; Sakamoto, A.; Park, H.R.; Shin-Ya, K.; Tsuruo, T.; Tomida, A. Chemical Genomics Identifies the Unfolded Protein Response as a Target for Selective Cancer Cell Killing During Glucose Deprivation. *Cancer Res.* **2009**, *69*, 4225–4234. [CrossRef]

61. Schroder, M.; Kaufman, R.J. The Mammalian Unfolded Protein Response. *Annu. Rev. Biochem.* **2005**, *74*, 739–789. [CrossRef] [PubMed]

62. Limonta, P.; Moretti, R.M.; Marzagalli, M.; Fontana, F.; Raimondi, M.; Marelli, M.M. Role of Endoplasmic Reticulum Stress in the Anticancer Activity of Natural Compounds. *Int. J. Mol. Sci.* **2019**, *20*, 961. [CrossRef] [PubMed]

63. Haga, N.; Saito, S.; Tsukumo, Y.; Sakurai, J.; Furuno, A.; Tsuruo, T.; Tomida, A. Mitochondria Regulate the Unfolded Protein Response Leading to Cancer Cell Survival under Glucose Deprivation Conditions. *Cancer Sci.* **2010**, *101*, 1125–1132. [CrossRef] [PubMed]

64. Xu, Z.; Bu, Y.; Chitnis, N.; Koumenis, C.; Fuchs, S.Y.; Diehl, J.A. Mir-216b Regulation of C-Jun Mediates Gadd153/Chop-Dependent Apoptosis. *Nat. Commun.* **2016**, *7*, 11422. [CrossRef] [PubMed]

65. Verfaillie, T.; Garg, A.D.; Agostinis, P. Targeting Er Stress Induced Apoptosis and Inflammation in Cancer. *Cancer Lett.* **2013**, *332*, 249–264. [CrossRef] [PubMed]

66. Salis, O.; Bedir, A.; Ozdemir, T.; Okuyucu, A.; Alacam, H. The Relationship between Anticancer Effect of Metformin and the Transcriptional Regulation of Certain Genes (Chop, Cav-1, Ho-1, Sgk-1 and Par-4) on Mcf-7 Cell Line. *Eur. Rev. Med. Pharmacol. Sci.* **2014**, *18*, 1602–1609.

67. Yang, J.; Wei, J.; Wu, Y.; Wang, Z.; Guo, Y.; Lee, P.; Li, X. Metformin Induces Er Stress-Dependent Apoptosis through Mir-708-5p/Nnat Pathway in Prostate Cancer. *Oncogenesis* **2015**, *4*, e158. [CrossRef]

68. Ma, L.; Wei, J.; Wan, J.; Wang, W.; Wang, L.; Yuan, Y.; Yang, Z.; Liu, X.; Ming, L. Low Glucose and Metformin-Induced Apoptosis of Human Ovarian Cancer Cells Is Connected to Ask1 Via Mitochondrial and Endoplasmic Reticulum Stress-Associated Pathways. *J. Exp. Clin. Cancer Res.* **2019**, *38*, 77. [CrossRef]

69. Li Volti, G.; Tibullo, D.; Vanella, L.; Giallongo, C.; di Raimondo, F.; Forte, S.; di Rosa, M.; Signorelli, S.S.; Barbagallo, I. The Heme Oxygenase System in Hematological Malignancies. *Antioxid. Redox Signal.* **2017**, *27*, 363–377. [CrossRef]

70. Tibullo, D.; Barbagallo, I.; Giallongo, C.; Vanella, L.; Conticello, C.; Romano, A.; Saccone, S.; Godos, J.; di Raimondo, F.; Volti, G.L. Heme Oxygenase-1 Nuclear Translocation Regulates Bortezomibinduced Cytotoxicity and Mediates Genomic Instability in Myeloma Cells. *Oncotarget* **2016**, *7*, 28868–28880. [CrossRef]

71. Onyango, P.; Celic, I.; McCaffery, J.M.; Boeke, J.D.; Feinberg, A.P. Sirt3, a Human Sir2 Homologue, Is an Nad-Dependent Deacetylase Localized to Mitochondria. *Proc. Natl. Acad. Sci. USA* **2002**, *99*, 13653–13658. [PubMed]

72. Liu, R.; Fan, M.; Candas, D.; Qin, L.; Zhang, X.; Eldridge, A.; Zou, J.X.; Zhang, T.; Juma, S.; Jin, C.; et al. Cdk1-Mediated Sirt3 Activation Enhances Mitochondrial Function and Tumor Radioresistance. *Mol. Cancer Ther.* **2015**, *14*, 2090–2102. [CrossRef] [PubMed]

73. Chen, X.; Xu, Z.; Zeng, S.; Wang, X.; Liu, W.; Qian, L.; Wei, J.; Yang, X.; Shen, Q.; Gong, Z.; et al. Sirt5 Downregulation Is Associated with Poor Prognosis in Glioblastoma. *Cancer Biomark.* **2019**, *24*, 449–459. [CrossRef]

74. Deng, C.X. Sirt1, Is It a Tumor Promoter or Tumor Suppressor? *Int. J. Biol. Sci.* **2009**, *5*, 147–152.

75. Wang, C.; Chen, L.; Hou, X.; Li, Z.; Kabra, N.; Ma, Y.; Nemoto, S.; Finkel, T.; Gu, W.; Cress, W.D.; et al. Interactions between E2f1 and Sirt1 Regulate Apoptotic Response to DNA Damage. *Nat. Cell Biol.* **2006**, *8*, 1025–1031. [CrossRef]

76. Liu, T.F.; McCall, C.E. Deacetylation by Sirt1 Reprograms Inflammation and Cancer. *Genes Cancer* **2013**, *4*, 135–147. [CrossRef] [PubMed]

77. Yi, J.; Luo, J. Sirt1 and P53, Effect on Cancer, Senescence and Beyond. *Biochim. Biophys. Acta* **1804**, *2010*, 1684–1689. [CrossRef]

78. Zhang, Y.G.; Cai, X.Q.; Chai, N.; Gu, Y.; Zhang, S.; Ding, M.L.; Cao, H.C.; Sha, S.M.; Yin, J.P.; Li, M.B.; et al. Sirt1 Is Reduced in Gastric Adenocarcinoma and Acts as a Potential Tumor Suppressor in Gastric Cancer. *Gastrointest. Tumors* **2015**, *2*, 109–123.

79. Gerthofer, V.; Kreutz, M.; Renner, K.; Jachnik, B.; Dettmer, K.; Oefner, P.; Riemenschneider, M.J.; Proescholdt, M.; Vollmann-Zwerenz, A.; Hau, P.; et al. Combined Modulation of Tumor Metabolism by Metformin and Diclofenac in Glioma. *Int. J. Mol. Sci.* **2018**, *19*, 2586.

80. Tseng, H.W.; Li, S.C.; Tsai, K.W. Metformin Treatment Suppresses Melanoma Cell Growth and Motility through Modulation of Microrna Expression. *Cancers* **2019**, *11*, 209. [CrossRef]

81. Michishita, E.; Park, J.Y.; Burneskis, J.M.; Barrett, J.C.; Horikawa, I. Evolutionarily Conserved and Nonconserved Cellular Localizations and Functions of Human Sirt Proteins. *Mol. Biol. Cell* **2005**, *16*, 4623–4635. [CrossRef]

82. Matsushita, N.; Yonashiro, R.; Ogata, Y.; Sugiura, A.; Nagashima, S.; Fukuda, T.; Inatome, R.; Yanagi, S. Distinct Regulation of Mitochondrial Localization and Stability of Two Human Sirt5 Isoforms. *Genes Cells* **2011**, *16*, 190–202. [CrossRef]

83. Jung-Hynes, B.; Nihal, M.; Zhong, W.; Ahmad, N. Role of Sirtuin Histone Deacetylase Sirt1 in Prostate Cancer. *A Target for Prostate Cancer Management Via Its Inhibition?" J. Biol. Chem.* **2009**, *284*, 3823–3832. [CrossRef]

84. Barbagallo, I.; Parenti, R.; Zappala, A.; Vanella, L.; Tibullo, D.; Pepe, F.; Onni, T.; Volti, G.L. Combined Inhibition of Hsp90 and Heme Oxygenase-1 Induces Apoptosis and Endoplasmic Reticulum Stress in Melanoma. *Acta Histochem.* **2015**, *117*, 705–711. [CrossRef]

85. Gozzelino, R.; Jeney, V.; Soares, M.P. Mechanisms of Cell Protection by Heme Oxygenase-1. *Annu. Rev. Pharmacol. Toxicol.* **2010**, *50*, 323–354. [CrossRef]

86. Sacerdoti, D.; Singh, S.P.; Schragenheim, J.; Bellner, L.; Vanella, L.; Raffaele, M.; Meissner, A.; Grant, I.; Favero, G.; Rezzani, R.; et al. Development of Nash in Obese Mice Is Confounded by Adipose Tissue Increase in Inflammatory Nov and Oxidative Stress. *Int. J. Hepatol.* **2018**, *2018*, 3484107. [CrossRef]

87. Poljsak, B.; Milisav, I. Clinical Implications of Cellular Stress Responses. *Bosn J. Basic Med. Sci.* **2012**, *12*, 122–126. [CrossRef]

88. Choi, A.M.; Alam, J. Heme Oxygenase-1: Function, Regulation, and Implication of a Novel Stress-Inducible Protein in Oxidant-Induced Lung Injury. *Am. J. Respir. Cell Mol. Biol.* **1996**, *15*, 9–19. [CrossRef]

89. Bahmani, P.; Hassanshahi, G.; Halabian, R.; Roushandeh, A.M.; Jahanian-Najafabadi, A.; Roudkenar, M.H. The Expression of Heme Oxygenase-1 in Human-Derived Cancer Cell Lines. *Iran. J. Med. Sci.* **2011**, *36*, 260–265.

90. Chiang, S.K.; Chen, S.E.; Chang, L.C. A Dual Role of Heme Oxygenase-1 in Cancer Cells. *Int. J. Mol. Sci.* **2018**, *20*, 39. [CrossRef]

91. Ryter, S.W.; Choi, A.M. Targeting Heme Oxygenase-1 and Carbon Monoxide for Therapeutic Modulation of Inflammation. *Transl. Res.* **2016**, *167*, 7–34. [CrossRef]

92. Sena, P.; Mancini, S.; Benincasa, M.; Mariani, F.; Palumbo, C.; Roncucci, L. Metformin Induces Apoptosis and Alters Cellular Responses to Oxidative Stress in Ht29 Colon Cancer Cells: Preliminary Findings. *Int. J. Mol. Sci.* **2018**, *19*, 1478. [CrossRef]

93. Li, Y.; Wang, M.; Zhi, P.; You, J.; Gao, J.Q. Metformin Synergistically Suppress Tumor Growth with Doxorubicin and Reverse Drug Resistance by Inhibiting the Expression and Function of P-Glycoprotein in Mcf7/Adr Cells and Xenograft Models. *Oncotarget* **2018**, *9*, 2158–2174. [CrossRef]

94. Teh, J.T.; Zhu, W.L.; Newgard, C.B.; Casey, P.J.; Wang, M. Respiratory Capacity and Reserve Predict Cell Sensitivity to Mitochondria Inhibitors: Mechanism-Based Markers to Identify Metformin-Responsive Cancers. *Mol. Cancer Ther.* **2019**, *18*, 693–705. [CrossRef] [PubMed]

95. Wheaton, W.W.; Weinberg, S.E.; Hamanaka, R.B.; Soberanes, S.; Sullivan, L.B.; Anso, E.; Glasauer, A.; Dufour, E.; Mutlu, G.M.; Budigner, G.S.; et al. Metformin Inhibits Mitochondrial Complex I of Cancer Cells to Reduce Tumorigenesis. *Elife* **2014**, *3*, e02242. [CrossRef]

96. Wang, H.; Cheng, Q.; Li, X.; Hu, F.; Han, L.; Zhang, H.; Li, L.; Ge, J.; Ying, X.; Guo, X.; et al. Loss of Tigar Induces Oxidative Stress and Meiotic Defects in Oocytes from Obese Mice. *Mol. Cell Proteom.* **2018**, *17*, 1354–1364. [CrossRef] [PubMed]

97. Du, Z.X.; Zhang, H.Y.; Meng, X.; Guan, Y.; Wang, H.Q. Role of Oxidative Stress and Intracellular Glutathione in the Sensitivity to Apoptosis Induced by Proteasome Inhibitor in Thyroid Cancer Cells. *BMC Cancer* **2009**, *9*, 56. [CrossRef]

98. Wong, E.Y.; Wong, S.C.; Chan, C.M.; Lam, E.K.; Ho, L.Y.; Lau, C.P.; Au, T.C.; Chan, A.K.; Tsang, C.M.; Tsao, S.W.; et al. Tp53-Induced Glycolysis and Apoptosis Regulator Promotes Proliferation and Invasiveness of Nasopharyngeal Carcinoma Cells. *Oncol. Lett.* **2015**, *9*, 569–574. [CrossRef]

99. Bensaad, K.; Tsuruta, A.; Selak, M.A.; Vidal, M.N.; Nakano, K.; Bartrons, R.; Gottlieb, E.; Vousden, K.H. Tigar, a P53-Inducible Regulator of Glycolysis and Apoptosis. *Cell* **2006**, *126*, 107–120. [CrossRef]

100. Winkler, B.S.; DeSantis, N.; Solomon, F. Multiple Nadph-Producing Pathways Control Glutathione (Gsh) Content in Retina. *Exp. Eye Res.* **1986**, *43*, 829–847. [CrossRef]

101. Zhou, J.H.; Zhang, T.T.; Song, D.D.; Xia, Y.F.; Qin, Z.H.; Sheng, R. Tigar Contributes to Ischemic Tolerance Induced by Cerebral Preconditioning through Scavenging of Reactive Oxygen Species and Inhibition of Apoptosis. *Sci. Rep.* **2016**, *6*, 27096. [CrossRef] [PubMed]

102. Heiss, K.; Raffaele, M.; Vanella, L.; Murabito, P.; Prezzavento, O.; Marrazzo, A.; Arico, G.; Castracani, C.C.; Barbagallo, I.; Zappala, A.; et al. (+)-Pentazocine Attenuates Sh-Sy5y Cell Death, Oxidative Stress and Microglial Migration Induced by Conditioned Medium from Activated Microglia. *Neurosci. Lett.* **2017**, *642*, 86–90. [CrossRef]

103. Salerno, L.; Amata, E.; Romeo, G.; Marrazzo, A.; Prezzavento, O.; Floresta, G.; Sorrenti, V.; Barbagallo, I.; Rescifina, A.; Pittala, V. Potholing of the Hydrophobic Heme Oxygenase-1 Western Region for the Search of Potent and Selective Imidazole-Based Inhibitors. *Eur. J. Med. Chem.* **2018**, *148*, 54–62. [CrossRef] [PubMed]

104. Lanteri, R.; Acquaviva, R.; di Giacomo, C.; Sorrenti, V.; Destri, G.L.; Santangelo, M.; Vanella, L.; di Cataldo, A. Rutin in Rat Liver Ischemia/Reperfusion Injury: Effect on Ddah/Nos Pathway. *Microsurgery* **2007**, *27*, 245–251. [CrossRef]

International Journal of
Molecular Sciences

MDPI

Article

High-Pressure Carbon Monoxide and Oxygen Mixture is Effective for Lung Preservation

Atsushi Fujiwara [1,2], Naoyuki Hatayama [2], Natsumi Matsuura [1], Naoya Yokota [1], Kaori Fukushige [2], Tomiko Yakura [2], Shintaro Tarumi [1], Tetsuhiko Go [1], Shuichi Hirai [2,*], Munekazu Naito [2] and Hiroyasu Yokomise [1]

[1] Department of General Thoracic, Breast and Endocrinological Surgery, Kagawa University, Kagawa 761-0793, Japan; a24fujiwara@gmail.com (A.F.); nmori1130@gmail.com (N.M.); naoe@med.kagawa-u.ac.jp (N.Y.); shintarotarumi@gmail.com (S.T.); g-tetsu@wa3.so-net.ne.jp (T.G.); yokomise@med.kagawa-u.ac.jp (H.Y.)
[2] Department of Anatomy, Aichi Medical University, Aichi 480-1195, Japan; nhatayama416@gmail.com (N.H.); kaori.fukushige@gmail.com (K.F.); tomi.tomi105@gmail.com (T.Y.); munekazunaito@gmail.com (M.N.)
* Correspondence: shinamon611@gmail.com; Tel.: +81-561-62-3311

Received: 11 April 2019; Accepted: 30 May 2019; Published: 3 June 2019

Abstract: (1) Background: Heme oxygenase-1 (HO-1) degrades heme and generates carbon monoxide (CO), producing various anti-inflammatory, anti-oxidative, and anti-apoptotic effects. This study aimed to confirm the effects of CO on the ischemia–reperfusion injury (IRI) of donor lungs using a high-pressure gas (HPG) preservation method. (2) Methods: Donor rat and canine lungs were preserved in a chamber filled with CO (1.5 atm) and oxygen (O_2; 2 atm) and were ventilated with either CO and O_2 mixture (CO/O_2 group) or air (air group) immediately before storage. Rat lungs were subjected to heterotopic cervical transplantation and evaluated after reperfusion, whereas canine lungs were subjected to allogeneic transplantation and evaluated. (3) Results: Alveolar hemorrhage in the CO/O_2 group was significantly milder than that in the air group. mRNA expression levels of HO-1 remained unchanged in both the groups; however, inflammatory mediator levels were significantly lower in the CO/O_2 group than in the air group. The oxygenation of graft lungs was comparable between the two groups, but lactic acid level tended to be higher in the air group. (4) Conclusions: The HO-1/CO system in the HPG preservation method is effective in suppressing IRI and preserving donor lungs.

Keywords: carbon monoxide; lung preservation; ischemia–reperfusion injury; high-pressure gas

1. Introduction

Carbon monoxide (CO), a colorless and odorless gas, produces carboxyhemoglobin (CO-Hb) and inhibits the P450 enzyme of the mitochondrial electron transfer system [1,2], resulting in tissue oxygen deficiency. Moreover, endogenous CO is biosynthesized when heme oxygenase-1 (HO-1) is involved in the degradation of hemoglobin to bilirubin [3]. CO plays anti-inflammatory, anti-oxidative, vasodilatory, and anti-apoptotic roles. The p38 mitogen-activated protein kinase pathway is reportedly one of the mechanisms underlying the cytoprotective effects of CO [4–7]. Furthermore, exogenous CO administration therapeutically affects transplantation by suppressing ischemia–reperfusion injury (IRI) [8–10]. Therefore, the HO-1/CO system has protective effects against stress reactions.

Lung transplantation is an important treatment approach for end-stage lung disease. However, ischemia–reperfusion injury (IRI), which is one of the most important complications of lung transplantation, hinders its utility and efficacy [11]. Mortality rates after transplantation are higher in cases with severe IRI, which is difficult to control. Successful IRI suppression is critical for maintaining good preservation conditions. Although the simple immersion method using an organ preservation solution is the globally used and clinically applied approach for lung preservation during transplantation, it cannot completely prevent damage due to IRI [12–14]. To suppress IRI in donor lungs, several recent methods, including ex vivo lung perfusion, have been explored, and the development of new organ preservation methods that can satisfactorily preserve lungs for extended periods is urgent.

We recently developed a high-pressure gas (HPG) preservation method using a mixture of CO and oxygen (O_2) and succeeded in resuscitating rat hearts after long-term preservation [15,16], revealing the utility of this method through histological analysis and metabolic evaluation [17]. Furthermore, this preservation method was reportedly useful for the preservation of rat kidneys [18] and limbs [19]. In addition, the lung is a solid organ similar to the kidney and liver, although it can be filled with gas from the inside, which is its decisively unique character. Appropriate delivery of CO and O_2 in their gaseous forms to the alveoli, in addition to their simple diffusion from the serosal side in a hyperbaric chamber, is anticipated to maximize their organ-preserving effects. Therefore, we hypothesized that the lung is a suitable organ to test the advantages of the HO-1/CO system in the HPG preservation method and investigated the utility of this approach using rat and canine models of lung transplantation.

2. Results

2.1. Experiment 1. Comparison of the Preservation Status of Rat Donor Lungs Filled and Preserved with Mixture of CO and O_2 or Air

We evaluated the preservation status of rat donor lungs preserved using the HPG preservation method. The donor lungs were filled with a mixture of CO and O_2 or air (Figure 1).

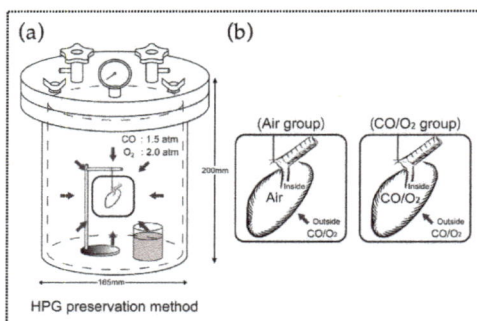

Figure 1. Schematic representation of the preservation method (Experiment 1). (**a**) Donor lungs were preserved in a chamber with high-pressure carbon monoxide (CO; 1.5 atm) and oxygen (O_2; 2.0 atm). (**b**) Before preservation, donor lungs were filled with either room air (containing nitrogen (N_2) and O_2 at a ratio of 4:1; air group) or a mixture of CO and O_2 at the same ratio as that outside (CO:O_2, 3:4; CO/O_2 group).

Mean weight of the donor lungs in the normal group (0.44 ± 0.01 g) was not significantly different from that of the donor lungs in the control (0.81 ± 0.35 g, $p = 0.509$) and CO/O_2 (0.97 ± 0.09 g, $p = 0.212$) groups. Mean weight of the donor lungs of the air group (1.77 ± 0.53 g) was significantly greater than that of the donor lungs in the normal ($p < 0.001$), control ($p = 0.005$), and CO/O_2 groups ($p = 0.011$; Figure 2a). Light microscopic evaluation of alveolar hemorrhage, which is an indicator of IRI, revealed that similar to the result observed in terms of lung weight, there was no significant difference between the control and CO/O_2 groups ($p = 0.999$). However, there was significantly more severe alveolar hemorrhage in the air group than in the CO/O_2 group ($p = 0.005$). A similar trend was observed among the air, control, and CO/O_2 groups (Figure 2b,c).

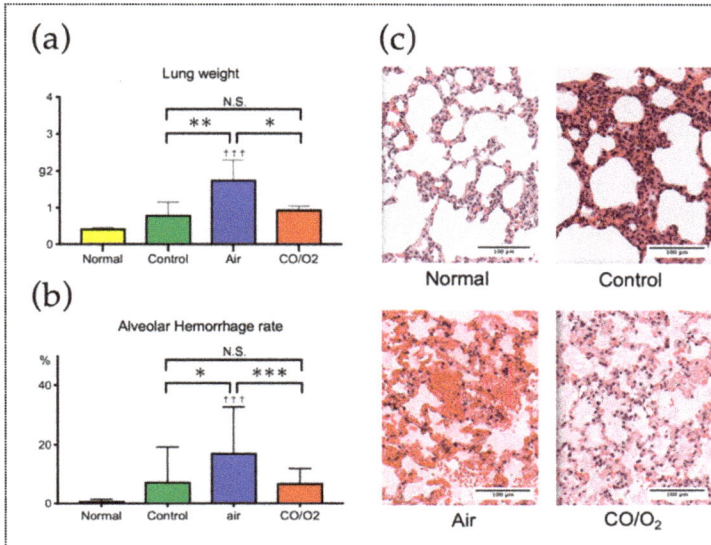

Figure 2. Assessment of rat lungs in the 24-h preservation model. (**a**) Comparison of the lung weights among the four groups. (**b**) Comparison of the severity of alveolar hemorrhage determined by histological assessment among the four groups. (**c**) Comparison of changes observed by light microscopy among the four groups. Each data of bars are expressed as the means ± standard deviations. N.S.—not significant, *** $p < 0.001$, ** $p < 0.01$, * $p < 0.05$, [†††] $p < 0.001$ compared with the normal group. Black bar = 100 μm

Real-time reverse transcription (RT)–polymerase chain reaction (PCR) indicated that the gene expression levels of HO-1 remained unchanged in all groups; however, expression levels of inflammatory mediators, particularly interleukin (IL)-6 and IL-1β, were significantly higher in the air group than in the CO/O$_2$ (IL-6: $p = 0.015$ and IL-1b: $p < 0.001$) and control (IL-6: $p = 0.020$ and IL-1b: $p = 0.023$) groups. In addition, there were no significant differences in expression levels of anti-inflammatory (transforming growth factor (TGF)-β, inducible nitric oxide synthase (iNOS), and granulocyte-macrophage colony-stimulating factor (GM-CSF)) and apoptotic (caspase 3) mediators (Figure 3).

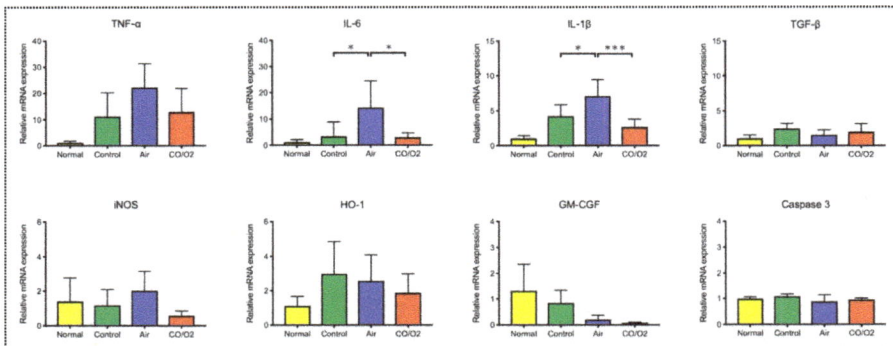

Figure 3. Changes in relative mRNA expression levels of mediators in donor rat lungs after 90 min of reperfusion. Each data of bars are expressed as the means ± standard deviations. *** $p < 0.001$, * $p < 0.05$.

2.2. Experiment 2. Comparison of Microscopic Findings and Arterial Blood Gas (ABG) of Canine Donor Lungs Ventilated From the Inside With Different Gases and Preserved in a High-Pressure CO/O$_2$ Gas Mixture

Canine lung transplantation was compared between the CO/O$_2$ and air groups. Since the canine lung was larger than the rat lung, a donor canine lung was placed on a net (Figure 4a). Before preservation, each lung was ventilated five times either with a mixture of CO and O$_2$ at a ratio of 3:4 (CO/O$_2$ group) or with air (air group). The size of the canine lungs after preservation reduced to about half of that before preservation because of the high pressure. (Figure 4b).

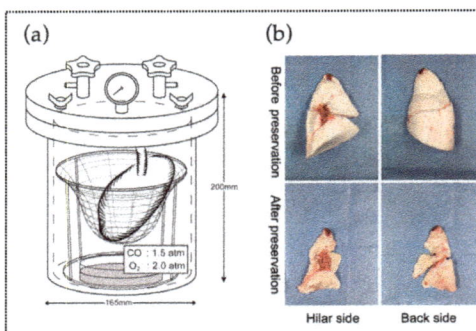

Figure 4. Schematic representation of the preservation method used in this study (Experiment 2). (a) The chamber was filled with a mixture of CO and O$_2$ [PCO (partial carbon monoxide pressure), 1.5 atm; PO$_2$ (partial oxygen pressure), 2 atm], and a flask with 50 mL distilled water was placed inside to maintain humidity for 20 h. The lung was gently placed on the net placed in the container. Before preservation, each lung was ventilated five times either with a mixture of CO and O$_2$ at a ratio of 3:4 (CO/O$_2$ group) or air (air group). (b) In the chamber, the trachea was kept open. After preservation, the lung shrunk due to high pressure.

Light microscopic evaluation revealed that the normal lung structure was well preserved in the air and CO/O$_2$ groups (Figure 5a,b).

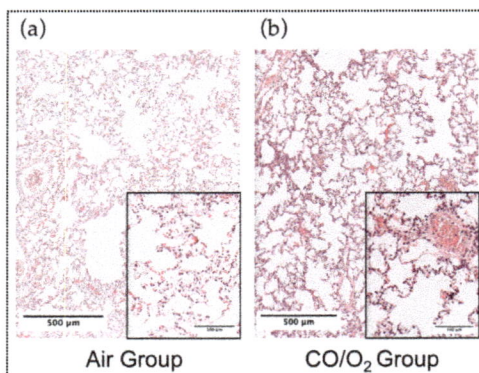

Figure 5. Light microscopic findings of the donor canine lungs after reperfusion by hematoxylin and eosin staining in the air (a) and CO/O$_2$ groups (b).

Results of the arterial blood gas (ABG) assessment at multiple time points after reperfusion are presented in Figure 6. Briefly, there was no noticeable increase in CO-Hb concentration in any of the groups before or after reperfusion. Partial arterial oxygen pressure (PaO$_2$) in the CO/O$_2$ and air groups was 288.3 ± 94.7 and 464 torr, respectively, at 120 min after reperfusion and 336.3 ± 70.9 and 275 torr, respectively, at 180 min after reperfusion. These results indicated that a good O_2 supply capacity after reperfusion was maintained in both the groups. In the CO/O_2 group, the partial arterial carbon dioxide pressure ($PaCO_2$) at 30, 60, 120, and 180 min after reperfusion was 35.7 ± 7.54, 27.9 ± 0.64, 41.8 ± 11.6, and 42.2 ± 11.6 torr, respectively. Conversely, $PaCO_2$ at 30, 60, 120, and 180 min after reperfusion was 25.7, 52.1, 55.6, and 71.3 torr, respectively, in the air group, indicating the tendency for CO_2 retention over time in the air group. Additionally, arterial blood lactate levels were high in the air group at 30 min after reperfusion. In both groups, peak inspiratory pressure (PIP) tended to increase starting at 60 min after reperfusion. PIP levels showed similar trends between the CO/O_2 and air groups at other time points. In addition, other parameters (pH, Na, K, and Ca levels) showed similar trends between the CO/O_2 and air groups at all time points (pre-operation, 30, 60, 120, and 180 min after reperfusion, Figure 7).

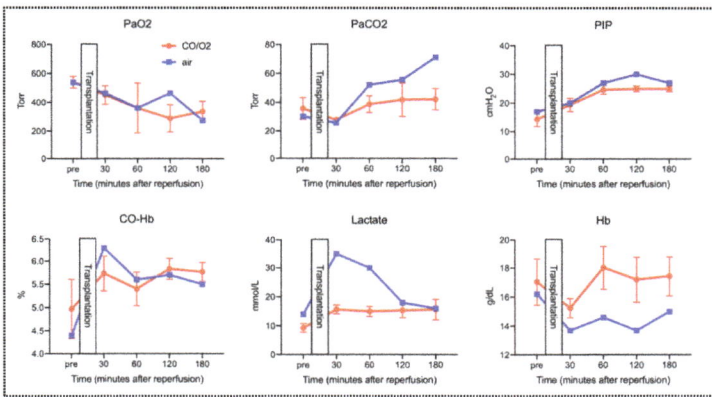

Figure 6. Assessment of chorological changes in parameters of arterial blood gases (ABG). The red line shows parameters in the CO/O_2 group, and the blue line shows parameters in the air group. Each data of red lines are expressed as the means ± standard deviations.

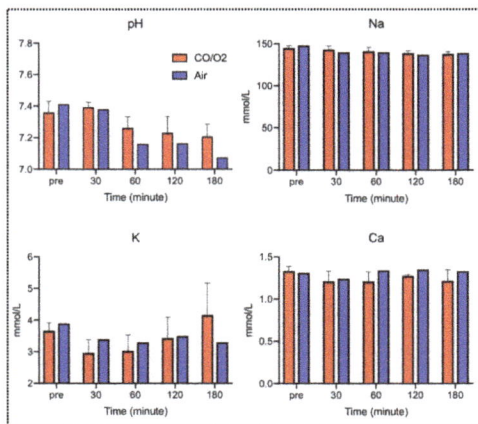

Figure 7. Arterial blood gas analysis for pH, Na, K, and Ca levels in the CO/O_2 and air groups. The red bars show parameters in the CO/O_2 group, and the blue bars show parameters in the air group. Each data of red bars are expressed as the means ± standard deviations.

3. Discussion

This study demonstrated that filling rat lungs with a CO and O_2 mixture via the HPG preservation method suppressed IRI. Furthermore, filling canine lungs with a CO and O_2 mixture maintained PaO_2 but suppressed the increase in lactic acid levels in the recipient.

Reportedly, HO-1 generates endogenous CO, and the HO-1/CO system confers tissue protective anti-inflammatory, anti-oxidative, and anti-apoptotic effects [20,21]. Recently, several trials have employed exogenous CO administration via inhalation or pharmacologic delivery [22–24]. However, the human body cannot be exposed to high concentrations of CO because it causes hypoxemia by generating CO-Hb [25]. Meanwhile, donor organs for transplantation may benefit from the effects of CO because they are preserved after removing blood, including hemoglobin (Hb), by the reflux of preservation solution. A previous study showed that rat heart transplantation using the HPG preservation method did not lead to increased CO-Hb concentration in recipients despite the exposure of the donor organs to high CO concentrations [17]. Similarly, the present study showed that CO-Hb concentration in the canine recipient remained unchanged until 180 min after reperfusion (Figure 6). These results indicate that the HPG preservation method, which does not affect the CO-Hb concentration in the recipient, is safe and could be one of the methods that can maximize the effects of the HO-1/CO system.

The lung, one of the most commonly transplanted organs, characteristically includes abundant alveolar spaces that are constantly exposed to gases, such as O_2, N_2, and CO_2, during gas exchange [26]. Therefore, the lungs, unlike other organs, can be filled with gases, such as CO, both from the surface and from the inside in the HPG preservation method. This study demonstrated that the preserved rat lungs filled with air tended to be heavier and had more extensive and severe alveolar hemorrhage than those filled with the CO and O_2 mixture when examined histologically. mRNA expression levels of HO-1 were not significantly different among the control, air, and CO/O_2 groups. A previous study showed that exposure to exogenous CO increased HO-1 expression in the body [22]. However, this study suggested that exogenous CO administration during organ preservation using the HPG method did not activate the endogenous HO-1/CO system. Moreover, after reperfusion of the excised rat lungs, expression levels of various inflammatory mediators (IL-6 and IL-1β) were significantly increased in the lungs filled with air, but their levels were suppressed in the lungs filled with the CO and O_2 mixture (Figure 3). Furthermore, western blotting was performed to assess IL-6 and IL-1β levels in the lungs. Similar to the results of real-time RT–PCR, IL-6 and IL-1β protein levels in the CO/O_2 groups were lower than those in the air group (Figure S1).

These results indicate that exposure to the CO and O_2 mixture from inside the lung is important and effective in reducing IRI during the HPG preservation method. Further examination revealed more severe alveolar hemorrhage in the lungs filled with CO or O_2 alone than in the lungs filled with their mixture (Figure S2); therefore, both CO and O_2 are essential. Further studies are warranted to determine more optimal gas combinations and ratios.

The characteristic structure of the lungs indicates the feasibility of the HPG preservation of large solid organs because the lungs can be filled with gases by delivery to every alveolus from the inside. Therefore, we examined the differences in the effects of HPG preservation based on organ size by comparing canine lungs, which are approximately half the size of adult human lungs, with rat lungs. Compared with the air group, the CO/O_2 group demonstrated PaO_2 preservation even after 180 min of reperfusion and exhibited a tendency of suppressed lactic acid production (Figure 5). These results indicate that the lungs may be one of the ideal organs that could benefit the greatest from the HPG preservation method because of their unique characteristics. We propose that adaptation of this method to human lungs, which have larger volumes, should be considered in the future.

Although the HPG preservation method can take full advantage of the effects of gases, pressure injury is one of the critical issues. To verify the pressure damage, we examined whether the trachea should be kept open when the lungs were exposed to the HPG mixture following their inflation. We found that there was considerable histological damage to the alveolar structure when the trachea was closed, which was evidently worse than that observed in the inflated lungs with the trachea kept open (Figure S3). Pressure trauma is a recognized injury that occurs in paranasal sinuses and the middle ear during diving and is believed to be due to a pressure gradient between the external environment and the closed lumen, resulting in relative negative pressure in the lumen and edema/bleeding of mucosal surfaces lining the lumen [27]. In the present study, there was a pressure gradient of 2.5 atm between the outside and inside of the lungs during HPG preservation in the group with the trachea closed at the time of preservation. Relative negative pressure on the alveolar tissue might have contributed, at least partially, to the alveolar damage. Therefore, to maximize the effect of inflation by gas for lung preservation while preventing pulmonary injury due to high pressure, the gas mixture should be delivered for lung inflation only before storage and the trachea should be kept open at the time of preservation.

The HPG preservation method used in this study can be applied to human lungs in a similar manner as that for canine lungs. However, during clinical lung transplantation, it is often necessary to transfer donor lungs from the donation institute to the transplantation institute; therefore, we think that some improvements in the storage chamber are necessary. First, because the chamber used in this study is not large enough to preserve the human lung, it is necessary to build a larger chamber. Second, the storage chamber should have a refrigeration function to perform cryopreservation, thereby preventing tissue necrosis during delivery. Finally, a small CO or O_2 bomb should be attached in case CO or O_2 needs to be refilled into the chamber during transport.

There are some possible limitations in this study. To focus on animal protection and minimize the number of experimental animals used, the efficacy of the HPG preservation method was first demonstrated in a technically simple rat ectopic transplant model. Second, the safety of this method was verified in an orthotopic transplant canine model with large organs. Therefore, the usefulness and safety of the method warrant evaluation in an animal model. Moreover, to protect animals, only one sample was included in the control group and no statistical analysis could thus be conducted in terms of the functional aspect of the donor lung in the canine model. To demonstrate the usefulness of the HPG preservation method for the lungs in large animals, additional experiments with canine model should be conducted. Finally, comparative studies with the existing simple immersion method were not conducted. Such a comparison is warranted prior to the clinical application the HPG preservation method in the future.

In conclusion, this study demonstrated that the physiological effects of the HO-1/CO system were employed for preserving donor lungs with unique characteristics via the HPG preservation method. This approach has significant potential to be used as a new preservation method for lungs.

4. Materials and Methods

This study was performed in accordance with the Guide for the Care and Use of Laboratory Animals prepared by the Institute of Laboratory Resources at the National Research Council (http: //nap.edu/catalog/12910.html). In experiment 1, the Regulations for Experimental Animals, prepared by Aichi Medical University (2010), was followed after the approval of the study design by the Committee for Animal Experiments at Aichi Medical University (Authorization number: 2016-11, Authorization date: 3 June 2016). Similarly, experiment 2 followed the Rules on Animal Experiments in Kagawa University prepared by Kagawa University (2016) following the approval of the study design by the Committee for Animal Experiments at Kagawa University (Authorization number: 15105-01, Authorization date: 10 November 2015).

4.1. Experiment 1. Rat Lung Preservation for 24 h Using the HPG Preservation Method

4.1.1. Animals

Ten-week-old male rats from the LEW/SsN Slc inbred line, with an average weight of 230 g (range, 220–245 g), were purchased from Shizuoka Laboratory Animal Center (Shizuoka, Japan). All rats were maintained under standard conditions and fed rodent food and water in accordance with the Regulations for Experimental Animals by Aichi Medical University (Authorization number: 2016-11).

4.1.2. Anesthesia, Extraction, and Preservation of Donor Lungs, and Ectopic Lung Transplantation

Previous studies have achieved the long-term preservation of the rat heart using the HPG preservation method [15–17]. In the rat model used in this study, ectopic cervical lung transplantation was performed and IRI severity was evaluated according to previous studies.

The left lungs were extracted from rats under deep anesthesia using pentobarbital (50 mg/kg; Kyoritsu Seiyaku, Tokyo, Japan), and blood was removed by the organ preservation solution (ET-K solution, Otsuka Pharmaceutical, Tokyo, Japan) via retrograde reflux from the ascending aorta to the pulmonary artery. A custom-built 7 air pressure-resistant chamber (length, 165 mm; width, 165 mm; height, 200 mm; material, iron; Nakamura Iron Works, Tokyo, Japan) was cooled to 4 °C in advance. Before undergoing HPG preservation, donor lungs were filled with either room air (air group, $n = 5$) or a mixture of CO and O_2 (CO/O_2 group, $n = 5$). Rat lungs that were immediately transplanted after extraction were used as the control group ($n = 4$), whereas the lungs that were immediately analyzed after extraction were used as the normal group ($n = 3$). Next, the rat lungs were placed in the chamber filled with a CO and O_2 mixture (3.5 atm; PCO, 1.5 atm, PO_2, 2.0 atm) and contained a flask with 50 mL distilled water to maintain humidity for 24 h. Lung preservation was performed with the trachea kept open (Figure 1). After preservation, the lungs were removed from the chamber, and ectopic lung transplantation to recipient rats was performed with anastomoses of the pulmonary artery with the internal carotid artery and the pulmonary vein with the external jugular vein. Follow-up observation was performed for 90 min after the reperfusion, followed by excision of the transplanted lungs and weight measurement.

4.1.3. Light Microscopy

Extracted rat lungs were processed with extended fixation in formalin solution for 24 h. The samples were washed, dehydrated in a series of graded ethanol, and embedded in paraffin. Serial sections (thickness, 6 μg) were cut using a microtome and stained with Gill's hematoxylin III and 2% eosin Y. Tissue slides were digitized and stored using an image storage device (Aperio AT, Leika Biosystems, Wetzlar, Germany). The digitized slides were analyzed using imaging software (Aperio esilde manager, Leika Biosystems, Wetzlar, Germany). To assess IRI, alveolar hemorrhage rate was evaluated using for ImageJ (National Institutes of Health, Bethesda, MD, USA). For each tissue sample, five fields of view were randomly selected to be captured as images. The percent area occupied by red blood cells relative to the area of the entire image was defined as the alveolar hemorrhage rate.

4.1.4. Analysis for Gene Expression Levels of Mediators

Gene expression levels of inflammatory mediators (TNF-α, IL-6, and IL-1β), anti-inflammatory mediators (HO-1, iNOS, and TGF-β), and apoptotic mediators (caspase 3 and GM-CSF) were quantified in duplicates using a two-step SYBR Green-based real-time RT–PCR protocol.

4.1.5. Statistical Analysis

Values were presented as means and standard deviations. Significance in differences was determined using an analysis of variation (ANOVA) and post hoc tests. A p value of <0.05 was considered statistically significant.

4.2. *Experiment 2. Canine Lung Preservation for 24 h Using the HPG Preservation Method*

4.2.1. Animals

Four pairs of weight-matched (8.9–11.0 kg) adult Narc Beagle canines (Kitayama Labes, Ina, Japan) were evaluated. All canines were maintained under standard conditions and fed dog chow and water in accordance with the Rules on Animal Experiments at Kagawa University (Authorization number: 15105-01).

4.2.2. Anesthesia

All canines were anesthetized with intramuscular injection of ketamine (10 mg/kg; Daiichi Sankyo, Tokyo, Japan), xylazine (0.25 mg/kg; Bayer Yakuhin, Osaka, Japan), and atropine sulfate (0.05 mg/kg; Mitsubishi Tanabe Pharma, Osaka, Japan) and intubated for mechanical ventilation. Anesthesia was maintained via the inhalation of 2–3% sevoflurane (Pfizer, New York, NY, USA) and O_2. All donor and recipient canines were ventilated with an inspired O_2 fraction of 1.0 at a tidal volume of 20 mL/kg, a respiratory rate of 15 breaths/min, and a positive end-expiratory pressure of 5 cmH_2O.

4.2.3. Extraction of the Donor Lung and Preservation

We regarded the canine model as an intermediate stage for the clinical application of the HPG preservation method and emphasized the evaluation in a more clinical form. Therefore, the orthotopic transplantation of the donor lungs was performed for evaluation.

Catheters were introduced into the right femoral artery and the right external jugular vein. After the left thoracotomy, the left pulmonary artery and vein as well as the left main bronchus were encircled. After systemic heparinization (250 IU/kg; MOCHIDA PHARMACEUTICAL, Tokyo, Japan), bolus prostaglandin E1 (25 μg/kg; ONO PHARMACEUTICAL, Osaka, Japan) was injected into the right external jugular vein. After confirming that the atrial blood pressure (ABP) was at ≤90 mmHg, the left main pulmonary artery was cannulated with a 16-Fr-diameter catheter. After ligation of the proximal pulmonary artery and incision of the left atrium, the lung was flushed with 70 mL/kg ET-K solution to remove blood, with anterograde perfusion of the solution from the pulmonary artery and flushing with 30 cmH_2O pressure. After perfusion, the left main bronchus and left atrium were incised, and the left lung was removed from inside the thoracic cavity. Donor lungs were preserved in a custom-built 7 air pressure-resistant chamber (Figure 4a).

4.2.4. Allogeneic Lung Transplantation of the Preserved Lung and Functional Evaluation after Reperfusion

Recipient canines were anesthetized using the same approach for donor canines, and Swan-Ganz catheters were introduced to the right external jugular vein. Before surgery, heart rate, ABP, percutaneous O_2 saturation, PIP, ABG, CO-binding hemoglobin (CO-Hb) concentration in blood, and systemic cardiac functions (pulmonary artery pressure, cardiac output, and pulmonary capillary wedge pressure) were evaluated. Left pneumonectomy was performed using the same approach for donor canines, and left single-lung transplantation was performed. Anastomosis was performed in the following order: Main bronchus, left atrium, and pulmonary artery. After 30 min of reperfusion, right main pulmonary artery and right main bronchus were ligated to evaluate the donor lung function. The ventilator settings were changed to achieve 2/3 of the tidal volume before reperfusion and a respiratory rate of 20 breaths/min. Heart rate, ABP, percutaneous O_2 saturation, and PIP were measured at five time points: Pre-operation and 30, 60, 120, and 180 min after reperfusion. After 180 min of reperfusion, the transplanted lungs were excised for histological evaluation.

4.2.5. Assessment of ABG and CO-Hb Concentration

Arterial blood obtained from the recipient canines at the indicated five time points were analyzed to measure partial arterial O_2 pressure (PaO_2), partial arterial CO_2 pressure ($PaCO_2$), CO-Hb concentration, lactate, Na, K, and Ca using a radiometer (ABL800 FLEX blood gas analyzer, Copenhagen, Denmark). Furthermore, ABG data mentioned above were considered to be more accurate because the absence of any increase in CO-Hb concentration was reflected in the minimal effect on the PaO_2 value by ABG analysis.

4.2.6. Light Microscopy

Canine lungs were processed by extended fixation in 10% formalin solution for 24 h. The samples were washed, dehydrated in a series of graded ethanol, and embedded in paraffin. Serial sections (thickness, 6 μm) were cut with a microtome and stained with Gill's hematoxylin III and 2% eosin Y. Tissue slides were digitized and stored using an image storage device (Aperio AT, Leika Biosystems, Wetzlar, Germany). The digitized slides were analyzed using imaging software (Aperio esilde manager, Leika Biosystems, Wetzlar, Germany).

4.2.7. Statistical Analysis

Values are presented as means and standard deviation.

Supplementary Materials: Supplementary materials can be found at http://www.mdpi.com/1422-0067/20/11/2719/s1.

Author Contributions: Conceptualization, A.F., N.H., S.H., M.N. and H.Y.; Data curation, A.F.; Formal analysis, A.F.; Funding acquisition, M.N.; Investigation, A.F., N.H., N.M., N.Y., K.F., T.Y., S.T. and T.G.; Project administration, S.H.; Supervision, H.Y.; Validation, S.H.; Writing—original draft, A.F.; Writing—review & editing, S.H.

Funding: This work was supported by MEXT KAKENHI [Grant Numbers JP 17911763, JP 19091362, JP 19090214].

Acknowledgments: The authors would like to thank ENAGO (www.enago.jp) for the English language review.

Conflicts of Interest: The authors declare no conflict of interest.

Abbreviations

CO	Carbon monoxide
CO-Hb	Carboxyhemoglobin
HO-1	Heme oxygenase-1
MAPK	Mitogen-activated protein kinase
IRI	Ischemia–reperfusion injury
HPG	High-pressure gas
O_2	Oxygen
N_2	Nitrogen
RT–PCR	Reverse transcription–polymerase chain reaction
IL	Interleukin
TGF	Transforming growth factor
iNOS	Inducible nitric oxide synthase
GM-CSF	Granulocyte-macrophage colony-stimulating factor
ABG	Arterial blood gas
PaO_2	Partial arterial oxygen pressure
$PaCO_2$	Partial arterial carbon dioxide pressure
PCO	Partial carbon monoxide pressure
PO_2	Partial oxygen pressure
CO_2	Carbon dioxide
PIP	Peak inspiratory pressure
Hb	Hemoglobin
TNF	Tumor necrosis factor
ANOVA	Analysis of variation
ABP	Arterial blood pressure
RIPA	Radioimmunoprecipitation assay
SDS-PAGE	SDS-polyacrylamide gel electrophoresis
PVDF	Polyvinylidene di fluoride

References

1. Bauer, I.; Pannen, B.H. Bench-to-bedside review: Carbon monoxide—From mitochondrial poisoning to therapeutic use. *Crit. Care (Lond. Engl.)* **2009**, *13*, 220. [CrossRef] [PubMed]
2. Ryter, S.W.; Otterbein, L.E. Carbon monoxide in biology and medicine. *Bioessays News Rev. Mol. Cell. Dev. Biol.* **2004**, *26*, 270–280. [CrossRef] [PubMed]
3. Ryter, S.W.; Choi, A.M. Targeting heme oxygenase-1 and carbon monoxide for therapeutic modulation of inflammation. *Transl. Res. J. Lab. Clin. Med.* **2016**, *167*, 7–34. [CrossRef] [PubMed]
4. Zhang, X.; Shan, P.; Otterbein, L.E.; Alam, J.; Flavell, R.A.; Davis, R.J.; Choi, A.M.; Lee, P.J. Carbon monoxide inhibition of apoptosis during ischemia-reperfusion lung injury is dependent on the p38 mitogen-activated protein kinase pathway and involves caspase 3. *J. Biol. Chem.* **2003**, *278*, 1248–1258. [CrossRef] [PubMed]
5. Zhang, X.; Shan, P.; Alam, J.; Fu, X.Y.; Lee, P.J. Carbon monoxide differentially modulates stat1 and stat3 and inhibits apoptosis via a phosphatidylinositol 3-kinase/akt and p38 kinase-dependent stat3 pathway during anoxia-reoxygenation injury. *J. Biol. Chem.* **2005**, *280*, 8714–8721. [CrossRef]
6. Kohmoto, J.; Nakao, A.; Stolz, D.B.; Kaizu, T.; Tsung, A.; Ikeda, A.; Shimizu, H.; Takahashi, T.; Tomiyama, K.; Sugimoto, R.; et al. Carbon monoxide protects rat lung transplants from ischemia-reperfusion injury via a mechanism involving p38 mapk pathway. *Am. J. Transplant. Off. J. Am. Soc. Transplant. Am. Soc. Transpl. Surg.* **2007**, *7*, 2279–2290. [CrossRef]
7. Mishra, S.; Fujita, T.; Lama, V.N.; Nam, D.; Liao, H.; Okada, M.; Minamoto, K.; Yoshikawa, Y.; Harada, H.; Pinsky, D.J. Carbon monoxide rescues ischemic lungs by interrupting mapk-driven expression of early growth response 1 gene and its downstream target genes. *Proc. Natl. Acad. Sci. USA* **2006**, *103*, 5191–5196. [CrossRef]
8. Fujita, T.; Toda, K.; Karimova, A.; Yan, S.F.; Naka, Y.; Yet, S.F.; Pinsky, D.J. Paradoxical rescue from ischemic lung injury by inhaled carbon monoxide driven by derepression of fibrinolysis. *Nat. Med.* **2001**, *7*, 598–604. [CrossRef]
9. Kohmoto, J.; Nakao, A.; Kaizu, T.; Tsung, A.; Ikeda, A.; Tomiyama, K.; Billiar, T.R.; Choi, A.M.; Murase, N.; McCurry, K.R. Low-dose carbon monoxide inhalation prevents ischemia/reperfusion injury of transplanted rat lung grafts. *Surgery* **2006**, *140*, 179–185. [CrossRef]
10. Meng, C.; Ma, L.; Niu, L.; Cui, X.; Liu, J.; Kang, J.; Liu, R.; Xing, J.; Jiang, C.; Zhou, H. Protection of donor lung inflation in the setting of cold ischemia against ischemia-reperfusion injury with carbon monoxide, hydrogen, or both in rats. *Life Sci.* **2016**, *151*, 199–206. [CrossRef]
11. Meyers, B.F.; de la Morena, M.; Sweet, S.C.; Trulock, E.P.; Guthrie, T.J.; Mendeloff, E.N.; Huddleston, C.; Cooper, J.D.; Patterson, G.A. Primary graft dysfunction and other selected complications of lung transplantation: A single-center experience of 983 patients. *J. Thorac. Cardiovasc. Surg.* **2005**, *129*, 1421–1429. [CrossRef] [PubMed]
12. Ikeda, M.; Bando, T.; Yamada, T.; Sato, M.; Menjyu, T.; Aoyama, A.; Sato, T.; Chen, F.; Sonobe, M.; Omasa, M.; et al. Clinical application of et-kyoto solution for lung transplantation. *Surg. Today* **2015**, *45*, 439–443. [CrossRef]
13. Kosaka, S.; Ueda, M.; Bando, T.; Liu, C.J.; Hitomi, S.; Wada, H. Ultrastructural damage to the preserved lung and its function after reperfusion. *Jpn. J. Thorac. Cardiovasc. Surg. Off. Publ. Jpn. Assoc. Thorac. Surg. Nihon Kyobu Geka Gakkai Zasshi* **2002**, *50*, 6–14. [CrossRef]
14. Wada, H.; Liu, C.J.; Hirata, T.; Bando, T.; Kosaka, S. Effective 30-hour preservation of canine lungs with modified et-kyoto solution. *Ann. Thorac. Surg.* **1996**, *61*, 1099–1105. [CrossRef]
15. Hatayama, N.; Yoshida, Y.; Seki, K. Seventy-two-hour preservation, resuscitation, and transplantation of an isolated rat heart with high partial pressure carbon monoxide gas (pCO = 400 hpa) and high partial pressure carbon dioxide (pCO$_2$ = 100 hpa). *Cell Transplant.* **2012**, *21*, 623. [CrossRef]
16. Hatayama, N.; Naito, M.; Hirai, S.; Yoshida, Y.; Kojima, T.; Seki, K.; Li, X.K.; Itoh, M. Preservation by desiccation of isolated rat hearts for 48 hours using carbon monoxide (pCO = 4,000 hpa) and oxygen (pO$_2$ = 3,000 hpa). *Cell Transplant.* **2012**, *21*, 609–615. [CrossRef]
17. Hatayama, N.; Inubushi, M.; Naito, M.; Hirai, S.; Jin, Y.N.; Tsuji, A.B.; Seki, K.; Itoh, M.; Saga, T.; Li, X.K. Functional evaluation of rat hearts transplanted after preservation in a high-pressure gaseous mixture of carbon monoxide and oxygen. *Sci. Rep.* **2016**, *6*, 32120. [CrossRef]

18. Abe, T.; Yazawa, K.; Fujino, M.; Imamura, R.; Hatayama, N.; Kakuta, Y.; Tsutahara, K.; Okumi, M.; Ichimaru, N.; Kaimori, J.Y.; et al. High-pressure carbon monoxide preserves rat kidney grafts from apoptosis and inflammation. *Lab. Investig. J. Tech. Methods Pathol.* **2017**, *97*, 468–477. [CrossRef] [PubMed]

19. Hatayama, N.; Hirai, S.; Naito, M.; Terayama, H.; Araki, J.; Yokota, H.; Matsushita, M.; Li, X.K.; Itoh, M. Preservation of rat limbs by hyperbaric carbon monoxide and oxygen. *Sci. Rep.* **2018**, *8*, 6627. [CrossRef] [PubMed]

20. Brouard, S.; Otterbein, L.E.; Anrather, J.; Tobiasch, E.; Bach, F.H.; Choi, A.M.; Soares, M.P. Carbon monoxide generated by heme oxygenase 1 suppresses endothelial cell apoptosis. *J. Exp. Med.* **2000**, *192*, 1015–1026. [CrossRef]

21. Hayashi, S.; Takamiya, R.; Yamaguchi, T.; Matsumoto, K.; Tojo, S.J.; Tamatani, T.; Kitajima, M.; Makino, N.; Ishimura, Y.; Suematsu, M. Induction of heme oxygenase-1 suppresses venular leukocyte adhesion elicited by oxidative stress: Role of bilirubin generated by the enzyme. *Circ. Res.* **1999**, *85*, 663–671. [CrossRef]

22. Sawle, P.; Foresti, R.; Mann, B.E.; Johnson, T.R.; Green, C.J.; Motterlini, R. Carbon monoxide-releasing molecules (co-rms) attenuate the inflammatory response elicited by lipopolysaccharide in raw264.7 murine macrophages. *Br. J. Pharmacol.* **2005**, *145*, 800–810. [CrossRef]

23. Musameh, M.D.; Green, C.J.; Mann, B.E.; Fuller, B.J.; Motterlini, R. Improved myocardial function after cold storage with preservation solution supplemented with a carbon monoxide-releasing molecule (corm-3). *J. Heart Lung Transplant. Off. Publ. Int. Soc. Heart Transplant.* **2007**, *26*, 1192–1198. [CrossRef]

24. Babu, D.; Motterlini, R.; Lefebvre, R.A. Co and co-releasing molecules (co-rms) in acute gastrointestinal inflammation. *Br. J. Pharmacol.* **2015**, *172*, 1557–1573. [CrossRef]

25. Rodkey, F.L.; O'Neal, J.D.; Collison, H.A.; Uddin, D.E. Relative affinity of hemoglobin s and hemoglobin a for carbon monoxide and oxygen. *Clin. Chem.* **1974**, *20*, 83–84.

26. Knudsen, L.; Ochs, M. The micromechanics of lung alveoli: Structure and function of surfactant and tissue components. *Histochem. Cell Biol.* **2018**, *150*, 661–676. [CrossRef]

27. Livingstone, D.M.; Smith, K.A.; Lange, B. Scuba diving and otology: A systematic review with recommendations on diagnosis, treatment and post-operative care. *Diving Hyperb. Med.* **2017**, *47*, 97–109. [CrossRef]

International Journal of
Molecular Sciences

MDPI

Article

Betula etnensis Raf. (Betulaceae) Extract Induced HO-1 Expression and Ferroptosis Cell Death in Human Colon Cancer Cells

Giuseppe Antonio Malfa [1,†], Barbara Tomasello [1,†], Rosaria Acquaviva [1,*], Carlo Genovese [2], Alfonsina La Mantia [1], Francesco Paolo Cammarata [3,4], Monica Ragusa [5], Marcella Renis [1] and Claudia Di Giacomo [1]

[1]   Department of Drug Science, Section of Biochemistry, University of Catania, Viale A. Doria 6, 95125 Catania, Italy; g.malfa@unict.it (G.A.M.); btomase@unict.it (B.T.); alfy.lamantia@gmail.com (A.L.M.); renis@unict.it (M.R.); cdigiaco@unict.it (C.D.G.)
[2]   Department of Biomedical and Biotechnological Sciences, Microbiology Section University of Catania, 95125 Catania, Italy; gnv.carlo@gmail.com
[3]   Institute of Bioimaging and Molecular Physiology, National Council of Research (IBFM-CNR), 90015 Cefalù (PA), Italy; fracammarata@gmail.com
[4]   National Institute of Nuclear Physics, South National Laboratory (LNS-INFN), 95125 Catania, Italy
[5]   Department of Experimental and Clinical Medicine, University Magna Graecia of Catanzaro, 88100 Catanzaro, Italy; m.ragusa@unicz.it
*   Correspondence: racquavi@unict.it; Tel.: +39-957384219; Fax: +39-957384220
†   These authors contributed equally to this work.

Received: 30 April 2019; Accepted: 31 May 2019; Published: 3 June 2019

Abstract: *Betula etnensis* Raf. (Birch Etna) belonging to the Betulaceae family grows on the eastern slope of Etna. Many bioactive compounds present in Betula species are considered promising anticancer agents. In this study, we evaluated the effects of *B. etnensis* Raf. bark methanolic extract on a human colon cancer cell line (CaCo2). In order to elucidate the mechanisms of action of the extract, cellular redox status, cell cycle, and heme oxygenase-1 (HO-1) expression in ferroptosis induction were evaluated. Cell viability and proliferation were tested by tetrazolium (MTT) assayand cell cycle analysis, while cell death was evaluated by annexin V test and lactic dehydrogenase (LDH) release. Cellular redox status was assessed by measuring thiol groups (RSH) content, reactive oxygen species (ROS) production, lipid hydroperoxide (LOOH) levels and (γ-glutamylcysteine synthetase) γ-GCS and HO-1 expressions. The extract significantly reduced cell viability of CaCo2, inducing necrotic cell death in a concentration-depending manner. In addition, an increase in ROS levels and a decrease of RSH content without modulation in γ-GCS expression were detected, with an augmentation in LOOH levels and drastic increase in HO-1 expression. These results suggest that the *B. etnensis* Raf. extract promotes an oxidative cellular microenvironment resulting in CaCo2 cell death by ferroptosis mediated by HO-1 hyper-expression.

Keywords: Colon cancer; *Betula etnensis* Raf.; oxidative stress; heme oxigenase-1; ferroptosis; thiol groups

1. Introduction

After prostate cancer in men, breast cancer in women and lung cancer, the colorectal cancer (CRC) represents the second leading cause of death in the Western world in both males and females. The association between nutrition and colon cancer has extensively been investigating by many studies but is controversial because of the diet shows a causal and protective role in the CRC development [1,2].

It is well known that the consumption of vegetables is correlated with a low incidence of cancer and, in particular, is effective in the prevention and reduction of CRC risk [3,4]. A plausible reason might be plant foodstuff is a good source of fibers, folate vitamins, various antioxidants, and other bioactive compounds including polyphenols, terpenoids, saponins, and carotenoids [5–7]. Besides their well-known anti-oxidant activities, they have been reported to be anti-mutagenic and/or anti-carcinogenic and to possess several other biological activities. In addition, natural antioxidants are known to have a dual face, behaving as pro-oxidant compounds after reacting directly with reactive oxygen species (ROS) in the presence of transition metal ions, such as copper and iron. This increase of cellular ROS to cytotoxic level may generate secondary oxidative damage and induce a selective killing of cancer cells by a variety of ways including ferroptosis, a non-apoptotic form of cell death characterized by the high expression of heme oxygenase-1 (HO-1) and accumulation of lipid hydroperoxides (LOOH) [8]. The role of heme oxygenase-1 (HO-1) in cancer biology is poorly understood. In fact, HO-1 has been described as survival molecule because of its anti-apoptotic and pro-angiogenic effects in several cancer types and its modulation can be induced by several natural compounds such as polyphenols and terpenoids [5].

Today potential usefulness of natural compounds as anti-cancer agents has to be ascribed prevalently to improve patient quality of life and to support conventional chemotherapy or radiotherapy. In the genus *Betula* (Betulaceae) there belongs common trees and shrubs of the boreal and north temperate zones. The bioactivities of Betula species are well documented, in fact the Betula was largely used in human and veterinary medicine to treat various diseases. Betula compounds have a wide variety of properties including anticancer, anti-inflammatory, and immunomodulatory beyond being antioxidant [9–11].

Currently, of particular interest are anticancer activities showed in vitro by some constituents isolated from these plants, which displayed cytotoxic effects on neuroblastoma, melanoma, medulloblastoma, and Ewing's sarcoma cells [10,12]. In particular, it has been reported that betulinic acid may be considered a promising anticancer agent [10,13]. Commonly known as Birch Etna, *B. etnensis* Raf. is a medium-sized deciduous tree typically reaching 5–20 m tall, which belongs to the family Betulaceae. It grows on Mt. Etna volcano, at an altitude between 1200 and 2000 m and its ivory colored bark, in particular in the young branches, is rich of resinous chemicals secreted by the numerous glands present [14]. Polyphenols, terpenoids, betulin, betulinic acid, and ursolic acid are the main constituents present in the *B. etnensis* Raf. bark. As for as our knowledge no literature data are present on biological activities of *B. etnensis* Raf.

In the present study, we evaluated, for the first time, the effects of *B. etnensis* Raf. bark methanolic extract on a human colon cancer cell line (CaCo2). In order to elucidate the mechanisms of action of this extract, cell viability, annnexin, lactic dehydrogenase (LDH) release, ROS production, thiol groups (RSH) content, lipid peroxidation (LOOH) levels, cell cycle, HO-1, and γ-GCS protein expression were evaluated.

2. Results

2.1. MTT Bioassay

Methanolic extract of bark from *B. etnensis* Raf. (5, 50, 250, or 500 µg/mL) was able to significantly reduce in a dose dependent manner cell viability (Figure 1); because 250 and 500 µg/mL of methanolic extract showed similar effects, in the other subsequent experiments we omitted 500 µg/mL of extract.

2.2. LDH Release

As shown in Figure 2, after 72 h of incubation with methanolic extract of bark from *B. etnensis* Raf. a statistically significant increase in LDH release was observed in CaCo2 cells treated with 50 and 250 µg/mL of extract.

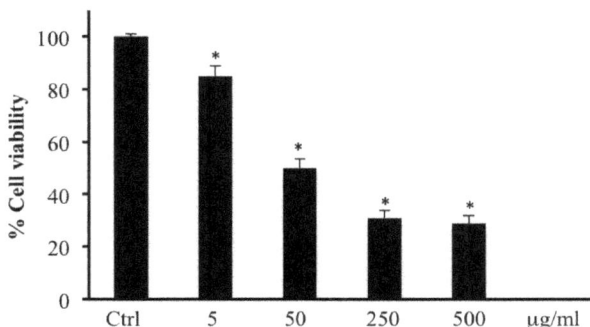

Figure 1. Cell viability in human colon cancer cell line (CaCo2) cells untreated and treated for 72 h with methanolic extract of *B. etnensis* Raf. at different concentrations (5–500 µg/mL). Values are the mean ± SD of four experiments in triplicate. * Significant vs. untreated control cells: $p < 0.001$.

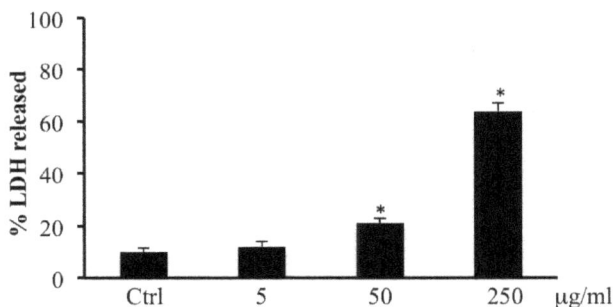

Figure 2. Lactic dehydrogenase (LDH) released in CaCo2 cells untreated and treated for 72 h with methanolic extract of *B. etnensis. Raf.* at different concentrations (5–250 µg/mL). Values are the mean ± SD of four experiments in triplicate. * Significant vs. untreated control cells: $p < 0.001$.

2.3. Annexin V and Dead Cell Evaluation

A slight non-significant increase in the percentage of total apoptotic cells from 9.20 ± 1.25% to 11.75 ± 0.40% after treatment with 50–250 µg/mL of the extract was observed (Table 1). The annexin-V/7-AAD results confirm that most of antiproliferative activity of *B. etnensis* Raf. observed by viability assay is mediated by necrosis.

Table 1. Annexin V in CaCo2 cells untreated and treated for 72 h with methanolic extract of *B. etnensis* Raf. at different concentrations (5–250 µg/mL). Values are the mean ± SD of four experiments in triplicate. * Significant vs. untreated control cells: $p < 0.001$.

	Live	Early Apoptosis	Late Apoptosis	Dead Cells	Total Apoptosis
CTRL	96.0 ± 0.045%	0%	1.64 ± 0.09%	2.36 ± 0.06%	1.64 ± 0.05%
5 µg/mL	92.30 ± 0.055%	0%	2.1 ± 0.035%	4.9 ± 0.065%	2.1 ± 0.08%
50 µg/mL	84.70 ± 0.08% *	0%	9.20 ± 0.09%	6.10 ± 0.1%	9.20 ± 1.25%
250 µg/mL	4.0 ± 0.084% *	1.10 ± 0.085%	10.65 ± 0.075%	84.25 ± 0.5%	11.75 ± 0.40%

2.4. ROS Levels

Data reported in Figure 3 demonstrate that exposure of CaCo2 cells to several concentrations of *B. etnensis* Raf. methanolic extracts resulted in a significant increase in radical species, as revealed by fluorescence intensity.

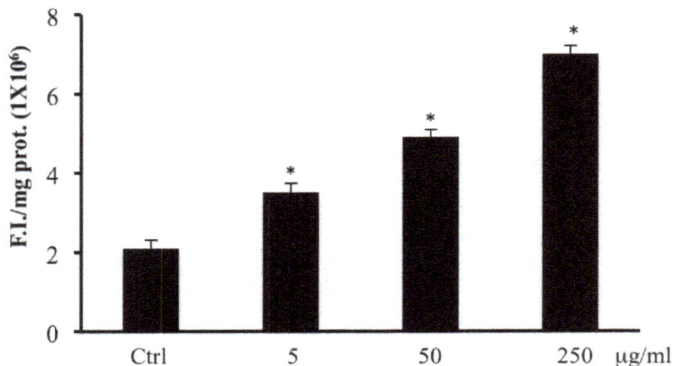

Figure 3. Intracellular oxidants in CaCo2 cells untreated and treated for 72 h with methanolic extract of *B. etnensis* Raf. at different concentrations (5–250 µg/mL). Values are the mean ± SD of four experiments in triplicate. * Significant vs. untreated control cells: $p < 0.001$.

2.5. LOOH Levels

Figure 4 shows that the addition of methanolic extract of *B. etnensis* Raf. at 5, 50, and 250 µg/mL for 72 h induced a significant and dose-dependent increase in LOOH levels with respect to untreated CaCo2 cells.

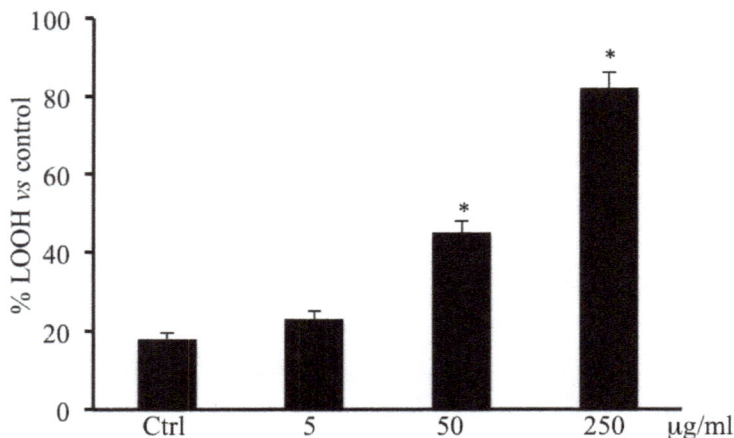

Figure 4. LOOH levels in CaCo2 cells untreated and treated for 72 h with methanolic extract of *B. etnensis* Raf. at different concentrations (5–250 µg/mL). Values are the mean ± SD of four experiments in triplicate. * Significant vs. untreated control cells: $p < 0.001$.

2.6. Total Thiol Groups

Treatment of cells with 50–250 μg/mL methanolic extract of *B. etnensis* Raf. resulted in a significant reduction in RSH levels (Figure 5). Instead, the lowest concentration of methanolic extract of *B. etnensis* Raf. did not alter the levels of RSH groups with respect to the control.

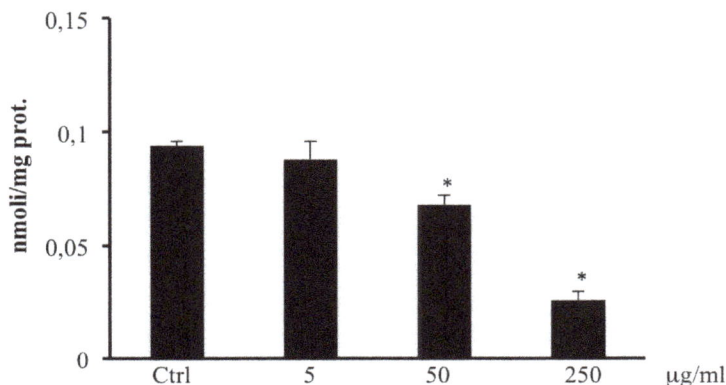

Figure 5. Thiol groups in CaCo2 cells untreated and treated for 72 h with methanolic extract of *B. etnensis* Raf. at different concentrations (5–250 μg/mL). Values are the mean ± SD of four experiments in triplicate. * Significant vs. untreated control cells: $p < 0.001$.

2.7. γ-GCS Determination

No significant change in γ-GCS expression was observed in CaCo2 cells treated with the extract of *B. etnensis* Raf. with respect to untreated cells (Figure 6).

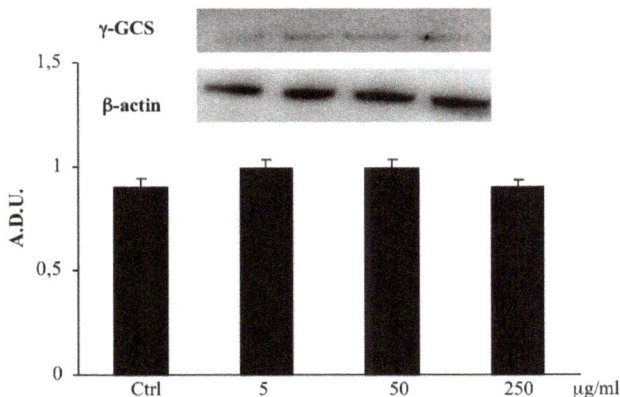

Figure 6. Immunoblotting of (γ-glutamylcysteine synthetase) γ-GCS levels in CaCo2 cells untreated and treated for 72 h with methanolic extract of *B. etnensis* Raf. at different concentrations (5–250 μg/mL). Values are the mean ± SD of four experiments performed in triplicate. * Significant vs. untreated control cells: $p < 0.001$.

2.8. HO-1 by ELISA

Results reported in Figure 7 show that the addition of 250 μg/mL of extract to CaCo2 cells, caused a significant increase in HO-1 protein expression (Figure 7).

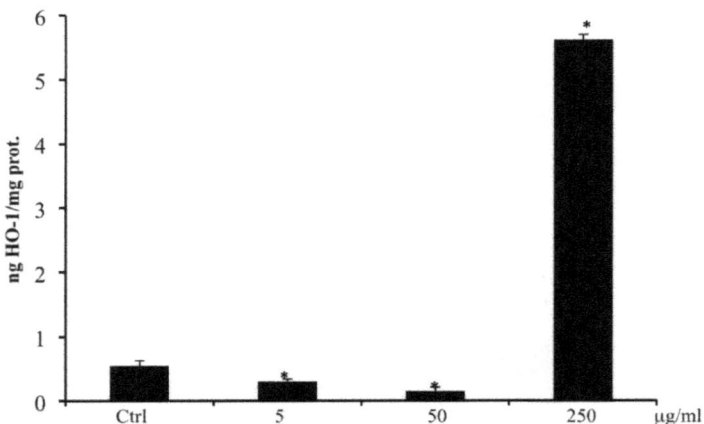

Figure 7. Heme oxygenase-1 (HO-1) levels in CaCo2 cells untreated and treated for 72 h with methanolic extract of *B. etnensis* Raf. at different concentrations (5–250 µg/mL). Values are the mean ± SD of four experiments in triplicate. * Significant vs. untreated control cells: $p < 0.001$.

2.9. Cell Cycle Analysis

When CaCo2 cells were treated with increasing concentrations of *B. etnensis* Raf. for 72 h, a dose dependent cell cycle arrest at G0/G1 and S phases were evident with concomitant marked decrease in the G2/M phase. After the treatment with 50 µg/mL of *B. etnensis* Raf., cell population in G0/G1 and S phases increased by ~5.14% and ~16% respectively, whereas cells in G2/M were decreased by ~21% (Figure 8). The same trend was observed with 250 µg/mL of extract.

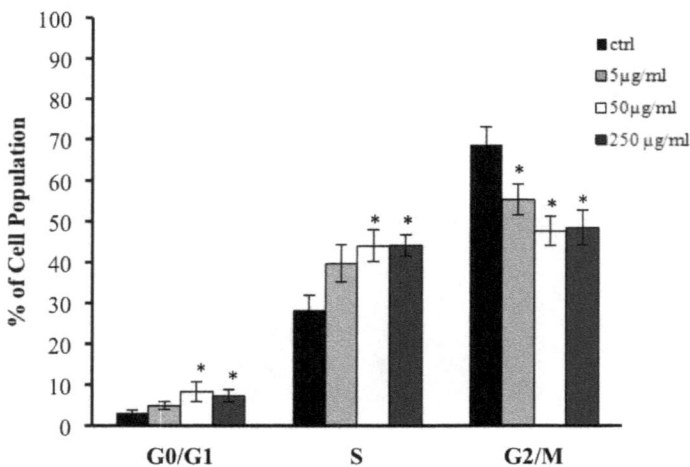

Figure 8. *B. etnensis* Raf. induces a cell cycle arrest at both G0/G1 and S phases in CaCo2 cells. The panel shows bar graph representing quantified values of the flow cytometry data. The graphs show the mean ± SD of four independent cell cycle experiments. (*) For values that are significantly different from the untreated control. * Significant vs. untreated control cells: $p < 0.001$.

3. Discussion

Tumors of the digestive tract, particularly CRC, are among the most common forms of cancer, with thousands of deaths worldwide per year. Chemotherapy and radiation therapy are the main treatments with significant side effects. Among cancers, CRC is the most responsive to dietary modification, in fact, several studies demonstrated that approximatively 75% of all sporadic cases of CRC are directly influenced by diet and that dietary modification is a feasible strategy for reducing CRC risk [5,15,16]. The consumption of fruit and vegetables and herbs is associated with a low incidence of cancer [7,17] and this may be partly due to the presence of several bioactive natural compounds in plants [6].

Previous studies have shown that betulinic acid is cytotoxic for different types of human cancer cells [10,13,18]. In this study, we evaluated the anti-cancer activity of *B. etnensis* Raf. methanolic extract in CaCo2 cells.

In order to evaluate cytotoxic effects of the methanolic extract, we first conducted an MTT test. The results, showed in Figure 1, indicate that cell viability is significantly reduced by the extract. Cell viability reaches a plateau at 250 µg/mL with a 70% decrease and the same result was obtained with 500 µg/mL so the highest concentration of treatment used in next experiments was 250 µg/mL of *B. etnensis* Raf. methanolic extract.

In consideration of the results obtained on cell viability, we performed LDH assay in medium to evaluate cell membrane disruption which is accompanied by necrotic cells death. Results showed a dose-dependent necrotic effect with a maximum at 250 µg/mL in line with results obtained by MTT assay (Figure 2). These results were confirmed by annexin V levels, in fact, treatment with *B. etnensis* Raf. methanolic extract did not induce apoptotic cell death (Table 1). In order to verify the role of oxidative stress in the mechanism of cell death, the determination of ROS and LOOH was performed. Clearly, the treatments with *B. etnensis* Raf. extract induced an increase in ROS levels in CaCo2 tumor cell line, indicating that the extract does not act as an antioxidant agent, but rather exerts its cytotoxic effect as a pro-oxidant with concomitant increase lipid peroxidation particularly evident at 250 µg/mL (Figures 3 and 4).

This result suggests that the extract may exert its activity by destabilizing the cellular redox balance so involving intracellular factors probable targets of ROS. Certainly the high production of ROS is a key factor that contributes to carcinogenesis. Within a neoplastic cell the persistent oxidative stress may be responsible for the activation of growth factor pathways and the increased resistance to apoptosis [19]. In fact, ROS have been shown to act as second messengers [20,21], stimulating the transduction of intracellular signals [22,23].

Recently, it has been suggested that GSH, as well as the main thiol responsible for maintaining the intracellular redox state, through thiol/disulfide exchange reactions, may be involved in the redox regulation of this type of signal [20–24].

In our study, the treatment of CaCo2 cells with the extract causes a dose-dependent reduction of intracellular RSH levels (Figure 5). To further confirm the involvement of the antioxidant defense systems in the mechanism of action of the extract, we also evaluated the expression of γ-GCS, a key enzyme in glutathione synthesis the expression of which in cancer cells is involved in tumor aggressiveness and chemo-resistance. The Western blot analysis of γ-GCS showed that the extract does not significantly influence the expression of the protein at all concentrations tested, suggesting that decreased RSH levels are most due to an inhibition of GSH synthesis and to a depletion for excessive ROS reactions (Figure 6). All these results confirms the persistence of a pro-oxidative imbalance induced by *B. etnensis* Raf. methanolic extract.

To further understand the mechanism of action of the extract, under the same experimental conditions, the expression of HO-1, one of the main cellular defense mechanisms against oxidative stress was evaluated. The role of HO-1 in tumors is still not very clear [25]. Results obtained in the present study show that the *B. etnensis* Raf. extract may exert a dual effect on HO-1 expression: at low concentration (5–50 µg/mL) it reduced HO-1 expression whereas, at higher concentrations (250 µg/mL) induced a significant increase in HO-1 expression (Figure 7). The reported inverse correlation between GSH levels and HO-1 expression is only confirmed at the highest concentration of the extract [26].

The HO-1 effect in cancer cells is not yet clear, but it is wildly documented that an HO-1 over-expression confers resistance to chemotherapy and radiation therapy. This protective effect could be due to its reaction products such as CO or biliverdin/bilirubin [27,28]. In spite of its implication in tumor initiation, angiogenesis, and metastasis, excessively increased expression of HO-1 in tumor cells may lead to cell death through a process called ferroptosis [29–32] In fact, a lot of evidence showed that HO-1 induces ferroptosis through an increase of ROS production mediated by iron accumulation [33–35] and accompanied by augmentation lipid peroxidation and glutathione depletion. Results obtained in the present study, at the highest concentration of the extract, may suggest a ferroptotic cell death.

The antiproliferative effect of extract was also demonstrated by the analysis of the cytometric flow, in fact the treatment induced cell cycle arrest at G0/G1 and S phases and a concomitant decrease in the G2/M phase.

Results obtained in the present research demonstrated that methanolic extract of *B. etnensis* Raf. by inducing ROS production and decreasing antioxidant cellular defense, elicited an imbalance in intracellular redox status. This effect, in turn, leads to a higher lipid peroxidation and a drastic rise in HO-1 activity, letting us to hypothesize the induction of ferroptosis, a non-programmed cell death.

4. Materials and Methods

4.1. Chemicals

All solvents, chemicals and reference compounds were purchased from Sigma-Aldrich (Milano, Italy) except as otherwise specified.

4.2. Plant Collection and Preparation of Extract

B. etnensis Raf. bark was collected in the area around Linguaglossa (Catania, Italy) in November 2018. The specimen was obtained and authenticated by botanist R. Acquaviva, Department of Drug Science, Section of Biochemistry, University of Catania, Italy. A voucher specimen of the plant (No. 36/03) was deposited in the herbarium of the same Department. Dried bark from *B. etnensis* Raf. (50 g) was extracted at 80 °C in 70% methanol for 4 h. The extract was then filtered and evaporated to dryness under reduced pressure with a rotatory evaporator. The values of total phenolic and flavonoid content were 45 μM $\pm$ 0.05 gallic acid and 3.80 μM $\pm$ 0.09 catechin.

4.3. Cell Culture and Treatments

CaCo2 colon rectal cancer cells were obtained from ATCC cell bank (Rockville, MD, USA), were propagated in DMEM (Gibco BRL, Life Technologies, city, country) supplemented with 10% fetal calf serum, 1% sodium pyruvate, 1% L-glutamine solution, and 1% streptomycin/penicillin.

After 24 h of incubation in a humidified atmosphere of 5% CO_2 at 37 °C to allow cell attachment, the cells were treated for 72 h with different concentrations (5, 50, 100, 250 or 500 μg/mL) of the methanolic extract, previously dissolved in the minimum amount of dimethyl sulfoxide (DMSO) and diluted with medium.

4.4. MTT Bioassay

MTT assay was performed to assess cell viability on a 96 multiwell plate (8×10^3 cells/well). This assay measures the conversion of tetrazolium salt to yield colored formazan in the presence of metabolic activity. The amount of formazan is proportional to the number of living cells [36]. The optical density was measured with a microplate spectrophotometer reader (Titertek Multiskan, Flow Laboratories, Helsinki, Finland) at $\lambda = 570$ nm. Results are expressed as percentage cell viability respect to control (untreated cells).

4.5. Lactic Dehydrogenase Release

LDH release was measured to evaluate cell necrosis as a result of cell membrane disruption. LDH activity was measured spectrophotometrically in the culture medium and in the cellular lysates, at $\lambda = 340$ nm by analyzing NADH reduction [5]. The percentage of LDH release was calculated from the total sum of the enzymatic activity present in the cellular lysate and of that in the culture medium. Results are expressed as percentage in LDH released.

4.6. Annexin V and Dead Cell Evaluation

The percentage of cells undergoing apoptosis after treatment with *B. etnensis* Raf. was evaluated by Muse™ Annexin V & Dead cell kit (Catalog No. MCH100105, Millipore, Milan, Italy) according to the manufacture's protocol. Briefly for cells staining, 100 μL of the Muse™ Annexin V & Dead Cell Reagent was added to 100 μL of cell suspension, mixed thoroughly by vortexing and incubated for 20 min at room temperature in the dark. Then samples were analyzed by Muse™ Cell Analyzer (Millipore).

4.7. Reactive Oxygen Species Assay

Dichlorofluorescein diacetate (DCFH-DA) assay was performed to quantify ROS levels as previously described [37]. The fluorescence intensity was detected by fluorescence spectrophotometry (excitation, $\lambda = 488$ nm; emission, $\lambda = 525$ nm). Results are expressed as fluorescence intensity/mg protein and, for each sample the total protein content was determined using the Sinergy HTBiotech instrument by measuring the absorbance difference at $\lambda = 280$ and $\lambda = 260$. Protein content was determined using the Sinergy HTBiotech instrument by measuring the absorbance difference at $\lambda = 280$ and $\lambda = 260$.

4.8. Determination of Lipid Hydroperoxide Levels

LOOH levels were evaluated by oxidation of Fe^{+2} to Fe^{+3} that in the presence of xylenol orange forms a Fe^{3+}-xylenol orange complex which can be measured spectrophotometrically at $\lambda = 560$ nm [38]. Results are expressed as percentage increase respect to control (untreated cells).

4.9. Total Thiol Group Determination

RSH groups were measured by using a spectrophotometric assay as previously described [39]. Results are expressed in nmol/mg protein.

4.10. Western Blotting

CaCo2 cells were harvested using cell lysis buffer; cell lysates were collected for Western blot analysis and the level of γ-GCS expression was visualized by immunoblotting with primary antibody against γ-GCS as previously described [5]. The anti- γ-GCS was used in 1:1000 dilution in TBST solution containing 5% not fat milk and detected with secondary antibody conjugated to horseradish peroxidase and Pierce ECL Plus substrate solution (Thermo Scientific, Rockford, IL, USA). Beta-actin, was used as a loading control to normalize the expression level of γ-GCS protein. The result was expressed in arbitrary densitometric units (ADU).

4.11. HO-1 Protein Expression

HO-1 protein concentration in cellular lysates was measured using a commercial HO-1 ELISA kit, according to the manufacturers' instructions. A standard curve generated with purified HO-1 was used to calculate HO-1 concentration from the absorbance at $\lambda = 450$ nm detected for each sample [40]. Detection limits were 0.78–25 ng/mL as reported by the manufacturer. Results are expressed as ng/mg protein.

4.12. Cell Cycle Analysis

Distribution of cell cycle was evaluated by flow cytometry using Muse™ Cell Cycle Kit (Catalog No. MCH100106, Millipore, Milan, Italy) according to the manufacture's guidelines. Briefly, CaCo2 cells were treated with different concentration of *B. etnensis* Raf. (5, 50, 250 µg/mL) for 72 h. After incubation, the cells were harvest, washed with PBS and fixed with ice-cold 70% ethanol at −20 °C for 24 h. Then 200 µL of cell suspension (1×106 cells) was washed with PBS, stained with 200 µL of Muse™ Cell Cycle Reagent for 30 min in the dark at 37 °C and subsequently analyzed by Muse™ Cell Analyzer (Millipore).

4.13. Statistical Analysis

One-way analysis of variance (ANOVA) followed by Bonferroni's *t*-test was performed in order to estimate significant differences among groups. Data were reported as mean values ± SD and differences among groups were considered to be significant at $p < 0.005$.

5. Conclusions

These findings confirm the growing body of evidence on the bioactivities of *B. etnensis* Raf. and its potential impact on cancer therapy and human health. Our data demonstrate that exposure of CaCo2 cancer cells to *B. etnensis* Raf. extract decreased cell proliferation and induced necrotic death, suggesting that this extract may represent a subject for further studies on drug discovery, also helpful for cancer integrative medicine.

Moreover, our results show a prospective application of *B. etnensis* Raf. in the HO-1 mediated ferroptosis as a chemotherapeutic strategy against tumor. Further research should be carried out to better clarify the involvement of HO-1 in the mechanism of action of this promising anticancer medicinal plant.

Author Contributions: Conceptualization, writing original draft, R.A., G.A.M., B.T., and C.D.G.; Collect plant, authentication, preparation extract, R.A., G.A.M.; Methodology and investigation, B.T., G.A.M., A.L.M., F.P.C., and C.G.; Project administration, R.A. and G.A.M.; Formal analysis, M.R.; Writing–review & editing, C.D.G. and M.R.; Funding acquisition, R.A.; All authors read and approved the final manuscript.

Funding: Work was supported by University of Catania and Association Mani Amiche onlus.

Acknowledgments: The authors would like to thank M. Wilkinson (Research Assistant) for proofreading the manuscript and Associazione Mani Amiche Onlus.

Conflicts of Interest: The authors declare no conflict of interest.

Abbreviations

CRC	colorectal cancer
DCFH-DA	2′,7′-dichlorofluorescein diacetate
DMSO	dimethyl sulfoxide
HO-1	Heme oxygenase-1
LDH	Lactic dehydrogenase
ROS	Reactive oxygen species
LOOH	Lipid hydroperoxide
RSH	Thiol groups
γ-GCS	γ-glutamylcysteine synthetase

References

1. Thanikachalam, K.; Khan, G. Colorectal Cancer and Nutrition. *Nutrients* **2019**, *11*, 164. [CrossRef] [PubMed]
2. Haggar, F.A.; Boushey, R.P. Colorectal cancer epidemiology: Incidence, mortality, survival, and risk factors. *Clin. Colon Rectal. Surg.* **2009**, *22*, 191–197. [CrossRef] [PubMed]
3. Chao, A.; Thun, M.J.; Connell, C.J.; McCullough, M.L.; Jacobs, E.J.; Flanders, W.D.; Rodriguez, C.; Sinha, R.; Calle, E.E. Meat consumption and risk of colorectal cancer. *JAMA* **2005**, *293*, 172–182. [CrossRef] [PubMed]

4. Bultman, S.J. Interplay between diet, gut microbiota, epigenetic events, and colorectal cancer. *Mol. Nutr. Food Res.* **2017**, *61*. [CrossRef] [PubMed]

5. Acquaviva, R.; Sorrenti, V.; Santangelo, R.; Cardile, V.; Tomasello, B.; Malfa, G.; Vanella, L.; Amodeo, A.; Mastrojeni, S.; Pugliese, M.; et al. Effects of extract of *Celtis aetnensis* (Tornab.) Strobl twigs in human colon cancer cell cultures. *Oncol. Rep.* **2016**, *36*, 2298–2304. [CrossRef] [PubMed]

6. Weisburger, J.H. Nutritional approach to cancer prevention with emphasis on vitamins, antioxidants, and carotenoids. *Am. J. Clin. Nutr.* **1991**, *53*, S226–S237. [CrossRef] [PubMed]

7. Nakamura, Y.; Torikai, K.; Ohto, Y.; Murakami, A.; Tanaka, T.; Ohigashi, H. A simple phenolic antioxidant protocatechuic acid enhances tumor promotion and oxidative stress in female ICR mouse skin: Dose-and timing-dependent enhancement and involvement of bioactivation by tyrosinase. *Carcinogenesis* **2000**, *21*, 1899–1907. [CrossRef]

8. Chiang, S.K.; Chen, S.E.; Chang, L.C. A dual role of Heme Oxygenase-1 in cancer cells. *Int. J. Mol. Sci.* **2018**, *20*, 39. [CrossRef]

9. Acquaviva, R.; Menichini, F.; Ragusa, S.; Genovese, C.; Amodeo, A.; Tundis, R.; Loizzo, M.R.; Iauk, L. Antimicrobial and antioxidant properties of *Betula aetnensis* Rafin. (Betulaceae) leaves extract. *Nat. Prod. Res* **2013**, *27*, 475–479. [CrossRef]

10. Rastogi, S.; Pandey, M.M.; Kumar Singh Rawat, A. Medicinal plants of the genus Betula-traditional uses and a phytochemical-pharmacological review. *J. Ethnopharmacol.* **2015**, *159*, 62–83. [CrossRef]

11. Rapp, O.A.; Pashinskii, V.G.; Chuchalin, V.S. Comparative study of pharmacological activity of birch bark ethanol extracts. *Pharm. Chem. J.* **1999**, *33*, 1–3. [CrossRef]

12. Coates, E.M.; Popa, G.; Gill, C.I.; McCann, M.J.; McDougall, G.J.; Stewart, D.; Rowland, I. Colon-available raspberry polyphenols exhibit anti-cancer effects on in vitro models of colon cancer. *J. Carcinog.* **2007**, *6*, 4. [CrossRef]

13. Mshvildadze, V.; Legault, J.; Lavoie, S.; Gauthier, C.; Pichette, A. Anticancer diarylheptanoid glycosides from the inner bark of Betula papyrifera. *Phytochemistry* **2007**, *68*, 2531–2536. [CrossRef] [PubMed]

14. Pignatti, S. Flora d'Italia, EFIZIO, Edagricole, New Business Media Italia, 2017, Vol. II, pp. 676–677. Available online: https://books.google.it/books?id=AqFexQEACAAJ&hl=it&source=gbs_book_other_versions (accessed on 31 May 2019).

15. Johnson, I. New approaches to the role of diet in the prevention of cancers of the alimentary tract. *Mutat. Res.* **2004**, *551*, 9–28. [CrossRef] [PubMed]

16. Bruce, W.R.; Giacca, A.; Medline, A. Possible mechanisms relating diet and risk of colon cancer. *Cancer Epidemiol. Biomark. Prev.* **2000**, *9*, 1271–1279.

17. Shibata, A.; Paganini-Hill, A.; Ross, R.K.; Henderson, B.E. Intake of vegetables, fruits, beta-carotene, vitamin C and vitamin supplements and cancer incidence among the elderly: A prospective study. *Br. J. Cancer* **1992**, *66*, 673–679. [CrossRef] [PubMed]

18. Foo, J.B.; Saiful Yazan, L.; Tor, Y.S.; Wibowo, A.; Ismail, N.; How, C.W.; Armania, N.; Loh, S.P.; Ismail, I.S.; Cheah, Y.K.; et al. Induction of cell cycle arrest and apoptosis by betulinic acid-rich fraction from Dillenia suffruticosa root in MCF-7 cells involved p53/p21 and mitochondrial signalling pathway. *J. Ethnopharmacol.* **2015**, *166*, 270–278. [CrossRef]

19. Brown, N.S.; Bicknell, R. Hypoxia and oxidative stress in breast cancer. Oxidative stress: Its effects on the growth, metastatic potential and response to therapy of breast cancer. *Breast Cancer Res.* **2001**, *3*, 323–327. [CrossRef]

20. Bae, Y.S.; Kang, S.W.; Seo, M.S.; Baines, I.C.; Tekle, E.; Chock, P.B.; Rhee, S.G. Epidermal growth factor (EGF)-induced generation of hydrogen peroxide. Role in EGF receptor-mediated tyrosine phosphorylation. *J. Biol. Chem.* **1997**, *272*, 217–221. [CrossRef]

21. Hardwick, J.S.; Sefton, B.M. The activated form of the lck tyrosine protein kinase in cells exposed to hydrogen peroxide is phosphorylated at both Tyr-394 and Tyr-505. *J. Biol. Chem.* **1997**, *272*, 25429–25432. [CrossRef]

22. Sundersan, M.; Yu, Z.X.; Ferrans, V.J.; Irani, K.; Finkel, T. Requirement for generation of H2O2 for platelet-derived growth factor signal transduction. *Science* **1995**, *270*, 296–299. [CrossRef]

23. Devary, Y.; Gottlieb, R.A.; Smeal, T.; Karin, M. The mammalian ultraviolet response is triggered by activation of Src tyrosine kinases. *Cell* **1992**, *71*, 1081–1091. [CrossRef]

24. Lanceta, L.; Li, C.; Choi, A.M.; Eaton, J.W. Heme oxygenase-1 overexpression alters intracellular iron distribution. *Biochem. J.* **2013**, *449*, 189–194. [CrossRef]

25. Jozkowicz, A.; Was, H.; Dulak, J. Heme Oxygenase-1 in tumors: Is it a false friend? *Antioxid. Redox Signal.* **2007**, *9*, 2099–2118. [CrossRef] [PubMed]
26. Di Giacomo, C.; Acquaviva, R.; Piva, A.; Sorrenti, V.; Vanella, L.; Piva, G.; Casadei, G.; La Fauci, L.; Ritieni, A.; Bognanno, M.; et al. Protective effect of cyanidin 3-O-beta-D-glucoside on ochratoxin A-mediated damage in the rat. *Br. J. Nutr.* **2007**, *98*, 937–943. [CrossRef]
27. Nakasa, K.; Kitayama, M.; Fukuda, H.; Kimura, K.; Yanagawa, T.; Ishii, T.; Nakashima, K.; Yamada, K. Oxidative stress-related proteins A170 and heme oxygense-1 are differently induced in the rat cerebellum under kainate-mediated excitotoxicity. *Neurosci. Lett.* **2000**, *282*, 57–60. [CrossRef]
28. Busserolles, J.; Megias, J.; Terencio, M.C.; Alcaraz, M.J. Heme oxygenase-1 inhibits apoptosis in Caco-2 cells via activation of Akt pathway. *Int. J. Biochem. Cell. Biol.* **2006**, *38*, 1510–1517. [CrossRef]
29. Ryter, S.W.; Alam, J.; Choi, A.M. Heme oxygenase-1/carbon monoxide: From basic science to therapeutic applications. *Physiol. Rev.* **2006**, *86*, 583–650. [CrossRef] [PubMed]
30. Loboda, A.; Jozkowicz, A.; Dulak, J. HO-1/CO system in tumor growth, angiogenesis and metabolism-Targeting HO-1 as an anti-tumor therapy. *Vascul. Pharmacol.* **2015**, *74*, 11–22. [CrossRef]
31. Loboda, A.; Damulewicz, M.; Pyza, E.; Jozkowicz, A.; Dulak, J. Role of Nrf2/HO-1 system in development, oxidative stress response and diseases: An evolutionarily conserved mechanism. *Cell. Mol. Life Sci.* **2016**, *73*, 3221–3247. [CrossRef]
32. Nitti, M.; Piras, S.; Marinari, U.M.; Moretta, L.; Pronzato, M.A.; Furfaro, A.L. HO-1 induction in cancer progression: A matter of cell adaptation. *Antioxidants* **2017**, *6*, 29. [CrossRef] [PubMed]
33. Kwon, M.Y.; Park, E.; Lee, S.J.; Chung, S.W. Heme oxygenase-1 accelerates Erastin induced ferroptotic cell death. *Oncotarget* **2018**, *416*, 124–137. [CrossRef] [PubMed]
34. Chang, L.C.; Chiang, S.K.; Chen, S.E.; Yu, Y.L.; Chou, R.H.; Chang, W.C. Heme oxygenase-1 mediates BAY 11-7085 induced ferroptosis. *Cancer Lett.* **2018**, *416*, 124–137. [CrossRef] [PubMed]
35. Hassannia, B.; Wiernicki, B.; Ingold, I.; Qu, F.; Van Herck, S.; Tyurina, Y.Y.; Bayır, H.; Abhari, B.A.; Angeli, J.P.F.; Choi, S.M.; et al. Nano-targeted induction of dual ferroptotic mechanisms eradicates high-risk neuroblastoma. *J. Clin. Investig.* **2018**, *128*, 3341–3355. [CrossRef]
36. Chillemi, R.; Cardullo, N.; Greco, V.; Malfa, G.; Tomasello, B.; Sciuto, S. Synthesis of amphiphilic resveratrol lipoconjugates and evaluation of their anticancer activity towards neuroblastoma SH-SY5Y cell line. *Eur. J. Med. Chem.* **2015**, *96*, 467–481. [CrossRef]
37. Malfa, G.A.; Tomasello, B.; Sinatra, F.; Villaggio, G.; Amenta, F.; Avola, R.; Renis, M. Reactive response evaluation of primary human astrocytes after methylmercury exposure. *J. Neurosci. Res.* **2014**, *92*, 95–103. [CrossRef]
38. Acquaviva, R.; Di Giacomo, C.; Vanella, L.; Santangelo, R.; Sorrenti, V.; Barbagallo, I.; Genovese, C.; Mastrojeni, S.; Ragusa, S.; Iauk, L. Antioxidant activity of extracts of Momordica foetida Schumach. et Thonn. *Molecules* **2013**, *18*, 3241–3249. [CrossRef]
39. Acquaviva, R.; Di Giacomo, C.; Sorrenti, V.; Galvano, F.; Santangelo, R.; Cardile, V.; Gangia, S.; D'Orazio, N.; Abraham, N.G.; Vanella, L. Antiproliferative effect of oleuropein in prostate cell lines. *Int. J. Oncol.* **2012**, *41*, 31–38.
40. Li Volti, G.; Galvano, F.; Frigiola, A.; Guccione, S.; Di Giacomo, C.; Forte, S.; Tringali, G.; Caruso, M.; Adekoya, O.A.; Gazzolo, D. Potential immunoregulatory role of heme oxygenase-1 in human milk: A combined biochemical and molecular modeling approach. *J. Nutr. Biochem.* **2010**, *21*, 865–871. [CrossRef]

International Journal of
Molecular Sciences

MDPI

Article

Beneficial Role of HO-1-SIRT1 Axis in Attenuating Angiotensin II-Induced Adipocyte Dysfunction

Hari Vishal Lakhani [1], Mishghan Zehra [1], Sneha S. Pillai [1], Nitin Puri [1], Joseph I. Shapiro [1], Nader G. Abraham [2] and Komal Sodhi [1,*]

[1] Department of Surgery, Internal Medicine, and Biomedical Sciences, Marshall University Joan C. Edwards School of Medicine, Huntington, WV 25701, USA
[2] Department of Pharmacology, New York Medical College, Valhalla, NY 10595, USA
* Correspondence: sodhi@marshall.edu; Tel.: +1-304 691-1704; Fax: +1-0914 347-4956

Received: 29 April 2019; Accepted: 27 June 2019; Published: 29 June 2019

Abstract: Background: Angiotensin II (Ang II), released by the renin–angiotensin–aldosterone system (RAAS), contributes to the modulatory role of the RAAS in adipose tissue dysfunction. Investigators have shown that inhibition of AngII improved adipose tissue function and insulin resistance in mice with metabolic syndrome. Heme Oxygenase-1 (HO-1), a potent antioxidant, has been demonstrated to improve oxidative stress and adipocyte phenotype. Molecular effects of high oxidative stress include suppression of sirtuin-1 (SIRT1), which is amenable to redox manipulations. The mechanisms involved, however, in these metabolic effects of the RAAS remain incompletely understood. Hypothesis: We hypothesize that AngII-induced oxidative stress has the potential to suppress adipocyte SIRT1 via down regulation of HO-1. This effect of AngII will, in turn, upregulate mineralocorticoid receptor (MR). The induction of HO-1 will rescue SIRT1, hence improving oxidative stress and adipocyte phenotype. Methods and Results: We examined the effect of AngII on lipid accumulation, oxidative stress, and inflammatory cytokines in mouse pre-adipocytes in the presence and absence of cobalt protoporphyrin (CoPP), HO-1 inducer, tin mesoporphyrin (SnMP), and HO-1 inhibitor. Our results show that treatment of mouse pre-adipocytes with AngII increased lipid accumulation, superoxide levels, inflammatory cytokine levels, interleukin-6 (IL-6) and tumor necrosis factor α (TNFα), and adiponectin levels. This effect was attenuated by HO-1 induction, which was further reversed by SnMP, suggesting HO-1 mediated improvement in adipocyte phenotype. AngII-treated pre-adipocytes also showed upregulated levels of MR and suppressed SIRT1 that was rescued by HO-1. Subsequent treatment with CoPP and SIRT1 siRNA in mouse pre-adipocytes increased lipid accumulation and fatty acid synthase (FAS) levels, suggesting that beneficial effects of HO-1 are mediated via SIRT1. Conclusion: Our study demonstrates for the first time that HO-1 has the ability to restore cellular redox, rescue SIRT1, and prevent AngII-induced impaired effects on adipocytes and the systemic metabolic profile.

Keywords: angiotensin II; mineralocorticoid receptor; heme oxygenase 1; sirtuin 1; adipocytes; oxidative stress

1. Introduction

Classically, the RAAS has been described as a multi-organ endocrine axis that is governed by a negative feedback loop where elevated circulating AngII inhibits renal renin release. A large body of evidence, today, supports the notion that apart from regulating cardiac, vascular, and renal functions, the RAAS operates in nervous, immunological, and reproductive tissues [1]. Visceral adipose tissue (VAT), classified as an endocrine organ, is the latest to be shown to express and to be regulated by the RAAS [2]. VAT expresses all components of the RAAS, including angiotensinogen [3,4], renin [5], and aldosterone synthase (CYP11B2) [6]. Excess VAT underlies obesity, whose global burden is reflected in sheer numbers; 39% of the adult population is overweight, while 13% is obese [7]. Evidence suggests that RAAS blockade reduces obesity-related cardiovascular and renal complications, alleviates insulin resistance, and facilitates adipocyte differentiation [8–11]. AngII has been shown to reduce proliferation and increase lipid accumulation in mouse pre-adipocytes in culture. Recent studies have also shown that increased AngII in animals with Goldblatt's hypertension is associated with increased visceral adiposity and reduced plasma adiponectin levels [9]. Aldosterone synthase (CYP11B2) and MR [12], which are critical RAAS components, also affect adipocyte structure and function [13]. MR blockade reduces adiposity-related cardiovascular and hepatic complications and has been shown to facilitate adipocyte differentiation [8–11]. MR-dependent adipogenesis has been demonstrated in mouse pre-adipocytes in response to aldosterone [14]. A direct effect of the RAAS on VAT/adipocyte structure and function, however, remains unresolved and the mechanisms involved in the RAAS function are not clearly understood.

Increased VAT mass promotes inflammatory infiltration with compromised secretion of protective adipokines, like adiponectin. Thus, VAT hypertrophy launches a complex patho-physiological maladaptive response characterized by a systemic spillover of inflammatory and oxidative mediators, reduced secretion of protective adipokines, and increased circulating levels of glucose. A cumulative line of evidence suggests that redox imbalance and chronic oxidative stress increase adipogenesis and precipitate adipocyte dysfunction [15]. The RAAS has been reported to increase lipid accumulation [16], an effect attributed to RAAS-induced redox imbalance. Molecular effects of high oxidative stress include suppression of SIRT1, which is amenable to redox manipulations [17]. Studies have shown that SIRT1 regulates adipogenesis in murine adipocytes. In the model of oxidative stress, SIRT1 over-expression and/or an exogenous antioxidant prevents redox-induced adipocyte hypertrophy and dysfunction [18]. To this end, studies from our lab have documented that upregulation of the endogenous heme–heme oxygenase system (HO) reduces adipogenesis in murine pre-adipocytes and human mesenchymal stem cells [19–21]. Previous studies have documented the antioxidant properties of the HO system and HO-1 induction via CoPP reduces visceral adiposity and ablates metabolic imbalance in obese and diabetic mice [22–26].

Based on these observations, we propose to explore the mechanistic link between VAT-specific overactive RAAS and eventual development of adipocyte and metabolic dysfunction via increase in cellular oxidative stress. We hypothesize that AngII-induced oxidative stress has the potential to downregulate adipocyte SIRT1 via suppression of HO-1. We further hypothesize that HO-1 induction will provide an antioxidant setting, thus rescuing cellular SIRT1 and preventing adipocyte dysfunction. This study, for the first time, will demonstrate a negative regulatory effect of the HO-1-SIRT1 axis, which will improve the effects mediated by AngII on the adipocyte phenotype. Thus, the overall objective of this proposal is to uncover the molecular interplay between AngII, HO-1, and SIRT1 as it pertains to the regulation of adipocyte structure and function, eventually affecting the systemic metabolic homeostasis.

2. Results

2.1. Effects of HO-1 on AngII-Induced Alteration on Adipocyte Phenotype

The dose-dependent effect of AngII demonstrated that 10 uM was the optimal concentration in increasing lipogenesis in 3T3-L1 cells, as previously published [9]. Our RT-PCR analysis showed significantly reduced expression of HO-1 in response to AngII treatment, as compared to the control group treated with adipogenic media alone (Figure 1A). As expected, treatment with CoPP induced increased expression of HO-1, as compared to AngII-treated murine adipocytes. Interestingly, the expression of HO-1 was increased by treatment with SnMP alone and CoPP and SnMP together. However, these findings are not surprising, as SnMP, which induced a significant increase in HO-1 expression, remains a potent inhibitor of HO activity, as shown previously [27–29]. Lipid accumulation, measured as the relative absorbance of Oil Red O staining in murine adipocytes, demonstrated that AngII-induced increased lipogenesis, as compared to the control group treated with adipogenic media alone and 3T3-L1 cells treated with SnMP alone (HO-inhibitor) (Figure 1B). SnMP treatment alone induced significant lipid accumulation as compared to the control treated with adipogenic media alone. This effect, induced by SnMP or AngII treatment alone, was significantly alleviated by the induction of the antioxidant HO-1 system using CoPP, an effect reversed by the treatment with SnMP (HO-inhibitor) (Figure 1B). Apart from that, mRNA expression of fatty acid synthase (FAS), a marker of lipid accumulation, was significantly upregulated in SnMP-treated cells, which was even further increased by AngII treatment, as compared to the control (Figure 1C). This increase was attenuated by CoPP, which was reversed by concomitant treatment with SnMP (Figure 1C). Furthermore, our results demonstrated that treatment with AngII significantly increased triglyceride levels as compared to the control, which was improved by HO-1 induction (Figure 1D). Concurrent exposure to the HO-1 inhibitor (SnMP) reversed the effect, observed by CoPP causing an increase in triglycerides levels, similar to AngII-induced triglyceride increase. We next evaluated the mRNA expression of the marker of mitochondrial biogenesis, PGC-1α, which showed significant downregulation by the treatment with SnMP alone and by AngII treatment, as compared to control with adipogenic media alone (Figure 1E). The expression was improved by the induction of CoPP, which was further reversed when treated concurrently with SnMP (Figure 1E). Apart from that, our results demonstrated a significant increase in the mRNA expression of inflammatory marker IL-6 by treatment with SnMP alone, which was further exacerbated by AngII treatment as compared to the control (Figure 1F). This increase was attenuated by CoPP, which was reversed by concomitant treatment with SnMP (Figure 1F).

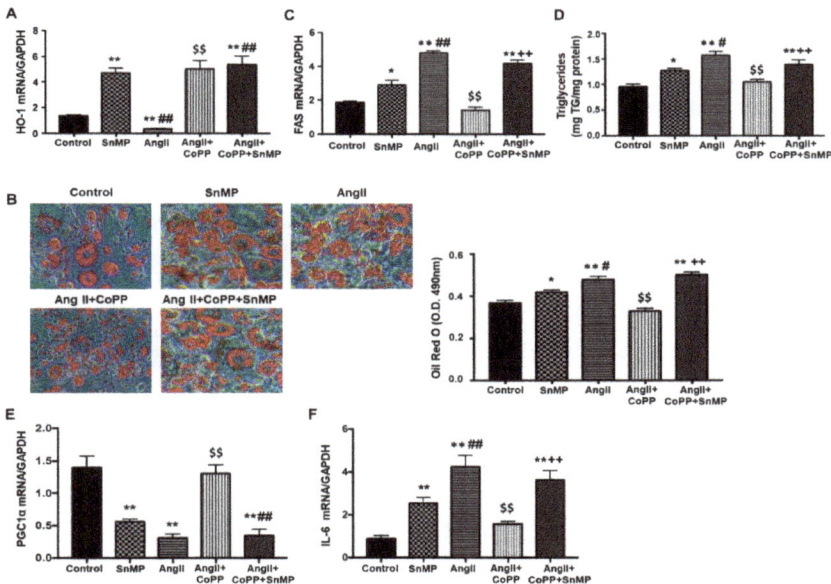

Figure 1. Effect of Angiotensin II (AngII) exposed to 3T3-L1 murine pre-adipocytes with or without Heme Oxygenase I (HO-1) induction. (**A**) RT-PCR analysis for the relative mRNA expression of HO-1; (**B**) representative images and quantitative data of lipid accumulation measured as the relative absorbance of Oil Red O staining. Images taken with 40× objective lens; (**C**) relative mRNA expression for marker of lipid accumulation, fatty acid synthase (FAS); (**D**) triglyceride levels measured by ELISA assay; RT-PCR analysis for the mRNA expression of (**E**) marker of mitochondrial biogenesis, PGC-1α, and (**F**) inflammatory marker, IL-6. Values represent means ± SEM. * $p < 0.05$ vs. control (CTR), ** $p < 0.01$ vs. CTR, # $p < 0.05$ vs. tin mesoporphyrin (SnMP), ## $p < 0.01$ vs. SnMP, $$ $p < 0.01$ vs. AngII, + $p < 0.05$ vs. AngII + cobalt protoporphyrin (CoPP), ++ $p < 0.01$ vs. AngII + CoPP ($n = 6$).

2.2. Effect of AngII on Mechanistic Interplay between HO-1/SIRT1 Axis in Mouse Adipocyte with or without HO-1 Induction

Our next set of experiments examined the AngII-induced molecular disruptions involved in causing altered adipocyte phenotype. Our Western blot analysis showed that treatment with CoPP induced increased expression of HO-1, as compared to AngII-treated murine adipocytes (Figure 2A). Interestingly, the protein levels of HO-1 were also increased by the treatment with CoPP that also received the SnMP (Figure 2A). However, these findings are not surprising, as SnMP, which induced a significant increase in HO-1 expression, remains a potent inhibitor of HO activity, as shown previously [27–29]. We next examined the effect of AngII treatment on the expression of SIRT1 in 3T3-L1 cells. Our Western blot analysis demonstrated that the treatment with AngII induced significantly reduced expression of SIRT1 as compared to controls, which was rescued by the induction of HO-1 (Figure 2B). However, the improved expression of SIRT1 was consequently decreased by the treatment with SnMP. Aldosterone synthase (CYP11B2), a critical RAAS component causing upregulation of MR, also affected adipocyte phenotype [30]. Our results further demonstrated that treatment with AngII stimulated the expression of CYP11B2, an effect negated by treatment with CoPP (Figure 2C). The expression of CYP11B2 was further increased by treatment of murine adipocytes with CoPP that were also exposed to SnMP. Apart from that, our results showed increased expression of MR, induced by the treatment with AngII, as compared to the control (Figure 2D). This increase was significantly negated by treatment with CoPP, which was again reversed by subsequent treatment with SnMP. Our Western blot analysis further demonstrated that AngII also significantly reduced insulin receptor-β (IR-β) expression, which was significantly improved by treatment with CoPP (Figure 2E). IR-β expression was suppressed significantly by treatment with CoPP that also received the SnMP. In concordance with these findings, we also performed RT-PCR analyses for the mRNA expression of SIRT1, CYP11B2, and MR in our murine pre-adipocytes. The findings were similar to our results from our Western blot analysis, with an addition of the SnMP-alone-treated experimental group, which showed significant upregulation in the expression of CYP11B2 and MR and also a significant reduction in SIRT1 expression, as compared to the control (Figure S1A–C). Our results also showed significant upregulation in expression of angiotensin II receptor type 1 (AT1R) by treatment with SnMP alone and more so by AngII treatment, as compared to the control (Figure S1D). The treatment with CoPP demonstrated significantly lower AT1R expression, which was reversed by consequent treatment with SnMP.

2.3. Effect of AngII in CM Obtained from Mouse Adipocytes with or without HO-1 Induction

The incubation of murine adipocytes with dihydroethidium (DHE) to measure superoxide levels (indicator of ROS) significantly showed upregulated levels by SnMP alone and a further significant increase in the AngII-treated group, as compared to the control treated with adipogenic media only (Figure 3A). The treatment with CoPP alone demonstrated significantly lower superoxide levels, which was consequently reversed by treatment with SnMP, indicating that upregulation of HO-1 is required for reduction in this oxidative marker. We next examined the effect of AngII on inflammatory markers in murine adipocytes. Conditioned media (CM) obtained from the treated 3T3-L1 cells flasks, demonstrated that the levels of key inflammatory markers, IL-6 and TNF-α, were significantly increased with AngII treatment, as compared to control and SnMP treatment alone (Figure 3C,D). The levels of these inflammatory markers were attenuated by treatment with CoPP, implying that the attenuation was dependent on HO-1 induction. Conversely, the levels of these inflammatory markers were upregulated in the CM of the cells treated with CoPP and exposed to SnMP. Furthermore, adiponectin levels were reduced significantly in CM in response to treatment with SnMP alone, which was further significantly decreased by AngII, as compared to the control (Figure 3D). However, treatment with CoPP improved adiponectin levels, an effect reversed by subsequent treatment with SnMP.

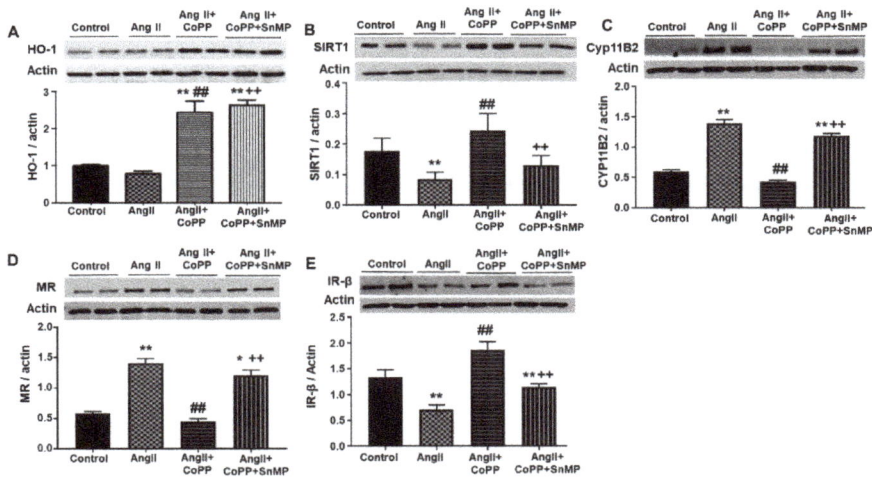

Figure 2. Effect of AngII exposed to 3T3-L1 murine pre-adipocytes by Western blot analysis for protein expression of (**A**) HO-1, (**B**) SIRT1, (**C**) CYP11B2, (**D**) MR, and (**E**) IR-β, shown as mean band densities normalized to β-actin. Values represent means ± SEM. * $p < 0.05$ vs. CTR, ** $p < 0.01$ vs. CTR, ## $p < 0.01$ vs. AngII, ++ $p < 0.01$ vs. AngII + CoPP ($n = 6$).

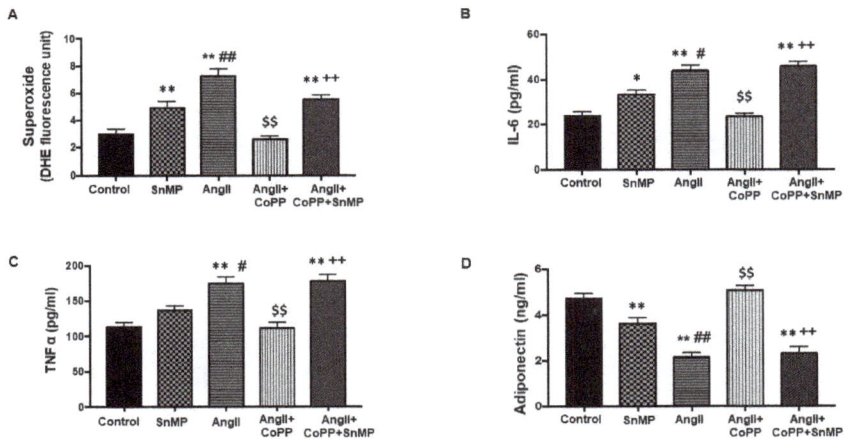

Figure 3. Effect of AngII exposed to 3T3-L1 murine pre-adipocytes on oxidative stress marker, inflammatory marker, and adiponectin levels. (**A**) Superoxide levels assessed with dihydroethidium (DHE) staining; (**B**,**C**) levels of inflammatory markers, IL-6 and TNFα, respectively; (**D**) adiponectin levels. Values represent means ± SEM. ** $p < 0.01$ vs. CTR, # $p < 0.05$ vs. SnMP, ## $p < 0.01$ vs. SnMP, $$ $p < 0.01$ vs. AngII, ++ $p < 0.01$ vs. AngII + CoPP. ($n = 6$).

2.4. Effect of AngII with or without HO-1 Induction and SIRT1 Knockdown on Lipogenesis and FAS Levels in Mouse Adipocytes

To assess whether HO-1 requires the participation of SIRT1 to mediate and/or amplify its actions, we studied the effect of SIRT1 siRNA and SIRT plasmid in 3T3-L1 cells treated with AngII. Our results showed that AngII increased lipid accumulation, measured as the relative absorbance of Oil Red O staining in murine adipocytes, and expression of FAS; this effect of AngII treatment was significantly negated by treatment with CoPP (Figures 4A and 4B, respectively). Interestingly, concurrent treatment with CoPP and SIRT1 siRNA increased lipid accumulation and FAS expression, suggesting that HO-1 is upstream of SIRT1 and that suppression of SIRT1 attenuates the beneficial effects of increased levels of HO-1. We also utilized plasmid SIRT to assess if increased expression of SIRT1 (in the absence of HO-1 upregulation) is sufficient to prevent the detrimental effects of AngII on lipid accumulation. Treatment of murine adipocytes with AngII, SnMP, and SIRT plasmid increased lipid accumulation and FAS levels, as compared to murine adipocytes treated with AngII, CoPP, and SIRT1 plasmid (Figures 4A and 4B, respectively). In concordance with our hypothesis, our results further showed that murine adipocytes treated with AngII, CoPP, and SIRT plasmid did not significantly decrease lipid accumulation and FAS levels as compared to cells treated with AngII and CoPP alone, indicating an HO-1-dependent activation of SIRT1 expression. We further demonstrated the mRNA expression of SIRT1, which was significantly reduced by treatment with AngII as compared to the control, was restored by the induction of CoPP (Figure 4C). Concurrent treatment with CoPP and SIRT1-siRNA decreased SIRT1 expression. The utilization of the SIRT plasmid in the group treated with AngII and CoPP significantly upregulated the expression of SIRT1, the effect of which was reversed by treatment with AngII, SnMP, and SIRT plasmid (Figure 4C).

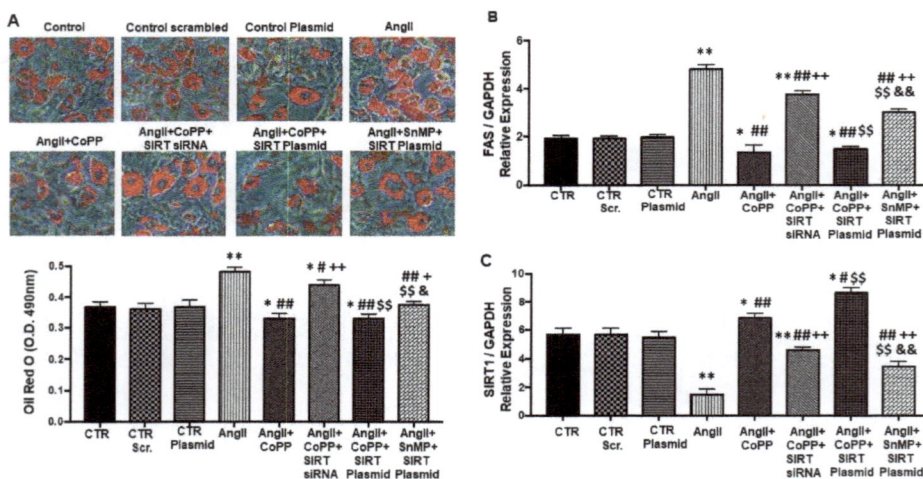

Figure 4. Effect of CoPP with and without SIRT1-siRNA and with and without SIRT plasmid on lipid accumulation and FAS expression in AngII-treated 3T3-L1 murine pre-adipocytes. (**A**) Representative images and quantitative data of lipid accumulation measured as the relative absorbance of Oil Red O staining. Images taken with 40× objective lens; RT-PCR analysis for mRNA expression of (**B**) FAS and (**C**) SIRT1. Values represent means ± SEM. * $p < 0.05$ vs. CTR, ** $p < 0.01$ vs. CTR, # $p < 0.05$ vs. AngII, ## $p < 0.01$ vs. AngII, + $p < 0.05$ vs. AngII + CoPP, ++ $p < 0.01$ vs. AngII+CoPP, \$\$ $p < 0.01$ vs. AngII + CoPP + SIRT siRNA, & $p < 0.05$ vs. AngII + CoPP + SIRT Plasmid, && $p < 0.01$ vs. AngII + CoPP + SIRT Plasmid ($n = 6$).

3. Discussion

This study demonstrates for the first time that AngII-induced increased adipocyte dysfunction is accompanied by suppression of cellular SIRT1; effects are reversed by concurrent exposure to CoPP, an inducer of the endogenous antioxidant HO-1. AngII stimulates lipogenesis, oxidative stress, release of inflammatory cytokines, and reduced adiponectin levels in vitro. Our results show the upregulation of MR and increased expression of aldosterone synthase (CYP11B2) by treatment with AngII, an effect that is attenuated by the HO-1-SIRT1 axis. We demonstrated that the alteration in adipocyte phenotype and molecular changes can be reversed in a setting where adipocyte redox balance and SIRT1 is restored by HO-1 upregulation. Hence, our results help establish "the proof-of-principle" of our hypothesis, confirming the beneficial role of the HO-1-SIRT1 axis in adipocytes, further providing a basis that HO-1 can improve adipocyte function by attenuating the activation of AngII expression in adipocytes.

Oxidative stress is known to be highly involved in the development and progression of clinical conditions like metabolic syndrome. Redox imbalance and chronic oxidative stress increase adipogenesis and cause adipocyte dysfunction [15,31]. We have shown that heme, a pro-oxidant molecule, increases lipid accumulation, promotes cell enlargement, induces over-expression of peroxisome proliferator-activated receptor gamma (PPARγ), CCAAT enhancer binding protein alpha (C/EBP-α), and adipocyte protein 2 (aP2), and downregulates adiponectin in mouse pre-adipocytes [18,31]. These effects of heme were prevented in cells treated with an antioxidant called tempol, a superoxide dismutase-mimetic. Additionally, in vivo administration of antioxidant reduces visceral adiposity, improves metabolic balance, and restores adipocyte function, as indicated by the recovery of adiponectin levels. Molecular effects of high oxidative stress include suppression of SIRT1, which is amenable to redox manipulations [17]. Our study shows that the upregulation of the antioxidant system by HO-1 induction attenuates AngII-mediated oxidative stress, as shown by the measurement of superoxide levels in our murine pre-adipocytes.

This study highlights the important role of upregulated MR expression in bringing about adipocyte dysfunction. MR is a member of superfamily of nuclear hormone receptors and belongs to the class of ligand-activated transcription factors [32]. Aldosterone and glucocorticoids have similarly high affinity for MR, with Kd values between 0.5 and 3 nM [33]. Inadvertent MR activation in epithelial tissues by circulating glucocorticoids (100- to 1000-fold higher than those of aldosterone) is prevented by the intracellular enzyme 11β-hydroxysteroid dehydrogenase, type 2 (HSD2). HSD2 catalyzes the inactivation of cortisol to cortisone; the latter has negligble affinity for the MR, thus protecting the MR from activation by excess levels of cortisol. Adipose tissues, however, in spite of high MR expression, have minimal HSD2 levels [34]. This sets a stage for aldosterone-mediated increased MR expression in the adipocytes. Congruent studies by other investigators have shown a central role of MR in the lipogenic process. These studies have demonstrated that in an altered pathophysiological condition, increased levels of circulating cortisol diffuse into the adipocytes and bind to MR, further activating it. The activation of ligand-bound MR, which is translocated in the nucleus, stimulates gene transcription, causing an increase in the downstream effectors of MR such as prostaglandin D2 synthase (PTGDS) and genes associated with adipocyte differentiation such as PPARγ and aP2 [35]. In line with these reports, we show that treatment of murine pre-adipocytes with AngII upregulates the expression of CYP11B2, subsequently causing increased expression of MR. Our results also show significant amelioration of these AngII-mediated effects by the induction of the HO-1 antioxidant system through the rescue of SIRT1, which establishes the mechanistic basis of our study.

Evidence has linked upregulated HO-1 expression results in increased insulin sensitivity and improvement in phosphorylation of insulin receptors and adipocyte function [32–34,36]. Additionally, induction of HO-1 in adipocyte cell cultures is associated with decreased pro-inflammatory cytokines TNFα and IL-6 [17]. Consistent with these findings, we can see that induction of HO-1 improved insulin sensitivity and decreased levels of TNFα and IL-6 in our murine adipocytes [37]. In concordance with our findings, previously published studies have shown the protective role of HO-1 in inhibiting the inflammatory effect of several mediators. Inflammatory and oxidative transcription factors like nuclear factor kappa-light-chain-enhancer of activated B cells (NF-κβ) and c-Jun N-terminal kinase (JNK) signaling mechanisms, which are primarily involved in inflammatory insult, stimulate inflammatory pathways like IL-signaling, creating a feedback mechanism of inflammation and further compromising pathophysiological conditions [32]. HO-1 have been demonstrated to be effective in suppressing these inflammatory mediators, hence, ameliorating the production of inflammatory cytokines. There is also evidence from the literature that suggests the presence of the HO-1-adiponectin regulatory axis in a murine model [23]. Our results corroborate the findings of the previous studies, as we have shown that HO-1 induction causes an upregulation of adiponectin levels in murine pre-adipocytes.

Adiponectin is a protective adipokine, the levels of which are often compromised in a state of excess VAT. A great emphasis has been given to adiponectin recently, due to its role associated with insulin sensitivity and positive effect on triglyceride levels [38–41]. Reports have also demonstrated that low adiponectin levels are associated with increased oxidative stress [42,43]. Our results show that AngII induces downregulated levels of adiponectin, which contributes toward altered adipocyte phenotype. However, the protective effect of HO-1 is able to increase adiponectin levels, leading to healthier adipocytes. Our study demonstrates that the upregulated HO-1 expression, induced by CoPP, has potential to reprogram the altered adipocyte phenotype to that of a healthy adipocyte by improving oxidative stress, decreasing release of inflammatory cytokines, decreasing lipid accumulation, and increasing adiponectin levels.

The mechanistic link of AngII exposure observed in our study is provided by SIRT1, which belongs to the family of nicotinamide adenine dinucleotide (NAD)-dependent deacetylases. SIRT1 is a class III protein deacetylase, a crucial cellular survival protein in combating metabolic imbalance [10,44]. SIRT1 is modulated by cellular redox. Resveratrol, an antioxidant, protects SIRT1 against oxidative stress and prolongs longevity in various animal models [45]. AngII activates NADPH oxidases and increases oxidative stress [46,47]. We demonstrate that the oxidative stress induced by AngII leads to attenuation of SIRT1 in adipocytes. SIRT1 rescue and MR-suppression in murine pre-adipocytes treated with CoPP alludes to the protective effect of HO-1 on cellular SIRT1.

Conclusively, AngII upregulation exerts its effects on oxidative and inflammatory pathways in adipocytes, further inducing phenotypic and molecular changes by inhibiting the HO-1-SIRT1 axis. Our study characterizes that the SIRT1 rescue is HO-1 dependent, which causes reversal of molecular and pathological effects of the AngII cascade in these adipocytes. Our findings are summarized in a schematic representation that demonstrates the AngII-mediated increase in aldosterone synthase, which causes an increase in inflammation, oxidative stress and lipogenesis (Figure 5). This effect is attenuated by the HO-1-induced rescue of SIRT1, hence improving the adipocyte phenotype and molecular changes. This study has far-reaching clinical implications for patients with RAS-dependent metabolic disorders and for patients with secondary RAS activation, systemic or local, as seen in obesity or diabetes. The study elucidates the potential for therapeutic application of HO-1-inducers as a complementary therapy toward abatement of adipose tissue dysfunction, reduction of systemic inflammation, enhancement of adiponectin, and restoration of metabolic balance in these patients.

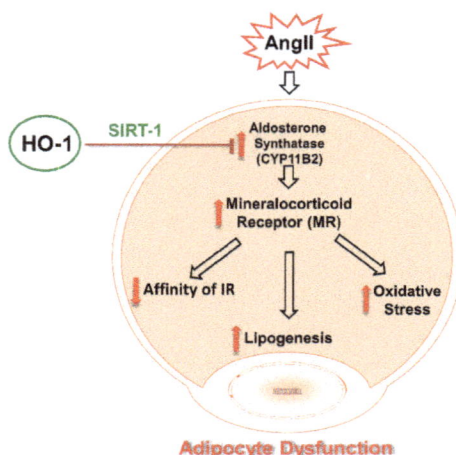

Figure 5. Schematic representation of AngII-mediated phenotypic alterations in adipocytes, reversed by the HO-1-dependent rescue of SIRT1 in 3T3-L1 murine adipocytes. Each arrow, shown in red, represents the upregulation or downregulation of the respective process.

4. Material and Methods

4.1. Experimental Design for In Vitro Experiments

Frozen mouse pre-adipocytes (3T3L-1) were purchased from ATTC (ATTC, Manassas, VA, USA). After thawing, 3T3L-1 cells were suspended in Dulbecco's Modified Eagle Medium (DMEM; Invitrogen, Carlsbad, CA, USA), supplemented with 10% heat-inactivated fetal bovine serum (FBS, Invitrogen, Carlsbad, CA, USA) and 1% antibiotic/antimycotic solution (Invitrogen, Carlsbad, CA, USA). The cells were plated at a density of 1×10^6 cells per 100 cm^2 dish. The cultures were maintained at 37 °C in a 5% CO_2 incubator and the medium was changed after 48 h and every ~3–4 days thereafter. Upon attaining confluence, 3T3L-1 cells were recovered by the addition of 0.25% trypsin/EDTA (Invitrogen, Carlsbad, CA, USA). 3T3L-1 cells were plated in 12-well dishes and 75 cm^2 flasks at a density of 1×10^4. Adipogenesis was initiated at confluence with the adipogenic media (DMEM high glucose with 0.5 mM dexamethasone, 10 µg/mL insulin, and 0.5 mM IBMX). Cells were treated every alternate day with AngII (10 µM), CoPP (5 µM), and SnMP (5 µM) for 7 days in mouse pre-adipocytes. Control groups were treated with adipogenic media alone to induce adipocyte differentiation only, without any treatment with AngII and/or other reagents. We also included an experimental group with our murine pre-adipocytes treated with SnMP alone to demonstrate its inhibitory effects, as previously published [37,48].

Commercially available (Ambion Silencer Select) siRNA and an appropriate scrambled RNA for SIRT1 was employed for "knockdown" studies [49]. For overexpression studies, we employed mouse SIRT1, full-length variant (isoform 1, Gene ID93759) synthesized into pJ603 vector, along with corresponding pJ603-GFP negative control by DNA 2.0 Inc. Transfection of cells was achieved using FuGENE HD transfection reagent, as described previously [10].

4.2. Oil Red O Staining

Lipid droplets were detected by Oil red O staining. For Oil Red O staining, 0.21% Oil Red O in 100% isopropanol (Sigma-Aldrich, St Louis, MO, USA) was used. Briefly, mouse pre-adipocytes were fixed in 10% formaldehyde, washed in Oil-red O for 10 min, rinsed with 60% isopropanol (Sigma-Aldrich), and the Oil red O was eluted by adding 100% isopropanol for 10 min. The optical density measured at 490 nm, for a 0.5-sec reading. Mouse adipocytes were measured by Oil red O staining (optical density = 490 nm) after day 7 [9].

4.3. Western Blot Analysis of IR-β, SIRT1, MR and CYP11B2

Frozen mouse adipocytes (3T3L-1) were pulverized under liquid nitrogen and placed in a homogenization buffer comprising (mmol/L): 10 phosphate buffer, 250 sucrose, 1 ethylenediaminetetraacetic acid (EDTA), 0.1 phenylmethanesulfonylfluoride, and 0.1% *v/v* tergitol, pH 7.5. Homogenates were centrifuged at 27,000 *g* for 10 min at 4°C. The supernatant was isolated and protein levels were assayed (Bradford Method). The supernatant was used for the determination of IR-β, SIRT1, MR, and CYP11B2. Immunoblotting was performed in mouse adipocytes. β-Actin was used to ensure adequate sample loading for all western blots.

4.4. Cytokines, Adiponectin, and Lipid Profile Measurements

Conditioned media (CM) was obtained from our cell culture The levels of interleukin (IL)-6, tumor necrosis factor α (TNFα) and the high molecular weight (HMW) form of adiponectin were determined using an enzyme-linked immunosorbent assay (ELISA) assay kit according to manufacturer's protocol (Abcam, Cambridge, MA, USA) [50]. Triglyceride levels were measured in CM using an ELISA assay (Assay Gate, Inc., Ijamsville, MD, USA).

4.5. Quantitative Real-Time PCR Analysis

Total RNA was extracted from mouse adipocytes using a 5-Prime PerfectPure RNA Tissue Kit (Fisher Scientific Company, LLC, Waltham, MA, USA). Total RNA was read on a NanoDrop 2000 Spectrophotometer (Thermo Scientific, Wilmington, DE, USA) and cDNA was synthesized using an iScript cDNA Synthesis kit (Bio-Rad, Hercules, CA, USA). PCR amplification of the cDNA was performed by quantitative real-time PCR using a qPCR Core kit for SYBR Green I (Applied Biosystems, Grand Island, NY, USA). The thermocycling protocol consisted of 10 min at 95 °C, 40 cycles of 15 s at 95 °C, 30 s at 60 °C, and finished with a melting curve ranging from 60 to 95 °C to allow distinction of specific products [9]. Primers were designed specific to each gene using Primer Express 3.0 software (Applied Biosystems). Normalization was performed in separate reactions with primers to GAPDH mRNA. Specific primers were used for HO-1 and FAS.

4.6. Measurement of Superoxide Levels for In Vitro Experiment

Mouse adipocytes were cultured on 96-well plates until they achieved approximately 70% confluence. After treatment with or without AngII (10 μM) in the absence and presence of CoPP (5 μM) and SnMP (5 μM) for 2 days, the cells were incubated with 10 μM dihydroethidium (DHE) for 30 min at 37 °C. Fluorescence intensity was measured using a Perkin–Elmer Luminescence Spectrometer at excitation/emission filters of 530/620 nm.

4.7. Statistical Analysis

Statistical significance between experimental groups was determined by the Fisher method of analysis of multiple comparisons ($p < 0.01$). For comparisons among treatment groups, the null hypothesis was tested by a two-factor ANOVA for multiple groups or unpaired t test for two groups. Data are presented as mean ± SE.

Supplementary Materials: Supplementary materials can be found at http://www.mdpi.com/1422-0067/20/13/3205/s1.

Author Contributions: Conceptualization—K.S.; Methodology, H.V.L., M.Z., S.S.P., N.P.; Supervision, K.S., J.I.S. and N.G.A.; Writing—original draft preparation, H.V.L.; Writing—review and editing, K.S.; Resources—J.I.S. and N.G.A.

Funding: This work was supported by National Institutes of Health Grants to JIS (HL109015, HL105649 and HL071556), and by the Brickstreet Foundation (J.I.S.).

Conflicts of Interest: The authors have declared that no competing interests exist.

References

1. Frederich, R.C., Jr.; Kahn, B.B.; Peach, M.J.; Flier, J.S. Tissue-specific nutritional regulation of angiotensinogen in adipose tissue. *Hypertension* **1992**, *19*, 339–344. [CrossRef] [PubMed]
2. Karlsson, C.; Lindell, K.; Ottosson, M.; Sjostrom, L.; Carlsson, B.; Carlsson, L.M. Human adipose tissue expresses angiotensinogen and enzymes required for its conversion to angiotensin II. *J. Clin. Endocrinol. Metab.* **1998**, *83*, 3925–3929. [CrossRef]
3. Vogel, J.; Bartels, V.; Tang, T.H.; Churakov, G.; Slagter-Jager, J.G.; Huttenhofer, A.; Wagner, E.G. RNomics in Escherichia coli detects new sRNA species and indicates parallel transcriptional output in bacteria. *Nucl. Acids Res.* **2003**, *31*, 6435–6443. [CrossRef] [PubMed]
4. Briones, A.M.; Nguyen Dinh Cat, A.; Callera, G.E.; Yogi, A.; Burger, D.; He, Y.; Correa, J.W.; Gagnon, A.M.; Gomez-Sanchez, C.E.; Gomez-Sanchez, E.P.; et al. Adipocytes produce aldosterone through calcineurin-dependent signaling pathways: implications in diabetes mellitus-associated obesity and vascular dysfunction. *Hypertension* **2012**, *59*, 1069–1078. [CrossRef] [PubMed]
5. Duvnjak, L.; Duvnjak, M. The metabolic syndrome—an ongoing story. *J. Physiol. Pharmacol.* **2009**, *60*, 19–24. [PubMed]

6.   Boustany, C.M.; Bharadwaj, K.; Daugherty, A.; Brown, D.R.; Randall, D.C.; Cassis, L.A. Activation of the systemic and adipose renin-angiotensin system in rats with diet-induced obesity and hypertension. *Am. J. Physiol. Regul. Integr. Comp. Physiol.* **2004**, *287*, R943–R949. [CrossRef] [PubMed]

7.   Togbo, I. Obesogenic factors influencing overweight among Asian children and youth. *J. Health Res. Rev.* **2018**, *5*, 111–116. [CrossRef]

8.   Yvan-Charvet, L.; Quignard-Boulange, A. Role of adipose tissue renin-angiotensin system in metabolic and inflammatory diseases associated with obesity. *Kidney Int.* **2011**, *79*, 162–168. [CrossRef]

9.   Sodhi, K.; Puri, N.; Kim, D.H.; Hinds, T.D.; Stechschulte, L.A.; Favero, G.; Rodella, L.; Shapiro, J.I.; Jude. D.; Abraham, N.G. PPARdelta binding to heme oxygenase 1 promoter prevents angiotensin II-induced adipocyte dysfunction in Goldblatt hypertensive rats. *Int. J. Obes.* **2014**, *38*, 456–465. [CrossRef]

10.   Puri, N.; Sodhi, K.; Haarstad, M.; Kim, D.H.; Bohinc, S.; Foglio, E.; Favero, G.; Abraham, N.G. Heme induced oxidative stress attenuates sirtuin1 and enhances adipogenesis in mesenchymal stem cells and mouse pre-adipocytes. *J. Cell Biochem.* **2012**, *113*, 1926–1935. [CrossRef]

11.   Guberman, C.; Jellyman, J.K.; Han, G.; Ross, M.G.; Desai, M. Maternal high-fat diet programs rat offspring hypertension and activates the adipose renin-angiotensin system. *Am. J. Obstet. Gynecol.* **2013**, *209*, e261–e268. [CrossRef] [PubMed]

12.   Abdul-Salam, V.B.; Ramrakha, P.; Krishnan, U.; Owen, D.R.; Shalhoub, J.; Davies, A.H.; Tang, T.Y.; Gillard, J.H.; Boyle, J.J.; Wilkins, M.R.; et al. Identification and assessment of plasma lysozyme as a putative biomarker of atherosclerosis. *Arterioscler. Thromb. Vasc. Biol.* **2010**, *30*, 1027–1033. [CrossRef] [PubMed]

13.   Hirata, A.; Maeda, N.; Hiuge, A.; Hibuse, T.; Fujita, K.; Okada, T.; Kihara, S.; Funahashi, T.; Shimomura, I. Blockade of mineralocorticoid receptor reverses adipocyte dysfunction and insulin resistance in obese mice. *Cardiovasc. Res.* **2009**, *84*, 164–172. [CrossRef] [PubMed]

14.   Boscaro, M.; Giacchetti, G.; Ronconi, V. Visceral adipose tissue: emerging role of gluco- and mineralocorticoid hormones in the setting of cardiometabolic alterations. *Ann. N. Y. Acad. Sci.* **2012**, *1264*, 87–102. [CrossRef] [PubMed]

15.   Chung, S.; Yao, H.; Caito, S.; Hwang, J.W.; Arunachalam, G.; Rahman, I. Regulation of SIRT1 in cellular functions: role of polyphenols. *Arch. Biochem. Biophys.* **2010**, *501*, 79–90. [CrossRef]

16.   Woo, Y.C.; Xu, A.; Wang, Y.; Lam, K.S. Fibroblast growth factor 21 as an emerging metabolic regulator: clinical perspectives. *Clin. Endocrinol.* **2013**, *78*, 489–496. [CrossRef]

17.   Kim, D.H.; Burgess, A.P.; Li, M.; Tsenovoy, P.L.; Addabbo, F.; McClung, J.A.; Puri, N.; Abraham, N.G. Heme oxygenase-mediated increases in adiponectin decrease fat content and inflammatory cytokines tumor necrosis factor-alpha and interleukin-6 in Zucker rats and reduce adipogenesis in human mesenchymal stem cells. *J. Pharmacol. Exp. Ther.* **2008**, *325*, 833–840. [CrossRef]

18.   Paul, M.; Poyan Mehr, A.; Kreutz, R. Physiology of local renin-angiotensin systems. *Physiol. Rev.* **2006**, *86*, 747–803. [CrossRef]

19.   Abraham, N.G.; Kappas, A. Pharmacological and clinical aspects of heme oxygenase. *Pharmacol. Rev.* **2008**, *60*, 79–127. [CrossRef]

20.   Abraham, N.G.; Kappas, A. Heme oxygenase and the cardiovascular-renal system. *Free Radic. Biol. Med.* **2005**, *39*, 1–25. [CrossRef]

21.   Gregoire, F.M.; Smas, C.M.; Sul, H.S. Understanding adipocyte differentiation. *Physiol. Rev.* **1998**, *78*, 783–809. [CrossRef] [PubMed]

22.   Kim, D.H.; Vanella, L.; Inoue, K.; Burgess, A.; Gotlinger, K.; Manthati, V.L.; Koduru, S.R.; Zeldin, D.C.; Falck, J.R.; Schwartzman, M.L.; et al. Epoxyeicosatrienoic acid agonist regulates human mesenchymal stem cell-derived adipocytes through activation of HO-1-pAKT signaling and a decrease in PPARgamma. *Stem Cells Dev.* **2010**, *19*, 1863–1873. [CrossRef] [PubMed]

23.   Li, M.; Kim, D.H.; Tsenovoy, P.L.; Peterson, S.J.; Rezzani, R.; Rodella, L.F.; Aronow, W.S.; Ikehara, S.; Abraham, N.G. Treatment of obese diabetic mice with a heme oxygenase inducer reduces visceral and subcutaneous adiposity, increases adiponectin levels, and improves insulin sensitivity and glucose tolerance. *Diabetes* **2008**, *57*, 1526–1535. [CrossRef] [PubMed]

24.   Bordone, L.; Guarente, L. Calorie restriction, SIRT1 and metabolism: understanding longevity. *Nat. Rev. Mol. Cell Biol.* **2005**, *6*, 298–305. [CrossRef] [PubMed]

25.   Kotronen, A.; Yki-Jarvinen, H. Fatty liver: a novel component of the metabolic syndrome. *Arterioscler. Thromb. Vasc. Biol.* **2008**, *28*, 27–38. [CrossRef] [PubMed]

26. Grundy, S.M.; Brewer, H.B., Jr.; Cleeman, J.I.; Smith, S.C., Jr.; Lenfant, C.; National Heart, L.; Blood, I.; American Heart, A. Definition of metabolic syndrome: report of the National Heart, Lung, and Blood Institute/American Heart Association conference on scientific issues related to definition. *Arterioscler. Thromb. Vasc. Biol.* **2004**, *24*, e13–e18. [CrossRef] [PubMed]

27. Cao, J.; Peterson, S.J.; Sodhi, K.; Vanella, L.; Barbagallo, I.; Rodella, L.F.; Schwartzman, M.L.; Abraham, N.G.; Kappas, A. Heme oxygenase gene targeting to adipocytes attenuates adiposity and vascular dysfunction in mice fed a high-fat diet. *Hypertension* **2012**, *60*, 467–475. [CrossRef] [PubMed]

28. Botros, F.T.; Schwartzman, M.L.; Stier, C.T., Jr.; Goodman, A.I.; Abraham, N.G. Increase in heme oxygenase-1 levels ameliorates renovascular hypertension. *Kidney Int.* **2005**, *68*, 2745–2755. [CrossRef]

29. Sodhi, K.; Puri, N.; Inoue, K.; Falck, J.R.; Schwartzman, M.L.; Abraham, N.G. EET agonist prevents adiposity and vascular dysfunction in rats fed a high fat diet via a decrease in Bach 1 and an increase in HO-1 levels. *Prostaglandins Other Lipid Mediat.* **2012**, *98*, 133–142. [CrossRef] [PubMed]

30. Nagase, M.; Fujita, T. Mineralocorticoid receptor activation in obesity hypertension. *Hypertens. Res.* **2009**, *32*, 649–657. [CrossRef] [PubMed]

31. Singh, S.P.; Bellner, L.; Vanella, L.; Cao, J.; Falck, J.R.; Kappas, A.; Abraham, N.G. Downregulation of PGC-1alpha Prevents the Beneficial Effect of EET-Heme Oxygenase-1 on Mitochondrial Integrity and Associated Metabolic Function in Obese Mice. *J. Nutr. Metab.* **2016**, *2016*, 9039754. [CrossRef] [PubMed]

32. Ndisang, J.F. Role of heme oxygenase in inflammation, insulin-signalling, diabetes and obesity. *Mediat. Inflamm.* **2010**, *2010*, 359732. [CrossRef] [PubMed]

33. Nicolai, A.; Li, M.; Kim, D.H.; Peterson, S.J.; Vanella, L.; Positano, V.; Gastaldelli, A.; Rezzani, R.; Rodella, L.F.; Drummond, G.; et al. Heme oxygenase-1 induction remodels adipose tissue and improves insulin sensitivity in obesity-induced diabetic rats. *Hypertension* **2009**, *53*, 508–515. [CrossRef] [PubMed]

34. Abraham, N.G.; Junge, J.M.; Drummond, G.S. Translational Significance of Heme Oxygenase in Obesity and Metabolic Syndrome. *Trends Pharmacol. Sci.* **2016**, *37*, 17–36. [CrossRef] [PubMed]

35. Gomez-Sanchez, C.E. What Is the Role of the Adipocyte Mineralocorticoid Receptor in the Metabolic Syndrome? *Hypertension* **2015**, *66*, 17–19. [CrossRef] [PubMed]

36. Vanella, L.; Sanford, C., Jr.; Kim, D.H.; Abraham, N.G.; Ebraheim, N. Oxidative stress and heme oxygenase-1 regulated human mesenchymal stem cells differentiation. *Int. J. Hypertens.* **2012**, *2012*, 890671. [CrossRef] [PubMed]

37. Singh, S.P.; Grant, I.; Meissner, A.; Kappas, A.; Abraham, N.G. Ablation of adipose-HO-1 expression increases white fat over beige fat through inhibition of mitochondrial fusion and of PGC1alpha in female mice. *Horm. Mol. Biol. Clin. Investig.* **2017**, *31*. [CrossRef]

38. Kim, J.Y.; van de Wall, E.; Laplante, M.; Azzara, A.; Trujillo, M.E.; Hofmann, S.M.; Schraw, T.; Durand, J.L.; Li, H.; Li, G.; et al. Obesity-associated improvements in metabolic profile through expansion of adipose tissue. *J. Clin. Investig.* **2007**, *117*, 2621–2637. [CrossRef]

39. Combs, T.P.; Pajvani, U.B.; Berg, A.H.; Lin, Y.; Jelicks, L.A.; Laplante, M.; Nawrocki, A.R.; Rajala, M.W.; Parlow, A.F.; Cheeseboro, L.; et al. A transgenic mouse with a deletion in the collagenous domain of adiponectin displays elevated circulating adiponectin and improved insulin sensitivity. *Endocrinology* **2004**, *145*, 367–383. [CrossRef]

40. Kubota, N.; Terauchi, Y.; Yamauchi, T.; Kubota, T.; Moroi, M.; Matsui, J.; Eto, K.; Yamashita, T.; Kamon, J.; Satoh, H.; et al. Disruption of adiponectin causes insulin resistance and neointimal formation. *J. Biol. Chem.* **2002**, *277*, 25863–25866. [CrossRef]

41. Berg, A.H.; Scherer, P.E. Adipose tissue, inflammation, and cardiovascular disease. *Circ. Res.* **2005**, *96*, 939–949. [CrossRef]

42. Vinatier, D.; Dufour, P.; Tordjeman-Rizzi, N.; Prolongeau, J.F.; Depret-Moser, S.; Monnier, J.C. Immunological aspects of ovarian function: role of the cytokines. *Eur. J. Obstet. Gynecol. Reprod. Biol.* **1995**, *63*, 155–168. [CrossRef]

43. Burgess, A.; Li, M.; Vanella, L.; Kim, D.H.; Rezzani, R.; Rodella, L.; Sodhi, K.; Canestraro, M.; Martasek, P.; Peterson, S.J.; et al. Adipocyte heme oxygenase-1 induction attenuates metabolic syndrome in both male and female obese mice. *Hypertension* **2010**, *56*, 1124–1130. [CrossRef] [PubMed]

44. Li, X. SIRT1 and energy metabolism. *Acta Biochim. Biophys. Sin.* **2013**, *45*, 51–60. [CrossRef] [PubMed]

45. Li, J.; Zhang, C.X.; Liu, Y.M.; Chen, K.L.; Chen, G. A comparative study of anti-aging properties and mechanism: resveratrol and caloric restriction. *Oncotarget* **2017**, *8*, 65717–65729. [CrossRef] [PubMed]

46. Nguyen Dinh Cat, A.; Montezano, A.C.; Burger, D.; Touyz, R.M. Angiotensin II, NADPH oxidase, and redox signaling in the vasculature. *Antioxid. Redox Signal.* **2013**, *19*, 1110–1120. [CrossRef] [PubMed]

47. Garrido, A.M.; Griendling, K.K. NADPH oxidases and angiotensin II receptor signaling. *Mol. Cell Endocrinol.* **2009**, *302*, 148–158. [CrossRef]

48. Vanella, L.; Sodhi, K.; Kim, D.H.; Puri, N.; Maheshwari, M.; Hinds, T.D.; Bellner, L.; Goldstein, D.; Peterson, S.J.; Shapiro, J.I.; et al. Increased heme-oxygenase 1 expression in mesenchymal stem cell-derived adipocytes decreases differentiation and lipid accumulation via upregulation of the canonical Wnt signaling cascade. *Stem Cell Res. Ther.* **2013**, *4*, 28. [CrossRef]

49. Sodhi, K.; Puri, N.; Favero, G.; Stevens, S.; Meadows, C.; Abraham, N.G.; Rezzani, R.; Ansinelli, H.; Lebovics, E.; Shapiro, J.I. Fructose Mediated Non-Alcoholic Fatty Liver Is Attenuated by HO-1-SIRT1 Module in Murine Hepatocytes and Mice Fed a High Fructose Diet. *PloS ONE* **2015**, *10*, e0128648. [CrossRef]

50. Bartlett, D.E.; Miller, R.B.; Thiesfeldt, S.; Lakhani, H.V.; Khanal, T.; Pratt, R.D.; Cottrill, C.L.; Klug, R.L.; Adkins, N.S.; Bown, P.C.; et al. Uremic Toxins Activates Na/K-ATPase Oxidant Amplification Loop Causing Phenotypic Changes in Adipocytes in In Vitro Models. *Int. J. Mol. Sci.* **2018**, *19*, 2685. [CrossRef]

International Journal of
Molecular Sciences

MDPI

Review

The Protective Role of Heme Oxygenase-1 in Atherosclerotic Diseases

Yoshimi Kishimoto [1,*], Kazuo Kondo [1,2] and Yukihiko Momiyama [3]

[1] Endowed Research Department "Food for Health", Ochanomizu University, 2-1-1 Otsuka, Bunkyo-ku, Tokyo 112-8610, Japan
[2] Institute of Life Innovation Studies, Toyo University, 1-1-1 Izumino, Itakura-machi, Ora-gun, Gunma 374-0193, Japan
[3] Department of Cardiology, National Hospital Organization Tokyo Medical Center, 2-5-1 Higashigaoka, Meguro-ku, Tokyo 152-8902, Japan
* Correspondence: kishimoto.yoshimi@ocha.ac.jp; Tel.: +81-3-5978-5810; Fax: +81-3-5978-2694

Received: 1 July 2019; Accepted: 22 July 2019; Published: 24 July 2019

Abstract: Heme oxygenase-1 (HO-1) is an intracellular enzyme that catalyzes the oxidation of heme to generate ferrous iron, carbon monoxide (CO), and biliverdin, which is subsequently converted to bilirubin. These products have anti-inflammatory, anti-oxidant, anti-apoptotic, and anti-thrombotic properties. Although HO-1 is expressed at low levels in most tissues under basal conditions, it is highly inducible in response to various pathophysiological stresses/stimuli. HO-1 induction is thus thought to be an adaptive defense system that functions to protect cells and tissues against injury in many disease settings. In atherosclerosis, HO-1 may play a protective role against the progression of atherosclerosis, mainly due to the degradation of pro-oxidant heme, the generation of anti-oxidants biliverdin and bilirubin and the production of vasodilator CO. In animal models, a lack of HO-1 was shown to accelerate atherosclerosis, whereas HO-1 induction reduced atherosclerosis. It was also reported that HO-1 induction improved the cardiac function and postinfarction survival in animal models of heart failure or myocardial infarction. Recently, we and others examined blood HO-1 levels in patients with atherosclerotic diseases, e.g., coronary artery disease (CAD) and peripheral artery disease (PAD). Taken together, these findings to date support the notion that HO-1 plays a protective role against the progression of atherosclerotic diseases. This review summarizes the roles of HO-1 in atherosclerosis and focuses on the clinical studies that examined the relationships between HO-1 levels and atherosclerotic diseases.

Keywords: heme oxygenase-1; atherosclerosis; coronary artery disease; peripheral artery disease; carotid plaque

---

1. Introduction

Atherosclerotic diseases are known to be the leading causes of death in the world. Atherosclerosis begins when the injured (or activated) artery wall creates chemical signals that cause certain types of leukocytes to attach to the endothelium [1]. These cells move into the wall of the artery and are transformed into foam cells by uptake of modified low-density lipoprotein (LDL) such as oxidized LDL, which collect cholesterol and other fatty materials and trigger smooth muscle cells to migrate from the media to the intima. They form atheromas, also called plaques, covered with a fibrous cap, and eventually the growing lesion begins to raise the endothelium and encroach on the lumen of the artery. When plaques rupture, the exposing material triggers blood clot formation, which can suddenly block blood flow through the artery, resulting in myocardial infarction or stroke.

Although heme serves key physiological functions and is tightly controlled, high levels of free heme, which may occur in various pathophysiological conditions, are toxic via pro-oxidant, pro-inflammatory, and cytotoxic effects [2]. Thus, the heme degradation pathway has been demonstrated to play a protective role against the development of atherosclerosis [3–5]. Heme oxygenase (HO) is the rate-limiting intracellular enzyme that catalyzes the oxidation of heme to generate biliverdin, carbon monoxide (CO), and ferrous iron. Biliverdin is subsequently reduced to bilirubin by biliverdin reductase; both of biliverdin and bilirubin have antioxidant properties. The endogenously produced CO can serve as a second messenger affecting several cellular functions, including proliferation, inflammation, and apoptosis [3,6,7]. Three isoforms in the HO family (HO-1, HO-2, and HO-3) are known to be the products of different genes and to be differently regulated. HO-1, also known as a 32-kDa heat shock protein, encoded by the gene *HMOX1*, is normally expressed at low levels in most tissues; however, HO-1 is highly inducible in response to various stresses/stimuli, including heme/hemoglobin, heavy metals, UV radiation, cytokines, and endotoxins [8–12]. In contrast, HO-2 is constitutively expressed in most tissues. HO-3 has a protein structure that is similar to that of HO-2 but has lower enzymatic activity and is less well characterized [13]. A variety of experimental studies have suggested that HO-1 is a stress-response protein that plays an important role in cell defense mechanisms against oxidative injury. In the pathogenesis of atherosclerosis, the ability of HO-1 to generate biliverdin and bilirubin, anti-oxidant molecules, and CO, a vasodilator and an anti-inflammatory and antiapoptotic molecule, is thought to play important roles in protecting the artery against oxidant-induced injury. This review documents the roles of HO-1 in atherosclerosis and focuses on the clinical significance as a potential therapeutic target in atherosclerotic diseases, such as coronary artery disease (CAD) and peripheral artery disease (PAD). Using a PubMed database, we reviewed the articles published by July 2019 only in English. The clinical studies included in this review that showed the relationships between HO-1 levels and atherosclerotic diseases are summarized in Table 1.

Table 1. Studies of the relationships between heme oxygenase (HO)-1 levels and atherosclerotic diseases.

Wang et al., 1998 [14]	Ascending and abdominal aortas	Patients undergoing surgery for CAD ($n = 3$) or abdominal aortic aneurysm ($n = 5$)	HO-1 was highly expressed in human atherosclerotic lesions
Chen et al., 2005 [15]	Blood leukocytes	Control ($n = 30$) SAP ($n = 30$) UAP ($n = 40$) AMI ($n = 35$)	HO-1 protein expression was higher in patients with CAD (AMI > UAP > SAP > Control)
Ameriso et al., 2005 [16]	Carotid endarterectomy specimens	Controls ($n = 7$) Patients with symptomatic plaques ($n = 25$) or asymptomatic plaques ($n = 23$)	HO-1 expression is highly prevalent in asymptomatic plaques
Ijas et al., 2007 [17]	Carotid plaques	(a) Patients with bilateral high-grade stenosis (one being symptomatic and the other asymptomatic) ($n = 4$) (b) Patients with ipsilateral stroke symptoms ($n = 22$) or without cerebrovascular symptoms ($n = 18$)	HO-1 and CD163 were overexpressed in symptomatic carotid plaques in both intra-individual and inter-individual comparison
Brydun et al., 2007 [18]	Blood mononuclear cells	110 patients undergoing coronary angiography	The capacity to upregulate *HMOX1* mRNA expression was inversely related to the degree of CAD
Cheng et al., 2009 [19]	Carotid endarterectomy specimens	112 CAD patients	HO-1 protein expression correlated with the vulnerability of atheromatous plaque
Idriss et al., 2010 [20]	Plasma	Healthy controls ($n = 50$) Stable CAD ($n = 70$) ACS ($n = 24$)	HO-1 levels were higher in stable CAD and ACS patients

Table 1. *Cont.*

Novo et al., 2011 [21]	Serum (or plasma)	Controls (n = 40) AMI (n = 40)	HO-1 levels in AMI patients were significantly higher than in controls, and showed an inverse association with the severity of CAD
Yunoki et al., 2013 [22]	Coronary atherectomy specimens	SAP (n = 33) UAP (n = 34)	HO-1-positive areas were significantly higher in UAP patients
Li et al., 2014 [23]	Serum	Stroke (n = 60) TIA (n = 50)	HO-1 levels were higher in patients with stroke than TIA
Signorelli et al., 2016 [24]	Serum	Controls (n = 27) PAD (n = 27)	HO-1 levels were lower in PAD patients
Kishimoto et al., 2018 [25]	Plasma	136 subjects undergoing carotid ultrasonography for medical check-up	HO-1 levels were high in subjects with carotid plaques
Kishimoto et al., 2018 [26]	Plasma	410 patients undergoing coronary angiography for suspected CAD	HO-1 levels were low in patients with PAD, in contrast to high levels in patients with CAD
Fiorelli et al., 2019 [27]	Monocyte-derived macrophages (MDMs)	Healthy controls (10) CAD patients undergoing coronary angiography (30)	HO-1 levels were higher in MDMs of CAD patients and were associated with rupture-prone coronary plaque

SAP, stable angina pectoris; UAP, unstable angina pectoris; AMI, acute myocardial infarction; CAD, coronary artery disease; ACS, acute coronary syndrome; PAD, peripheral artery disease; TIA, transient ischemic attack.

2. Important Role of HO-1 in Atherosclerosis

HMOX1 deficiency is very rare in humans [28,29]. In two reported cases, similar phenotypes characterized by generalized inflammation, nephropathy, asplenia, anemia, and tissue iron deposition were observed. Vascular injury and early atherosclerotic changes, as reflected by the presence of fatty streaks and fibrous plaques were also reported, suggesting the importance of HO-1 in vascular health. In HO-1-knockout (*Hmox1*$^{-/-}$) mice, growth retardation, anemia, iron deposition, and vulnerability to stressful injury were observed [30,31]. *Hmox1*$^{-/-}$ mice were also reported to develop severe aortitis and coronary arteritis with mononuclear cellular infiltration and fatty streak formation even on a standard chow diet [32].

In 1998, Wang et al. demonstrated that HO-1 expression was present throughout the development of human atherosclerotic lesions from early fatty streaks to advanced lesions [14]. Oxidized LDL, a major determinant in the pathogenesis of atherosclerosis, was identified to be a potent inducer of HO-1. The HO-1 expression in endothelial cells, monocytes, and macrophages was up-regulated by exposure to oxidized LDL [33–35]. HO-1 expression in atherosclerotic lesions is thus considered to be a protective response against the progression of atherosclerosis. HO-1 overexpression by pharmacological inducers or viral gene transfer successfully inhibited atherogenesis in hypercholesterolemic animal models [36–38]. In contrast, the genetic ablation of *Hmox1* in apolipoprotein E-knockout mice accelerated the development of atherosclerosis and exacerbated lesion formation [39]. These results thus suggest that HO-1 plays a protective role against the progression of atherosclerosis.

The genetic polymorphisms of the *HMOX1* gene in humans also indicate the potential importance of HO-1 in the pathogenesis of cardiovascular diseases (CVDs). Among the identified polymorphisms in the *HMOX1* gene, two have attracted the most attention: A (GT)n dinucleotide repeat length polymorphism and a common single-nucleotide polymorphism (SNP), T(-413)A (rs2071746) [40]. The (GT)n short allele (S, <25 repeats) and the A(-413) allele are reported to be associated with significantly increased *HMOX1* gene promoter activity compared to the long allele (L, ≥25 repeats) and the T(-413) allele, respectively [41,42]. The association of these polymorphisms with CAD has been discussed [18,43–47]. A meta-analysis by Qiao et al. [40] demonstrated that the (GT)n SS genotype was associated with a decreased risk of CAD after controlling for biases (age, sex, extent of coronary stenosis, ethnicity, and study quality). For the T(-413)A SNP, although a decreased CAD risk among individuals with the AA genotype was observed compared to individuals with the TT genotype, the authors mentioned that this effect was quite limited and should be interpreted cautiously [40].

3. Mechanistic Actions of HO-1 in Oxidative Stress and Inflammation

HO-1 is known to be regulated by the redox-sensitive transcription factor known as nuclear factor erythroid 2-related factor (Nrf2) [48]. Nrf2 is ubiquitously expressed and kept in a latent state through the interaction with its repressor protein, Kelch ECH associated protein 1 (Keap1). The exposure of cells to oxidative stimuli triggers a conformational change in Keap1 through a modification of its cysteine residues, which results in the release of Nrf2 from Keap1. Apart from this Keap1-dependent pathway, Nrf2 activation is also mediated by protein kinases such as glycogen synthase kinase-3β (GSK-3β), phosphatidylinositol-3-kinase (PI3K)/Akt, protein kinase C (PKC), and mitogen-activated protein kinase (MAPK) cascades via the phosphorylation of the serine or threonine residues of Nrf2. Stabilized cytosolic Nrf2 is translocated into the nucleus and binds to the antioxidant response element (ARE), thereby initiating the transcription of antioxidant and phase II detoxification enzymes, including HO-1, superoxide dismutase (SOD), catalase, and NAD(P)H quinone dehydrogenase 1 (NQO1) [49]. Additionally, Nrf2 has demonstrated anti-inflammatory properties through its ability to negatively regulate nuclear factor-kappaB (NF-κB), the transcription factor central to the inflammatory response [50].

The anti-oxidant activity of HO-1 is thought to be due to its byproducts biliverdin, bilirubin, and CO. Bilirubin strongly scavenges several oxygen free radicals including singlet oxygen, $O_2{}^-$, $ONOO^-$, and organic peroxy radicals [51,52]. Because of its lipophilic property, bilirubin is closely associated with cell membranes, and hence can protect them against lipid damage and also protect LDL against peroxidation. It was reported that the patients with Gilbert syndrome (i.e., unconjugated hyperbilirubinemia) had lower circulating levels of oxidized LDL [53] and also that total bilirubin levels were associated with oxidized LDL levels [54,55]. On the basis of the involvement of oxidized LDL in the formation of atherosclerotic plaques, it was suggested that increased physiological concentrations of plasma bilirubin may reduce atherogenic risk.

In 1994, Schwertner et al. [56] reported serum bilirubin levels to be low in patients with CAD. Since then, many studies showed bilirubin levels to be lower in patients with CAD than without CAD and to inversely correlate with the severity of CAD [57–59]. Low bilirubin levels in blood may thus play a promotive role in the development of CAD. Genetic variations in the UDP-glucuronosyltransferase 1A1 (*UGT1A1*) gene are known to be major determinants of serum bilirubin level [60]. However, Stender et al. [61] investigated the associations between the *UGT1A1* gene genotype and plasma bilirubin levels and between genetically elevated bilirubin levels and CAD in 67,068 subjects. They demonstrated that genetically elevated bilirubin levels were not associated with a decreased risk of CAD. They also performed a meta-analysis of 11 studies and showed no association between genetically elevated bilirubin levels and CAD. These findings thus suggest no causal relationship between elevated bilirubin levels and CAD. Since increased reactive oxygen species (ROS) and oxidative stress are involved in the pathogenesis of atherosclerosis, low bilirubin levels in patients with CAD may not be a cause of CAD but rather a result of increased oxidative stress, leading to the consumption of endogenous anti-oxidants [52].

It is suggested that CO can attenuate the production of intracellular ROS. Kobayashi et al. [62] demonstrated that low-dose exogenous CO exposure inhibited the activation of NADPH oxidase and effectively suppressed ROS generation in the heart tissues of angiotensin II–infused mice. This finding might support the previous in vitro observation that HO-1 inhibited NADPH oxidase activity in cultured cells [63]. Importantly, obesity enhances the activation of NADPH oxidase and the angiotensin II system, resulting in the development of diabetes and hypertension in part due to impairment of adipocyte function [64]. Hinds and colleagues demonstrated that the induction of HO-1 in obese mice resulted in the elevation of peroxisome proliferator-activated receptor-alpha (PPAR-α), reducing body weight and blood glucose [65]. Interestingly, they recently identified that bilirubin could bind directly to activate PPAR-α, which increased target genes to reduce adiposity [66,67].

Endothelial inflammation and dysfunction are key players in the initiation of atherosclerosis progression. The overexpression of endothelial HO-1 significantly attenuated the production of inflammatory mediators and improved the impaired vasodilatory responses of aortic segments treated with oxidized LDL [68]. Oxidized LDL increases the production of ROS, leading to NF-κB activation, which upregulates intercellular adhesion molecule (ICAM-1), vascular cell adhesion molecule (VCAM-1), and E-selectin expression in endothelial cells and increases the adhesion of monocytes [69]. The activation of Nrf2 was reported to suppress the endothelial cell activation by inactivating p38 MAPK activity, thereby suppressing VCAM-1 expression [70]. In vascular endothelium, atherosclerotic plaque is often observed in areas where disturbed blood flow is formed, whereas an atheroprotective region is found in areas where a steady laminar flow has developed. Kim et al. demonstrated that a laminar flow-induced activation of Nrf2 signaling pathway played a critical role in the anti-inflammatory and anti-apoptotic mechanisms in endothelial cells [71]. These observations thus suggest that the upregulation of HO-1 in vascular endothelial cells contributes significantly to the inhibition of atherosclerosis. Both cell-based and in vivo studies demonstrated that the induction of HO-1 protected the vessel walls from pathological remodeling and endothelial cell dysfunction [72,73]. In human endothelial cells, HO-1/CO also inhibited endoplasmic reticulum stress-induced apoptosis via p38 MAPK-dependent inhibition of the proapoptotic C/EBP homologous protein (CHOP) expression [74].

Atherosclerosis is regarded as a chronic inflammatory state in which macrophages play different and important roles. HO-1 in macrophages appears to be of critical importance for driving the resolution of inflammatory responses [75]. Orozco et al. reported that decreased HO-1 expression increased the expression of proinflammatory cytokines such as monocyte chemoattractant protein 1 (MCP-1) and interleukin 6 (IL-6) and the expression of scavenger receptor A (SR-A), and it also accelerated foam cell formation [76]. Ruotsalainen et al. [77] demonstrated that the peritoneal macrophages isolated from Nrf2-knockout (*Nfe2l2$^{-/-}$*) mice showed increased expressions of MCP-1, IL-6, and tumor necrosis factor-alpha (TNF-α). With the stimulation of Nrf2$^{-/-}$ peritoneal macrophages with oxidized LDL or lipopolysaccharide (LPS), the ROS production was increased with a concomitant induction of pro-inflammatory genes [78,79]. The anti-inflammatory effect of Nrf2 in macrophages is likely due to an improved antioxidant defense system. The cytoprotective action of bilirubin was reported to be partly related to its capacity to inhibit inducible NOS (iNOS), which leads to less production of the highly reactive and potent ONOO$^-$ free radical [80]. Bilirubin inhibited iNOS expression and NO production in response to endotoxin in murine macrophages and in rats. CO also mediates part of the antioxidant and anti-inflammatory effects of HO-1. The increase of CO-exposure, whether produced endogenously from induction of HO-1 or delivered exogenously via a CO-releasing molecule, inhibited the LPS-derived upregulation of iNOS expression and NO overproduction in macrophages [81].

4. HO-1 Expression in Atherosclerotic Diseases States (Animal Studies)

4.1. Myocardial Infarction

HO-1 is suggested to be a meaningful player in the maintenance of cardiac homeostasis and the subsequent cardiac damage. Sharma et al. [82] demonstrated that ischemia/reperfusion substantially enhanced HO-1 expression in the porcine heart, suggesting a potential role of HO-1 in the defense against pathophysiological stress. In HO-1-deficient (*Hmox1$^{-/-}$*) mice, hypoxia induced severe right ventricular dilatation and infarction [83]. The absence of *Hmox1* was reported to exacerbate ischemia/reperfusion-induced myocardial damage [84]. In contrast, a cardiac-specific overexpression of HO-1 reduced the myocardial infarct size and the inflammatory cell infiltration after ischemia/reperfusion [85]. The transfer of human HO-1 gene (*HMOX1*) before myocardial injury provided long-term myocardial protection from ischemia/reperfusion injury [86]. Tang et al.

demonstrated that *HMOX1* gene transfer improved the contractile and diastolic performance after myocardial infarction in mice [87,88]. Issan et al. [89] reported that HO-1 induction by cobalt-protoporphyrin (CoPP) improved the cardiac function and decreased the infarct size in diabetic mice subjected to myocardial infarction. They also demonstrated that HO-1 induction increased the activity of the Akt prosurvival pathway in cardiomyocytes and decreased the plasma TNF-α level.

4.2. Heart Failure

In heart failure model mice produced by coronary ligation, myocyte-specific HO-1 overexpression improved the postinfarction survival and alleviated left ventricular remodeling; it also promoted neovascularization and ameliorated apoptosis [90]. Cardiac HO-1 overexpression could be either protective or detrimental in the heart depending on the type of stress context. Allwood et al. [91] demonstrated that cardiac-specific HO-1 overexpression significantly attenuated cardiac dysfunction, interstitial fibrosis, and hypertrophy induced by isoproterenol, whereas HO-1 had detrimental effects on the development of cardiomyopathic heart failure induced by pressure overload or aging.

5. HO-1 Expression in Patients with Atherosclerotic Diseases (Clinical Studies)

5.1. Carotid Atherosclerosis

One of the major problems in CAD is related to the significant length of time between the start of subclinical atherosclerosis and the manifestation of the disease, highlighting the importance of identifying biomarkers that can be used to predict CVD progression at as early a stage as possible. Elevated blood levels of HO-1 have been reported in chronic diseases such as diabetes mellitus [92], chronic silicosis [93], Parkinson's disease [94], and hemophagocytic syndrome [95]. Although the precise mechanisms of secretion and the significance of the extracellular HO-1 remain to be determined, HO-1 is known to be released into the plasma by leukocytes, macrophages, smooth muscle cells, and endothelial cells that are activated or damaged by oxidative stress or inflammation [4]. Kishimoto et al. hypothesized that plasma HO-1 levels may be associated with the presence and severity of carotid atherosclerosis, and they measured the plasma HO-1 levels in 136 consecutive subjects (mean age 66 ± 9 years) who underwent carotid ultrasonography for a medical check-up to evaluate atherosclerosis [25]. The study was the first to reveal that the plasma HO-1 levels were significantly higher in the subjects with carotid plaque than in those without plaque (median 0.56 versus 0.44 ng/mL, $p < 0.05$), and the levels were stepwise increased depending on the severity of plaque, defined as the plaque score (Figure 1). Moreover, the plasma HO-1 levels were significantly correlated with the plaque score ($r = 0.23$, $p < 0.01$ by Spearman's rank correlation test). In a multivariate analysis, high HO-1 level (>0.50 ng/mL) was a significant factor associated with the presence of carotid plaque, independent of atherosclerotic risk factors (odds ratio: 2.33, 95% CI: 1.15–4.75, $p < 0.025$). Thus, the study reported plasma HO-1 levels to be high in subjects with carotid plaques and to be associated with the severity of carotid atherosclerosis. High plasma HO-1 levels may reflect an increased oxidative stress condition and may be aimed at protecting the body against the progression of atherosclerosis.

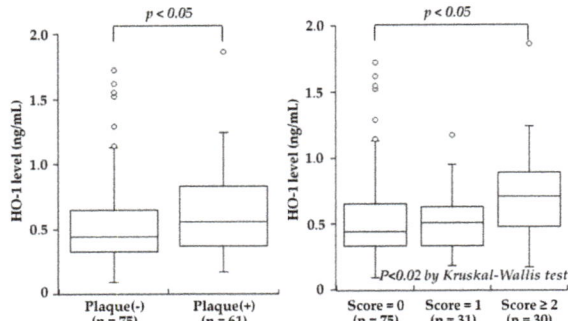

Figure 1. Plasma HO-1 levels and the presence of carotid plaque or the plaque score. Plasma HO-1 levels were significantly higher in subjects with carotid plaque than in those without plaque ($p < 0.05$) (**left**). A stepwise increase in HO-1 levels was found depending on the plaque score: 0.44 ng/mL in subjects with score = 0, 0.51 ng/mL in score = 1, and 0.70 ng/mL in score ≥ 2 ($p < 0.02$). HO-1 levels in score ≥ 2 were higher than those in score = 0 ($p < 0.05$) (**right**). The central line represents the median, and the box represents the 25th to 75th percentiles. The whiskers represent the lowest and highest value in the 25th percentile minus 1.5 IQR and 75th percentile plus 1.5 IQR, respectively. (Modified from Kishimoto et al. [25]).

5.2. Coronary Artery Disease (CAD) and Peripheral Artery Disease (PAD)

CAD and PAD are chronic progressive atherosclerotic diseases leading to thrombosis and ischemia. CAD is the most common type of atherosclerotic diseases, followed by stroke and PAD [96]. PAD is a common circulatory problem in which narrowed arteries reduce the blood flow, mainly to the lower limbs. Many patients with PAD have mild or no symptoms, but some may have leg pain when walking. Patients with PAD often suffer from multiple arterial co-morbidities leading to high CVD-related mortality or a poor prognosis within a short time frame.

Cheng et al. [19] reported that the HO-1 expression in carotid endarterectomy samples was higher in unstable plaques than in stable plaques. They showed that the HO-1 level was positively correlated with features of vulnerable human atheromatous plaque, such as macrophage and lipid accumulation, and that the HO-1 level was inversely correlated with stable plaque features like the presence of intra-plaque smooth muscle cells and collagen. Yunoki et al. [22] also reported that the majority of HO-1-positive cells were macrophages, and the percentage of HO-1-positive areas was significantly higher in coronary atherectomy samples from patients with unstable angina pectoris (UAP) compared to those from patients with stable angina pectoris (SAP). Ijas et al. [17] reported that symptomatic plaques overexpressed HO-1 and CD163 which is involved in the degradation of hemoglobin and such expressions were correlated with traditional markers of unstable carotid disease, i.e., the degree of carotid stenosis and plaque ulcerations. In contrast, Ameriso et al. [16] investigated HO-1 expression in relation to *Helicobacter pylori* (*H. pylori*) infection and stated that HO-1 expression was more frequent in infected and asymptomatic carotid plaques. They suggested a potential role of *H. pylori* in oxidative stress-mediated injury and a subsequent defense reaction represented by HO-1 expression. In blood leukocytes, high HO-1 expression was observed by Chen et al. in patients with CAD, especially those with acute myocardial infarction (AMI) or UAP [15]. Chen et al. also showed that the mRNA expression of *HMOX1* in leukocytes was associated with the severity of CAD [97]. Brydun et al. [18] demonstrated that the capacity to upregulate *HMOX1* mRNA expression in leukocytes was inversely related to the degree of CAD. More recently, Fiorelli et al. [27] detected higher levels of HO-1 and Nrf2 in monocyte-derived macrophages (MDMs) of their CAD patients compared to those of healthy subjects. Of note, the patients with high levels of HO-1 more frequently displayed a thin cap fibroatheroma, a ruptured plaque, and the presence of thrombi.

Several studies recently examined blood HO-1 levels in patients with atherosclerotic diseases. Idriss et al. [20] noted that the plasma HO-1 levels was raised in patients with stable CAD and increased further in those with acute coronary syndrome (ACS) compared to controls. Novo et al. [21] found that the serum HO-1 levels in patients with AMI were significantly higher compared to those of controls, and they revealed an inverse association with the severity of CAD. They also indicated that the HO-1 sequence was compatible with mechanisms of secretion and that therefore, its presence in the serum of patients might not necessarily be dependent on cell necrosis. Kishimoto et al. recently investigated the plasma HO-1 levels in 410 consecutive patients undergoing elective coronary angiography for suspected CAD who also had an ankle-brachial index (ABI) test to screen for PAD [26]. The plasma HO-1 levels did not differ between the patients with and without CAD (median 0.44 versus 0.35 ng/mL, p = NS). Notably, the HO-1 levels were significantly lower in the patients with PAD than in those without PAD (median 0.27 versus 0.41 ng/mL, p < 0.02) (Figure 2). However, the patients with PAD more often had CAD, especially three-vessel disease, compared to the patients without PAD (92% versus 51%, p < 0.01). After excluding the patients with PAD, the HO-1 levels were significantly higher in the patients with CAD than in those without CAD (0.45 versus 0.35 ng/mL, p < 0.05) and were highest in the patients with one-vessel disease among the four groups of CAD(-), one-vessel (1-VD), two-vessel (2-VD), and three-vessel disease (3-VD) (0.35, 0.49, 0.44, and 0.44 ng/mL, p < 0.05) (Figure 3). In a multivariate analysis, the odds ratios for CAD and PAD were 0.65 (95% CI: 0.42–0.99, p < 0.05) and 2.12 (95% CI: 1.03–4.37, p < 0.05) for low HO-1 level (<0.35 ng/mL), respectively. Therefore, plasma HO-1 levels were found to be low in patients with PAD, in contrast to high levels in patients with CAD. These results thus suggested that high plasma HO-1 levels in patients with CAD, especially one-vessel disease, may be aimed at protecting against the progression of CAD. In contrast, low plasma levels of HO-1 may be a marker reflecting the presence of PAD and may play a role in the development of PAD. This is in line with the results reported by Signorelli et al. [24], who noted that the serum HO-1 levels were lower in 27 patients with PAD compared to 27 controls. Although the mechanism of low plasma HO-1 levels in patients with PAD remains unclear, the HO-1 defensive response to oxidative stress was reported to be attenuated at advanced age [98] and at the late stage of diabetes mellitus [99]. A long duration of a severe stress condition may therefore cause some disruption of the HO-1 defense system. Gene and cell therapy with HO-1 were shown to be effective in animal models of limb ischemia [100,101]. Since patients with PAD have low HO-1 levels in blood, HO-1 inducers may be used to treat patients with PAD to inhibit the progression of PAD.

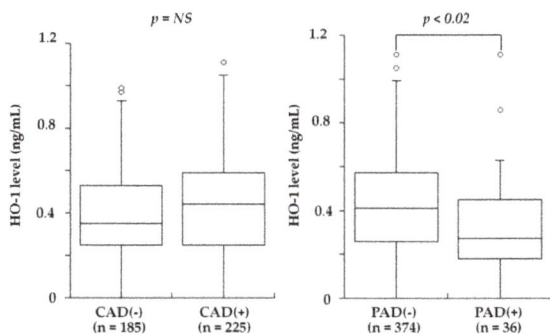

Figure 2. Plasma HO-1 levels and the presence of CAD or PAD. Plasma HO-1 levels tended to be higher in patients with CAD than in CAD(-) (median 0.44 versus 0.35 ng/mL), but this difference did not reach statistical significance (**left**). In contrast, HO-1 levels were significantly lower in patients with PAD than in PAD(-) (0.27 versus 0.41 ng/mL, p < 0.02) (**right**). The central line represents the median, and the box represents the 25th to 75th percentiles. The whiskers represent the lowest and highest value in the 25th percentile minus 1.5 IQR and 75th percentile plus 1.5 IQR, respectively. (Modified by Kishimoto et al. [26]).

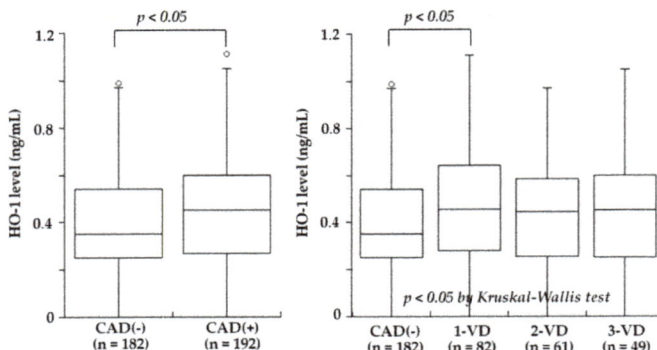

Figure 3. Plasma HO-1 levels and the presence of CAD or the number of stenotic coronary vessels among the 374 patients without PAD. After excluding the 36 patients with PAD, HO-1 levels were significantly higher in patients with CAD than in CAD(-) (median 0.45 versus 0.35 ng/mL, $p < 0.05$) (**left**). Furthermore, HO-1 levels in the 4 groups of CAD(-), 1-VD, 2-VD, and 3-VD were 0.35, 0.49, 0.44, and 0.44 ng/mL, respectively, and were highest in 1-VD ($p < 0.05$) (**right**). The central line represents the median, and the box represents the 25th to 75th percentiles. The whiskers represent the lowest and highest value in the 25th percentile minus 1.5 IQR and 75th percentile plus 1.5 IQR, respectively. (Modified by Kishimoto et al. [26]).

6. Conclusions

Taken together, the above studies' results strongly support the notion that HO-1 plays a protective role against the progression of atherosclerotic diseases, such as CAD and PAD. In the pathogenesis of atherosclerosis, the ability of HO-1 to generate bilirubin, an anti-oxidant molecule and an agonist for PPAR-α, and CO, a vasodilator and an anti-inflammatory and antiapoptotic molecule, is thought to play important roles. Although the relevance of pharmacological or gene therapy with HO-1 to atherosclerotic disease in humans has yet to be established, the overall outcome of the preclinical studies carried out clearly points to HO-1 as a potential therapeutic target in atherosclerotic diseases. It is also of interest that the anti-atherogenic effects of statins (HMG-CoA reductase inhibitors) and fibrates (PPAR ligands) are partly mediated through HO-1 induction [102–104]. A number of natural antioxidant compounds contained in foods and plants, such as curcumin and caffeic acid phenethyl ester (polyphenols), and sulforaphane (isothiocyanates), have been demonstrated to be effective inducers of HO-1 and exert defensive actions against oxidative stress-related diseases. [105–108]. Importantly, further prospective studies are needed to determine the precise association between plasma HO-1 levels and the progression of carotid atherosclerosis as well as CAD and PAD.

Funding: This work was supported in part by a grant from Honjo International Scholarship Foundation and JSPS KAKENHI (Grant No. 17K00847) from the Japan Society for the Promotion of Science.

Conflicts of Interest: The authors declare no conflict of interest.

Abbreviations

HO	Heme oxygenase
CO	Carbon monoxide
CAD	Coronary artery disease
PAD	Peripheral artery disease
LDL	Low-density lipoprotein
UGT1A1	UDP-glucuronosyltransferase 1A1

PPAR-α	Peroxisome proliferator-activated receptor-alpha
MCP-1	Monocyte chemoattractant protein-1
IL-6	Interleukin-6
SR-A	Scavenger receptor-A
AMI	Acute myocardial infarction
ABI	Ankle-brachial index
CI	Confidence interval
SAP	Stable angina pectoris
UAP	Unstable angina pectoris
ACS	Acute coronary syndrome
TIA	Transient ischemic attack

References

1. Libby, P.; Ridker, P.M.; Maseri, A. Inflammation and atherosclerosis. *Circulation* **2002**, *105*, 1135–1143. [PubMed]
2. Immenschuh, S.; Vijayan, V.; Janciauskiene, S.; Gueler, F. Heme as a target for therapeutic interventions. *Front. Pharmacol.* **2017**, *8*, 146. [PubMed]
3. Ryter, S.W.; Alam, J.; Choi, A.M. Heme oxygenase-1/carbon monoxide: From basic science to therapeutic applications. *Physiol. Rev.* **2006**, *86*, 583–650. [PubMed]
4. Abraham, N.G.; Kappas, A. Pharmacological and clinical aspects of heme oxygenase. *Pharmacol. Rev.* **2008**, *60*, 79–127. [PubMed]
5. Fredenburgh, L.E.; Merz, A.A.; Cheng, S. Haeme oxygenase signalling pathway: Implications for cardiovascular disease. *Eur. Heart J.* **2015**, *36*, 1512–1518. [PubMed]
6. Maines, M.D. The heme oxygenase system: A regulator of second messenger gases. *Ann. Rev. Pharmacol. Toxicol.* **1997**, *37*, 517–554.
7. Wu, M.L.; Ho, Y.C.; Lin, C.Y.; Yet, S.F. Heme oxygenase-1 in inflammation and cardiovascular disease. *Am. J. Cardiovasc. Dis.* **2011**, *1*, 150–158.
8. Yoshida, T.; Biro, P.; Cohen, T.; Muller, R.M.; Shibahara, S. Human heme oxygenase cdna and induction of its mrna by hemin. *Eur. J. Biochem.* **1988**, *171*, 457–461.
9. Alam, J.; Shibahara, S.; Smith, A. Transcriptional activation of the heme oxygenase gene by heme and cadmium in mouse hepatoma cells. *J. Biol. Chem.* **1989**, *264*, 6371–6375.
10. Keyse, S.M.; Tyrrell, R.M. Induction of the heme oxygenase gene in human skin fibroblasts by hydrogen peroxide and uva (365 nm) radiation: Evidence for the involvement of the hydroxyl radical. *Carcinogenesis* **1990**, *11*, 787–791.
11. Cantoni, L.; Rossi, C.; Rizzardini, M.; Gadina, M.; Ghezzi, P. Interleukin-1 and tumour necrosis factor induce hepatic haem oxygenase. Feedback regulation by glucocorticoids. *Biochem. J.* **1991**, *279*, 891–894. [PubMed]
12. Rizzardini, M.; Carelli, M.; Cabello Porras, M.R.; Cantoni, L. Mechanisms of endotoxin-induced haem oxygenase mrna accumulation in mouse liver: Synergism by glutathione depletion and protection by n-acetylcysteine. *Biochem. J.* **1994**, *304*, 477–483. [PubMed]
13. Hayashi, S.; Omata, Y.; Sakamoto, H.; Higashimoto, Y.; Hara, T.; Sagara, Y.; Noguchi, M. Characterization of rat heme oxygenase-3 gene. Implication of processed pseudogenes derived from heme oxygenase-2 gene. *Gene* **2004**, *336*, 241–250. [PubMed]
14. Wang, L.J.; Lee, T.S.; Lee, F.Y.; Pai, R.C.; Chau, L.Y. Expression of heme oxygenase-1 in atherosclerotic lesions. *Am. J. Pathol.* **1998**, *152*, 711–720. [PubMed]
15. Chen, S.M.; Li, Y.G.; Wang, D.M. Study on changes of heme oxygenase-1 expression in patients with coronary heart disease. *Clin. Cardiol.* **2005**, *28*, 197–201. [PubMed]
16. Ameriso, S.F.; Villamil, A.R.; Zedda, C.; Parodi, J.C.; Garrido, S.; Sarchi, M.I.; Schultz, M.; Boczkowski, J.; Sevlever, G.E. Heme oxygenase-1 is expressed in carotid atherosclerotic plaques infected by helicobacter pylori and is more prevalent in asymptomatic subjects. *Stroke* **2005**, *36*, 1896–1900.
17. Ijas, P.; Nuotio, K.; Saksi, J.; Soinne, L.; Saimanen, E.; Karjalainen-Lindsberg, M.L.; Salonen, O.; Sarna, S.; Tuimala, J.; Kovanen, P.T.; et al. Microarray analysis reveals overexpression of cd163 and ho-1 in symptomatic carotid plaques. *Arterioscler. Thromb Vasc. Biol* **2007**, *27*, 154–160.

18. Brydun, A.; Watari, Y.; Yamamoto, Y.; Okuhara, K.; Teragawa, H.; Kono, F.; Chayama, K.; Oshima, T.; Ozono, R. Reduced expression of heme oxygenase-1 in patients with coronary atherosclerosis. *Hypertens. Res.* **2007**, *30*, 341–348.

19. Cheng, C.; Noordeloos, A.M.; Jeney, V.; Soares, M.P.; Moll, F.; Pasterkamp, G.; Serruys, P.W.; Duckers, H.J. Heme oxygenase 1 determines atherosclerotic lesion progression into a vulnerable plaque. *Circulation* **2009**, *119*, 3017–3027.

20. Idriss, N.K.; Lip, G.Y.; Balakrishnan, B.; Jaumdally, R.; Boos, C.J.; Blann, A.D. Plasma haemoxygenase-1 in coronary artery disease. A comparison with angiogenin, matrix metalloproteinase-9, tissue inhibitor of metalloproteinase-1 and vascular endothelial growth factor. *Thromb. Haemost.* **2010**, *104*, 1029–1037.

21. Novo, G.; Cappello, F.; Rizzo, M.; Fazio, G.; Zambuto, S.; Tortorici, E.; Marino Gammazza, A.; Corrao, S.; Zummo, G.; De Macario, E.C.; et al. Hsp60 and heme oxygenase-1 (hsp32) in acute myocardial infarction. *Transl. Res.* **2011**, *157*, 285–292. [PubMed]

22. Yunoki, K.; Inoue, T.; Sugioka, K.; Nakagawa, M.; Inaba, M.; Wada, S.; Ohsawa, M.; Komatsu, R.; Itoh, A.; Haze, K.; et al. Association between hemoglobin scavenger receptor and heme oxygenase-1-related anti-inflammatory mediators in human coronary stable and unstable plaques. *Hum. Pathol.* **2013**, *44*, 2256–2265. [PubMed]

23. Li, X.; Song, G.; Jin, Y.; Liu, H.; Li, C.; Han, C.; Ren, S. Higher level of heme oxygenase-1 in patients with stroke than tia. *J. Thorac. Dis.* **2014**, *6*, 772–777. [PubMed]

24. Signorelli, S.S.; Li Volsi, G.; Fiore, V.; Mangiafico, M.; Barbagallo, I.; Parenti, R.; Rizzo, M.; Li Volti, G. Plasma heme oxygenase-1 is decreased in peripheral artery disease patients. *Mol. Med. Rep.* **2016**, *14*, 3459–3463. [PubMed]

25. Kishimoto, Y.; Sasaki, K.; Saita, E.; Niki, H.; Ohmori, R.; Kondo, K.; Momiyama, Y. Plasma heme oxygenase-1 levels and carotid atherosclerosis. *Stroke* **2018**, *49*, 2230–2232. [PubMed]

26. Kishimoto, Y.; Ibe, S.; Saita, E.; Sasaki, K.; Niki, H.; Miura, K.; Ikegami, Y.; Ohmori, R.; Kondo, K.; Momiyama, Y. Plasma heme oxygenase-1 levels in patients with coronary and peripheral artery diseases. *Dis. Markers* **2018**, *2018*, 6138124. [PubMed]

27. Fiorelli, S.; Porro, B.; Cosentino, N.; Di Minno, A.; Manega, C.M.; Fabbiocchi, F.; Niccoli, G.; Fracassi, F.; Barbieri, S.; Marenzi, G.; et al. Activation of nrf2/ho-1 pathway and human atherosclerotic plaque vulnerability:An in vitro and in vivo study. *Cells* **2019**, *8*, 356.

28. Yachie, A.; Niida, Y.; Wada, T.; Igarashi, N.; Kaneda, H.; Toma, T.; Ohta, K.; Kasahara, Y.; Koizumi, S. Oxidative stress causes enhanced endothelial cell injury in human heme oxygenase-1 deficiency. *J. Clin. Investig.* **1999**, *103*, 129–135. [PubMed]

29. Radhakrishnan, N.; Yadav, S.P.; Sachdeva, A.; Pruthi, P.K.; Sawhney, S.; Piplani, T.; Wada, T.; Yachie, A. Human heme oxygenase-1 deficiency presenting with hemolysis, nephritis, and asplenia. *J. Pediatr. Hematol. Oncol.* **2011**, *33*, 74–78. [PubMed]

30. Poss, K.D.; Tonegawa, S. Reduced stress defense in heme oxygenase 1-deficient cells. *Proc. Natl. Acad. Sci. USA* **1997**, *94*, 10925–10930. [PubMed]

31. Poss, K.D.; Tonegawa, S. Heme oxygenase 1 is required for mammalian iron reutilization. *Proc. Natl. Acad. Sci. USA* **1997**, *94*, 10919–10924. [PubMed]

32. Ishikawa, K.; Navab, M.; Lusis, A.J. Vasculitis, atherosclerosis, and altered hdl composition in heme-oxygenase-1-knockout mice. *Int. J. Hypertens.* **2012**, *2012*, 948203. [PubMed]

33. Agarwal, A.; Balla, J.; Balla, G.; Croatt, A.J.; Vercellotti, G.M.; Nath, K.A. Renal tubular epithelial cells mimic endothelial cells upon exposure to oxidized ldl. *Am. J. Physiol.* **1996**, *271*, F814–F823. [PubMed]

34. Yamaguchi, M.; Sato, H.; Bannai, S. Induction of stress proteins in mouse peritoneal macrophages by oxidized low-density lipoprotein. *Biochem. Biophys. Res. Commun.* **1993**, *193*, 1198–1201. [PubMed]

35. Ishikawa, K.; Navab, M.; Leitinger, N.; Fogelman, A.M.; Lusis, A.J. Induction of heme oxygenase-1 inhibits the monocyte transmigration induced by mildly oxidized ldl. *J. Clin. Investig.* **1997**, *100*, 1209–1216.

36. Ishikawa, K.; Sugawara, D.; Wang, X.; Suzuki, K.; Itabe, H.; Maruyama, Y.; Lusis, A.J. Heme oxygenase-1 inhibits atherosclerotic lesion formation in ldl-receptor knockout mice. *Circ. Res.* **2001**, *88*, 506–512.

37. Ishikawa, K.; Sugawara, D.; Goto, J.; Watanabe, Y.; Kawamura, K.; Shiomi, M.; Itabe, H.; Maruyama, Y. Heme oxygenase-1 inhibits atherogenesis in watanabe heritable hyperlipidemic rabbits. *Circulation* **2001**, *104*, 1831–1836.

38. Juan, S.H.; Lee, T.S.; Tseng, K.W.; Liou, J.Y.; Shyue, S.K.; Wu, K.K.; Chau, L.Y. Adenovirus-mediated heme oxygenase-1 gene transfer inhibits the development of atherosclerosis in apolipoprotein e-deficient mice. *Circulation* **2001**, *104*, 1519–1525.

39. Yet, S.F.; Layne, M.D.; Liu, X.; Chen, Y.H.; Ith, B.; Sibinga, N.E.; Perrella, M.A. Absence of heme oxygenase-1 exacerbates atherosclerotic lesion formation and vascular remodeling. *FASEB J.* **2003**, *17*, 1759–1761.

40. Qiao, H.; Sai, X.; Gai, L.; Huang, G.; Chen, X.; Tu, X.; Ding, Z. Association between heme oxygenase 1 gene promoter polymorphisms and susceptibility to coronary artery disease: A huge review and meta-analysis. *Am. J. Epidemiol.* **2014**, *179*, 1039–1048.

41. Yamada, N.; Yamaya, M.; Okinaga, S.; Nakayama, K.; Sekizawa, K.; Shibahara, S.; Sasaki, H. Microsatellite polymorphism in the heme oxygenase-1 gene promoter is associated with susceptibility to emphysema. *Am. J. Hum. Genet.* **2000**, *66*, 187–195. [PubMed]

42. Ono, K.; Goto, Y.; Takagi, S.; Baba, S.; Tago, N.; Nonogi, H.; Iwai, N. A promoter variant of the heme oxygenase-1 gene may reduce the incidence of ischemic heart disease in japanese. *Atherosclerosis* **2004**, *173*, 315–319. [PubMed]

43. Schillinger, M.; Exner, M.; Mlekusch, W.; Domanovits, H.; Huber, K.; Mannhalter, C.; Wagner, O.; Minar, E. Heme oxygenase-1 gene promoter polymorphism is associated with abdominal aortic aneurysm. *Thromb. Res.* **2002**, *106*, 131–136. [PubMed]

44. Kaneda, H.; Ohno, M.; Taguchi, J.; Togo, M.; Hashimoto, H.; Ogasawara, K.; Aizawa, T.; Ishizaka, N.; Nagai, R. Heme oxygenase-1 gene promoter polymorphism is associated with coronary artery disease in japanese patients with coronary risk factors. *Arterioscler. Thromb. Vasc. Biol.* **2002**, *22*, 1680–1685. [PubMed]

45. Funk, M.; Endler, G.; Schillinger, M.; Mustafa, S.; Hsieh, K.; Exner, M.; Lalouschek, W.; Mannhalter, C.; Wagner, O. The effect of a promoter polymorphism in the heme oxygenase-1 gene on the risk of ischaemic cerebrovascular events: The influence of other vascular risk factors. *Thromb. Res.* **2004**, *113*, 217–223. [PubMed]

46. Chen, Y.H.; Chau, L.Y.; Chen, J.W.; Lin, S.J. Serum bilirubin and ferritin levels link heme oxygenase-1 gene promoter polymorphism and susceptibility to coronary artery disease in diabetic patients. *Diabetes Care* **2008**, *31*, 1615–1620. [PubMed]

47. Chen, M.; Zhou, L.; Ding, H.; Huang, S.; He, M.; Zhang, X.; Cheng, L.; Wang, D.; Hu, F.B.; Wu, T. Short (gt) (n) repeats in heme oxygenase-1 gene promoter are associated with lower risk of coronary heart disease in subjects with high levels of oxidative stress. *Cell Stress Chaperones* **2012**, *17*, 329–338. [PubMed]

48. Itoh, K.; Chiba, T.; Takahashi, S.; Ishii, T.; Igarashi, K.; Katoh, Y.; Oyake, T.; Hayashi, N.; Satoh, K.; Hatayama, I.; et al. An nrf2/small maf heterodimer mediates the induction of phase ii detoxifying enzyme genes through antioxidant response elements. *Biochem. Biophys. Res. Commun.* **1997**, *236*, 313–322. [PubMed]

49. Ooi, B.K.; Goh, B.H.; Yap, W.H. Oxidative stress in cardiovascular diseases: Involvement of nrf2 antioxidant redox signaling in macrophage foam cells formation. *Int. J. Mol. Sci.* **2017**, *18*, 2336.

50. Sivandzade, F.; Prasad, S.; Bhalerao, A.; Cucullo, L. Nrf2 and nf-b interplay in cerebrovascular and neurodegenerative disorders: Molecular mechanisms and possible therapeutic approaches. *Redox Biol.* **2019**, *21*, 101059. [PubMed]

51. Stocker, R.; Yamamoto, Y.; McDonagh, A.F.; Glazer, A.N.; Ames, B.N. Bilirubin is an antioxidant of possible physiological importance. *Science* **1987**, *235*, 1043–1046. [PubMed]

52. Mayer, M. Association of serum bilirubin concentration with risk of coronary artery disease. *Clin. Chem.* **2000**, *46*, 1723–1727. [PubMed]

53. Boon, A.C.; Hawkins, C.L.; Bisht, K.; Coombes, J.S.; Bakrania, B.; Wagner, K.H.; Bulmer, A.C. Reduced circulating oxidized ldl is associated with hypocholesterolemia and enhanced thiol status in gilbert syndrome. *Free Radic. Biol. Med.* **2012**, *52*, 2120–2127. [PubMed]

54. Stojanov, M.; Stefanovic, A.; Dzingalasevic, G.; Ivanisevic, J.; Miljkovic, M.; Mandic-Radic, S.; Prostran, M. Total bilirubin in young men and women: Association with risk markers for cardiovascular diseases. *Clin. Biochem.* **2013**, *46*, 1516–1519. [PubMed]

55. Nascimento, H.; Alves, A.I.; Coimbra, S.; Catarino, C.; Gomes, D.; Bronze-da-Rocha, E.; Costa, E.; Rocha-Pereira, P.; Aires, L.; Mota, J.; et al. Bilirubin is independently associated with oxidized ldl levels in young obese patients. *Diabetol. Metab. Syndr.* **2015**, *7*, 4. [PubMed]

56. Schwertner, H.A.; Jackson, W.G.; Tolan, G. Association of low serum concentration of bilirubin with increased risk of coronary artery disease. *Clin. Chem.* **1994**, *40*, 18–23. [PubMed]

57. Turfan, M.; Duran, M.; Poyraz, F.; Yayla, C.; Akboga, M.K.; Sahinarslan, A.; Tavil, Y.; Pasaoglu, H.; Boyaci, B. Inverse relationship between serum total bilirubin levels and severity of disease in patients with stable coronary artery disease. *Coron. Artery Dis.* **2013**, *24*, 29–32.

58. Kang, S.J.; Kim, D.; Park, H.E.; Chung, G.E.; Choi, S.H.; Choi, S.Y.; Lee, W.; Kim, J.S.; Cho, S.H. Elevated serum bilirubin levels are inversely associated with coronary artery atherosclerosis. *Atherosclerosis* **2013**, *230*, 242–248.

59. Akboga, M.K.; Canpolat, U.; Sahinarslan, A.; Alsancak, Y.; Nurkoc, S.; Aras, D.; Aydogdu, S.; Abaci, A. Association of serum total bilirubin level with severity of coronary atherosclerosis is linked to systemic inflammation. *Atherosclerosis* **2015**, *240*, 110–114.

60. Lin, J.P.; Vitek, L.; Schwertner, H.A. Serum bilirubin and genes controlling bilirubin concentrations as biomarkers for cardiovascular disease. *Clin. Chem.* **2010**, *56*, 1535–1543.

61. Stender, S.; Frikke-Schmidt, R.; Nordestgaard, B.G.; Grande, P.; Tybjaerg-Hansen, A. Genetically elevated bilirubin and risk of ischaemic heart disease: Three mendelian randomization studies and a meta-analysis. *J. Intern. Med.* **2013**, *273*, 59–68. [PubMed]

62. Kobayashi, A.; Ishikawa, K.; Matsumoto, H.; Kimura, S.; Kamiyama, Y.; Maruyama, Y. Synergetic antioxidant and vasodilatory action of carbon monoxide in angiotensin ii - induced cardiac hypertrophy. *Hypertension* **2007**, *50*, 1040–1048. [PubMed]

63. Taille, C.; El-Benna, J.; Lanone, S.; Dang, M.C.; Ogier-Denis, E.; Aubier, M.; Boczkowski, J. Induction of heme oxygenase-1 inhibits nad(p)h oxidase activity by down-regulating cytochrome b558 expression via the reduction of heme availability. *J. Biol. Chem.* **2004**, *279*, 28681–28688. [PubMed]

64. Abraham, N.G.; Junge, J.M.; Drummond, G.S. Translational significance of heme oxygenase in obesity and metabolic syndrome. *Trends Pharmacol. Sci.* **2016**, *37*, 17–36. [PubMed]

65. Hinds, T.D., Jr.; Sodhi, K.; Meadows, C.; Fedorova, L.; Puri, N.; Kim, D.H.; Peterson, S.J.; Shapiro, J.; Abraham, N.G.; Kappas, A. Increased ho-1 levels ameliorate fatty liver development through a reduction of heme and recruitment of fgf21. *Obesity (Silver Spring)* **2014**, *22*, 705–712.

66. Stec, D.E.; John, K.; Trabbic, C.J.; Luniwal, A.; Hankins, M.W.; Baum, J.; Hinds, T.D., Jr. Bilirubin binding to pparalpha inhibits lipid accumulation. *PLoS ONE* **2016**, *11*, e0153427.

67. Gordon, D.M.; Blomquist, T.M.; Miruzzi, S.A.; McCullumsmith, R.; Stec, D.E.; Hinds, T.D., Jr. Rna sequencing in human hepg2 hepatocytes reveals ppar-alpha mediates transcriptome responsiveness of bilirubin. *Physiol. Genom.* **2019**, *51*, 234–240.

68. Kawamura, K.; Ishikawa, K.; Wada, Y.; Kimura, S.; Matsumoto, H.; Kohro, T.; Itabe, H.; Kodama, T.; Maruyama, Y. Bilirubin from heme oxygenase-1 attenuates vascular endothelial activation and dysfunction. *Arterioscler. Thromb. Vasc. Biol.* **2005**, *25*, 155–160. [PubMed]

69. Kim, J.A.; Territo, M.C.; Wayner, E.; Carlos, T.M.; Parhami, F.; Smith, C.W.; Haberland, M.E.; Fogelman, A.M.; Berliner, J.A. Partial characterization of leukocyte binding molecules on endothelial cells induced by minimally oxidized ldl. *Arterioscler. Thromb.* **1994**, *14*, 427–433.

70. Zakkar, M.; Van der Heiden, K.; Luong le, A.; Chaudhury, H.; Cuhlmann, S.; Hamdulay, S.S.; Krams, R.; Edirisinghe, I.; Rahman, I.; Carlsen, H.; et al. Activation of nrf2 in endothelial cells protects arteries from exhibiting a proinflammatory state. *Arterioscler. Thromb. Vasc. Biol.* **2009**, *29*, 1851–1857.

71. Kim, M.; Kim, S.; Lim, J.H.; Lee, C.; Choi, H.C.; Woo, C.H. Laminar flow activation of erk5 protein in vascular endothelium leads to atheroprotective effect via nf-e2-related factor 2 (nrf2) activation. *J. Biol. Chem.* **2012**, *287*, 40722–40731. [PubMed]

72. Duckers, H.J.; Boehm, M.; True, A.L.; Yet, S.F.; San, H.; Park, J.L.; Clinton Webb, R.; Lee, M.E.; Nabel, G.J.; Nabel, E.G. Heme oxygenase-1 protects against vascular constriction and proliferation. *Nat. Med.* **2001**, *7*, 693–698. [PubMed]

73. Li, T.; Tian, H.; Zhao, Y.; An, F.; Zhang, L.; Zhang, J.; Peng, J.; Zhang, Y.; Guo, Y. Heme oxygenase-1 inhibits progression and destabilization of vulnerable plaques in a rabbit model of atherosclerosis. *Eur. J. Pharmacol.* **2011**, *672*, 143–152. [PubMed]

74. Kim, K.M.; Pae, H.O.; Zheng, M.; Park, R.; Kim, Y.M.; Chung, H.T. Carbon monoxide induces heme oxygenase-1 via activation of protein kinase r-like endoplasmic reticulum kinase and inhibits endothelial cell apoptosis triggered by endoplasmic reticulum stress. *Circ. Res.* **2007**, *101*, 919–927. [PubMed]

75. Vijayan, V.; Wagener, F.; Immenschuh, S. The macrophage heme-heme oxygenase-1 system and its role in inflammation. *Biochem. Pharmacol.* **2018**, *153*, 159–167. [PubMed]

76. Orozco, L.D.; Kapturczak, M.H.; Barajas, B.; Wang, X.; Weinstein, M.M.; Wong, J.; Deshane, J.; Bolisetty, S.; Shaposhnik, Z.; Shih, D.M.; et al. Heme oxygenase-1 expression in macrophages plays a beneficial role in atherosclerosis. *Circ. Res.* **2007**, *100*, 1703–1711.

77. Ruotsalainen, A.K.; Inkala, M.; Partanen, M.E.; Lappalainen, J.P.; Kansanen, E.; Makinen, P.I.; Heinonen, S.E.; Laitinen, H.M.; Heikkila, J.; Vatanen, T.; et al. The absence of macrophage nrf2 promotes early atherogenesis. *Cardiovasc. Res.* **2013**, *98*, 107–115. [PubMed]

78. Barajas, B.; Che, N.; Yin, F.; Rowshanrad, A.; Orozco, L.D.; Gong, K.W.; Wang, X.; Castellani, L.W.; Reue, K.; Lusis, A.J.; et al. Nf-e2-related factor 2 promotes atherosclerosis by effects on plasma lipoproteins and cholesterol transport that overshadow antioxidant protection. *Arterioscler. Thromb. Vasc. Biol.* **2011**, *31*, 58–66.

79. Thimmulappa, R.K.; Scollick, C.; Traore, K.; Yates, M.; Trush, M.A.; Liby, K.T.; Sporn, M.B.; Yamamoto, M.; Kensler, T.W.; Biswal, S. Nrf2-dependent protection from lps induced inflammatory response and mortality by cddo-imidazolide. *Biochem. Biophys. Res. Commun.* **2006**, *351*, 883–889.

80. Wang, W.W.; Smith, D.L.; Zucker, S.D. Bilirubin inhibits inos expression and no production in response to endotoxin in rats. *Hepatology* **2004**, *40*, 424–433.

81. Srisook, K.; Han, S.S.; Choi, H.S.; Li, M.H.; Ueda, H.; Kim, C.; Cha, Y.N. Co from enhanced ho activity or from corm-2 inhibits both o2- and no production and downregulates ho-1 expression in lps-stimulated macrophages. *Biochem. Pharmacol.* **2006**, *71*, 307–318. [PubMed]

82. Sharma, H.S.; Maulik, N.; Gho, B.C.; Das, D.K.; Verdouw, P.D. Coordinated expression of heme oxygenase-1 and ubiquitin in the porcine heart subjected to ischemia and reperfusion. *Mol. Cell Biochem.* **1996**, *157*, 111–116. [PubMed]

83. Yet, S.F.; Perrella, M.A.; Layne, M.D.; Hsieh, C.M.; Maemura, K.; Kobzik, L.; Wiesel, P.; Christou, H.; Kourembanas, S.; Lee, M.E. Hypoxia induces severe right ventricular dilatation and infarction in heme oxygenase-1 null mice. *J. Clin. Investig.* **1999**, *103*, R23–R29.

84. Liu, X.; Wei, J.; Peng, D.H.; Layne, M.D.; Yet, S.F. Absence of heme oxygenase-1 exacerbates myocardial ischemia/reperfusion injury in diabetic mice. *Diabetes* **2005**, *54*, 778–784. [PubMed]

85. Yet, S.F.; Tian, R.; Layne, M.D.; Wang, Z.Y.; Maemura, K.; Solovyeva, M.; Ith, B.; Melo, L.G.; Zhang, L.; Ingwall, J.S.; et al. Cardiac-specific expression of heme oxygenase-1 protects against ischemia and reperfusion injury in transgenic mice. *Circ. Res.* **2001**, *89*, 168–173. [PubMed]

86. Melo, L.G.; Agrawal, R.; Zhang, L.; Rezvani, M.; Mangi, A.A.; Ehsan, A.; Griese, D.P.; Dell'Acqua, G.; Mann, M.J.; Oyama, J.; et al. Gene therapy strategy for long-term myocardial protection using adeno-associated virus-mediated delivery of heme oxygenase gene. *Circulation* **2002**, *105*, 602–607. [PubMed]

87. Tang, Y.L.; Tang, Y.; Zhang, Y.C.; Qian, K.; Shen, L.; Phillips, M.I. Protection from ischemic heart injury by a vigilant heme oxygenase-1 plasmid system. *Hypertension* **2004**, *43*, 746–751.

88. Tang, Y.L.; Qian, K.; Zhang, Y.C.; Shen, L.; Phillips, M.I. A vigilant, hypoxia-regulated heme oxygenase-1 gene vector in the heart limits cardiac injury after ischemia-reperfusion in vivo. *J. Cardiovasc. Pharmacol. Ther.* **2005**, *10*, 251–263. [PubMed]

89. Issan, Y.; Kornowski, R.; Aravot, D.; Shainberg, A.; Laniado-Schwartzman, M.; Sodhi, K.; Abraham, N.G.; Hochhauser, E. Heme oxygenase-1 induction improves cardiac function following myocardial ischemia by reducing oxidative stress. *PLoS ONE* **2014**, *9*, e92246.

90. Wang, G.; Hamid, T.; Keith, R.J.; Zhou, G.; Partridge, C.R.; Xiang, X.; Kingery, J.R.; Lewis, R.K.; Li, Q.; Rokosh, D.G.; et al. Cardioprotective and antiapoptotic effects of heme oxygenase-1 in the failing heart. *Circulation* **2010**, *121*, 1912–1925.

91. Allwood, M.A.; Kinobe, R.T.; Ballantyne, L.; Romanova, N.; Melo, L.G.; Ward, C.A.; Brunt, K.R.; Simpson, J.A. Heme oxygenase-1 overexpression exacerbates heart failure with aging and pressure overload but is protective against isoproterenol-induced cardiomyopathy in mice. *Cardiovasc. Pathol.* **2014**, *23*, 231–237. [PubMed]

92. Bao, W.; Song, F.; Li, X.; Rong, S.; Yang, W.; Zhang, M.; Yao, P.; Hao, L.; Yang, N.; Hu, F.B.; et al. Plasma heme oxygenase-1 concentration is elevated in individuals with type 2 diabetes mellitus. *PLoS ONE* **2010**, *5*, e12371.

93. Sato, T.; Takeno, M.; Honma, K.; Yamauchi, H.; Saito, Y.; Sasaki, T.; Morikubo, H.; Nagashima, Y.; Takagi, S.; Yamanaka, K.; et al. Heme oxygenase-1, a potential biomarker of chronic silicosis, attenuates silica-induced lung injury. *Am. J. Respir. Crit. Care Med.* **2006**, *174*, 906–914. [PubMed]

94. Mateo, I.; Infante, J.; Sanchez-Juan, P.; Garcia-Gorostiaga, I.; Rodriguez-Rodriguez, E.; Vazquez-Higuera, J.L.; Berciano, J.; Combarros, O. Serum heme oxygenase-1 levels are increased in parkinson's disease but not in alzheimer's disease. *Acta Neurol. Scand.* **2010**, *121*, 136–138. [PubMed]

95. Miyazaki, T.; Kirino, Y.; Takeno, M.; Hama, M.; Ushihama, A.; Watanabe, R.; Takase, K.; Tachibana, T.; Matsumoto, K.; Tanaka, M.; et al. Serum ho-1 is useful to make differential diagnosis of secondary hemophagocytic syndrome from other similar hematological conditions. *Int. J. Hematol.* **2010**, *91*, 229–237. [PubMed]

96. Fowkes, F.G.; Rudan, D.; Rudan, I.; Aboyans, V.; Denenberg, J.O.; McDermott, M.M.; Norman, P.E.; Sampson, U.K.; Williams, L.J.; Mensah, G.A.; et al. Comparison of global estimates of prevalence and risk factors for peripheral artery disease in 2000 and 2010: A systematic review and analysis. *Lancet* **2013**, *382*, 1329–1340. [PubMed]

97. Chen, S.M.; Li, Y.G.; Wang, D.M.; Zhang, G.H.; Tan, C.J. Expression of heme oxygenase-1, hypoxia inducible factor-1alpha, and ubiquitin in peripheral inflammatory cells from patients with coronary heart disease. *Clin. Chem. Lab. Med.* **2009**, *47*, 327–333.

98. Secher, N.; Ostergaard, L.; Tonnesen, E.; Hansen, F.B.; Granfeldt, A. Impact of age on cardiovascular function, inflammation, and oxidative stress in experimental asphyxial cardiac arrest. *Acta Anaesthesiol. Scand.* **2018**, *62*, 49–62.

99. Song, F.; Qi, X.; Chen, W.; Jia, W.; Yao, P.; Nussler, A.K.; Sun, X.; Liu, L. Effect of momordica grosvenori on oxidative stress pathways in renal mitochondria of normal and alloxan-induced diabetic mice. Involvement of heme oxygenase-1. *Eur. J. Nutr.* **2007**, *46*, 61–69.

100. Suzuki, M.; Iso-o, N.; Takeshita, S.; Tsukamoto, K.; Mori, I.; Sato, T.; Ohno, M.; Nagai, R.; Ishizaka, N. Facilitated angiogenesis induced by heme oxygenase-1 gene transfer in a rat model of hindlimb ischemia. *Biochem. Biophys. Res. Commun.* **2003**, *302*, 138–143.

101. Grochot-Przeczek, A.; Kotlinowski, J.; Kozakowska, M.; Starowicz, K.; Jagodzinska, J.; Stachurska, A.; Volger, O.L.; Bukowska-Strakova, K.; Florczyk, U.; Tertil, M.; et al. Heme oxygenase-1 is required for angiogenic function of bone marrow-derived progenitor cells: Role in therapeutic revascularization. *Antioxid. Redox Signal.* **2014**, *20*, 1677–1692. [PubMed]

102. Lee, T.S.; Chang, C.C.; Zhu, Y.; Shyy, J.Y. Simvastatin induces heme oxygenase-1: A novel mechanism of vessel protection. *Circulation* **2004**, *110*, 1296–1302. [PubMed]

103. Heeba, G.; Moselhy, M.E.; Hassan, M.; Khalifa, M.; Gryglewski, R.; Malinski, T. Anti-atherogenic effect of statins: Role of nitric oxide, peroxynitrite and haem oxygenase-1. *Br. J. Pharmacol.* **2009**, *156*, 1256–1266. [PubMed]

104. Wang, Y.; Yu, M.; Ma, Y.; Wang, R.; Liu, W.; Xia, W.; Guan, A.; Xing, C.; Lu, F.; Ji, X. Fenofibrate increases heme oxygenase 1 expression and astrocyte proliferation while limits neuronal injury during intracerebral hemorrhage. *Curr. Neurovasc. Res.* **2017**, *14*, 11–18. [PubMed]

105. Li Volti, G.; Sacerdoti, D.; Di Giacomo, C.; Barcellona, M.L.; Scacco, A.; Murabito, P.; Biondi, A.; Basile, F.; Gazzolo, D.; Abella, R.; et al. Natural heme oxygenase-1 inducers in hepatobiliary function. *World J. Gastroenterol.* **2008**, *14*, 6122–6132. [PubMed]

106. Pittala, V.; Vanella, L.; Salerno, L.; Romeo, G.; Marrazzo, A.; Di Giacomo, C.; Sorrenti, V. Effects of polyphenolic derivatives on heme oxygenase-system in metabolic dysfunctions. *Curr. Med. Chem.* **2018**, *25*, 1577–1595.

107. Pittala, V.; Vanella, L.; Salerno, L.; Di Giacomo, C.; Acquaviva, R.; Raffaele, M.; Romeo, G.; Modica, M.N.; Prezzavento, O.; Sorrenti, V. Novel caffeic acid phenethyl ester (cape) analogues as inducers of heme oxygenase-1. *Curr. Pharm Des.* **2017**, *23*, 2657–2664. [PubMed]

108. Pittala, V.; Salerno, L.; Romeo, G.; Acquaviva, R.; Di Giacomo, C.; Sorrenti, V. Therapeutic potential of caffeic acid phenethyl ester (cape) in diabetes. *Curr. Med. Chem.* **2018**, *25*, 4827–4836.

International Journal of
Molecular Sciences

MDPI

Review

Heme, Heme Oxygenase, and Endoplasmic Reticulum Stress—A New Insight into the Pathophysiology of Vascular Diseases

Tamás Gáll [1,2], György Balla [1,2] and József Balla [2,3,*]

[1] Department of Pediatrics, Faculty of Medicine, University of Debrecen, 4032 Debrecen, Hungary
[2] HAS-UD Vascular Biology and Myocardial Pathophysiology Research Group, Hungarian Academy of Sciences, 4032 Debrecen, Hungary
[3] Department of Internal Medicine, Faculty of Medicine, University of Debrecen, 4032 Debrecen, Hungary
* Correspondence: balla@belklinika.com; Tel.: +36-52-255-600/55004

Received: 18 June 2019; Accepted: 24 July 2019; Published: 26 July 2019

Abstract: The prevalence of vascular disorders continues to rise worldwide. Parallel with that, new pathophysiological pathways have been discovered, providing possible remedies for prevention and therapy in vascular diseases. Growing evidence suggests that endoplasmic reticulum (ER) stress is involved in a number of vasculopathies, including atherosclerosis, vascular brain events, and diabetes. Heme, which is released from hemoglobin or other heme proteins, triggers various pathophysiological consequence, including heme stress as well as ER stress. The potentially toxic free heme is converted by heme oxygenases (HOs) into carbon monoxide (CO), iron, and biliverdin (BV), the latter of which is reduced to bilirubin (BR). Redox-active iron is oxidized and stored by ferritin, an iron sequestering protein which exhibits ferroxidase activity. In recent years, CO, BV, and BR have been shown to control cellular processes such as inflammation, apoptosis, and antioxidant defense. This review covers our current knowledge about how heme induced endoplasmic reticulum stress (HIERS) participates in the pathogenesis of vascular disorders and highlights recent discoveries in the molecular mechanisms of HO-mediated cytoprotection in heme stress and ER stress, as well as crosstalk between ER stress and HO-1. Furthermore, we focus on the translational potential of HIERS and heme oxygenase-1 (HO-1) in atherosclerosis, diabetes mellitus, and brain hemorrhage.

Keywords: heme oxygenase; endoplasmic reticulum stress; hemoglobin; heme

---

1. Introduction

Hemoglobin (Hb) is not an innocent bystander of the pathophysiology in a number of diseases with extra- or intravascular hemorrhage/hemolysis. Free Hb, outside of red blood cells (RBCs), undergoes rapid oxidation from ferrous Hb (Fe^{2+}) to metHb (Fe^{3+}) that can be further oxidized to ferrylHb ($Fe^{4+=O2-}$). Hb oxidation is followed by rapid heme release, resulting in free labile heme pools in the extracellular spaces [1]. Organ/tissue injuries, inherited hemolytic syndromes, sepsis, surgical interventions, brain hemorrhages, atherosclerosis with ruptured plaques, kidney diseases with hematuria, rhabdomyolysis with kidney failure, hemolytic uremic syndromes, diabetic angiopathies, and neonatal retinopathy of prematurity are characteristic disorders for the demonstration of the pathophysiological role of free hemoglobin and heme. Excess free Hb and heme rapidly overwhelm the first line of the endogenous Hb and heme binding homeostatic protective system, the Hb scavenger haptoglobin, and the heme scavenger protein hemopexin and alpha-1-microglobulin. The labile unbound free heme and Hb initiate complex stress reactions, sensitizing cells and tissues towards reactive oxygens species (ROS), provoking cell damage or even cell death.

The second line of the protective system against Hb and heme stress are the intracellular heme oxygenases (HOs) and ferritin. HO enzymes catabolize heme into free ferrous iron, carbon monoxide (CO), and biliverdin (BV) converted to bilirubin (BR) by BV reductases. To eliminate the redox active free iron, cells rapidly express ferritin, an intracellular iron storage protein. The antioxidant character of ferritin depends on its ferroxidase activity and iron sequestering capability [2]. These products of heme catabolism possess a number of physiological functions. BR has been shown to possess remarkable antioxidant effects [3], while CO is an anti-inflammatory and anti-apoptotic gas molecule [4].

Among HOs, HO-1 and HO-2 are extensively characterized [5]. HO-1 is the inducible form, which presents a dramatic intracellular increase of mRNA and protein expressions in response to various environmental stimuli, such as radioactive and ultraviolet irradiation, heavy metals, reactive oxygen species, endotoxin, and several other agents, but most importantly to heme [6–8]. Importantly, the dramatic increase in HO-1 mRNA and protein levels does not ultimately correspond to HO-1 activity. In rat models, both Tin(IV)-protoporphyrin (SnPP) and Cobaltic(III)-protoporphyrin (CoPP) dramatically induce HO-1 expression in the liver. However, SnPP completely inhibits HO-1 activity, while CoPP leads to an overall increase in HO-1 catalytic activity [9,10]. The HO-2 enzyme is generally regarded as a constitutively expressed isoform; however, HO-2 expression also changes in response to hypoxia [11]. In addition, studies underline that HO-2 protects neurons against ischemia/reperfusion injury and oxidative damage [12,13].

Among cellular stress reactions, ER stress is one of the best characterized form of stress. The homeostasis of the ER is a finely tuned system; if newly synthesized misfolded or unfolded protein loads exceed the folding capacity of the ER, the pathways of unfolded protein response (UPR) are activated, leading to ER stress. Overwhelming evidence shows that ER stress is involved in a number diverse pathologies, such as diabetes [14], neurodegenerative diseases [15], rheumatic disorders [16], lung disease [17], and atherosclerosis [18]. Since heme is well known as a cell stressor, the hypothesis has to emerge that there must be a close relationship between ER and heme stress.

In the current article, we review the heme–heme oxygenase–ER stress relationship; the major mechanisms of their interactions by which ER stress contributes to the cell and organ damage in diabetes, atherosclerosis, and brain hemorrhage. Since HO-1 presents a unique Janus-faced character in brain pathologies, this issue has received special attention.

2. Heme Stress

From an evolutionary perspective, protoporphyrin ring is a unique metal chelator with outstanding properties. Chlorophyll, a magnesium-protoporphyrin, converts light energy to chemical energy and produces organic compounds. Heme, an amphipathic iron-protoporphyrin complex, is one of the most important prosthetic groups on Earth, which serves as an oxygen transporter and participates in various oxido/reductive processes in aerobic and anaerobic cell metabolism. However, heme released from the safe sanctuary area of heme proteins triggers a number of adverse effects. In '70s and '80s, several reports revealed the detrimental role of Hb in brain damage [19,20], the heme toxicity in malaria [21], and the potential role of heme in health and disease [22]. Ten years later, we were the first in the literature to be able to generate heme toxicity in cell cultures. We have shown that heme, liberated from heme proteins, is toxic and sensitizes vascular endothelial cells against oxidative stress [2]. Heme, due to its amphipathic nature, shows high affinity towards biological membranes, sensitizing them towards reactive oxygen species (ROS) and leading to the oxidative damage of membrane lipids [23,24], cell lysis [25], genomic [26], and mitochondrial DNA damage [27]. Free heme also triggers protein oxidative modifications [28] and activates the UPR pathways leading ER stress [29]. Moreover, both oxidized Hb and heme are endogenous pro-inflammatory agonists [30,31]. Heme also sensitizes cells to oxidative damage [32] and inflammatory cytokines [33]. Additionally, heme triggers oxidative modifications of lipid particles, such as low-density lipoprotein (LDL), promoting the progression of atherosclerosis [34], which will be discussed later in this review.

Considering the protective effect of HO-1, it is a logical explanation that end-products of heme degradation, BV/BR, and CO are responsible for the beneficial action. There are several experimental conditions where HO-1 provides defense for cells and tissues. Moreover, BR is considered to be an endogenous antioxidant in several clinical conditions. CO and its slow releasing agents (CO releasing molecules, CORMs) have been used as potential medicine, not only in cell culture models, but also in animal studies and clinical investigations. The scientific importance of CO is increasing, since quite a lot of work proves its anti-inflammatory effect.

The nuclear factor-E2-related factor-2 (Nrf2) is activated by diverse environmental stimuli such as oxidative stress, electrophilic, and xenobiotic compounds and plays a pivotal role in coping with oxidative stress [35]. Activation of Nrf2 triggers its dissociation from cytosolic Kelch ECH associating protein 1 (Keap-1) with the subsequent translocation of Nrf2 to the nucleus, where it binds to stress- or antioxidant-response elements (StRE/ARE) encoding a number of genes regulating redox homeostasis, such as HO-1, NAD(P)H:quinone oxidoreductase, glutathione S-transferases, glutamate-cysteine ligase, and glutathione oxidases [36]. On the other hand, HO-1 expression can also be induced by other transcription factors, such as Yin Yang 1 (YY1) [37], activator protein-1 (AP-1) [38], Bach 1 [39], and hypoxia inducible factor 1 [40]. Since HO-1 is abundantly induced by a broad spectrum of endogenous and exogenous stimuli, it is considered as an ideal cytoprotective enzyme.

3. Endoplasmic Reticulum Stress

The homeostasis of the endoplasmic reticulum (ER) is a finely tuned system. If newly synthesized misfolded or unfolded protein loads exceed the folding capacity of the ER, the pathways of unfolded protein response (UPR) are activated, leading to ER stress activating pancreatic ER kinase-like ER kinase (PERK), activating transcription factor-6 (ATF6), and inositol-requiring enzyme 1α (IRE1α), all of which are controlled by glucose-regulated protein 78 kDa (Grp78) [41]. PERK directly phosphorylates eukaryotic initiation factor 2α to reduce protein loads of the ER [42] and facilitates the expression of genes involved in the functional UPR, such as activating transcription factor 4 (ATF4), which controls genes involved in amino acid metabolism and redox homeostasis [43]. Severe or unresolved ER stress leads to cell death mediated by the pro-apoptotic transcription factor DNA-damage-inducible transcript 3 (CHOP), which is predominantly induced by the PERK/ATF4 pathway [44]. CHOP is a multifaceted transcription factor with various functions during ER stress. CHOP overexpression triggers apoptosis, whereas CHOP deficient cells are highly resistant to ER stress-induced cell death [45,46]. Furthermore, CHOP plays a key role in lipopolysaccharide-induced inflammation through the induction of caspase-11 [47]. Together, CHOP is an important factor in the pathogenesis of ER stress related diseases and largely determines the fate of cells in response to ER stress [48].

The second major effector protein of ER stress is IRE1α, which splices X-box binding protein (XBP1) mRNA, resulting in spliced XBP1 (XBP1s). XBP1s activates the machineries of endoplasmic reticulum-associated degradation (ERAD), which abolishes unfolded or misfolded proteins [49]. However, IRE1α can also induce cell death by apoptosis signal-regulating kinase (ASK1)-c-Jun amino-terminal kinase (JNK) [50].

Activating transcription factor-6 (ATF6), a transmembrane protein of the ER, activates the expression of ER chaperones, including Grp78, Grp94, protein disulfide isomerase, XBP1, and the components of ERAD [51].

Evidence suggests that ER stress pathways interact with Nrf2/HO-1 signaling (Figure 1).

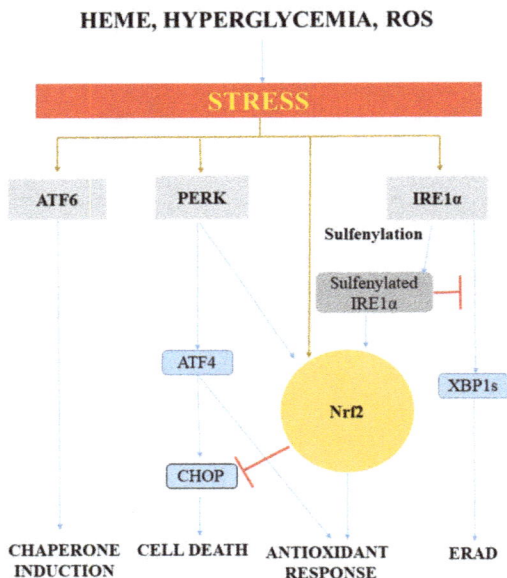

Figure 1. Crosstalk between ER stress and antioxidant response. ER stress in one of the best characterized stress response pathways induced by diverse stress stimuli such as heme, hyperglycemia, and reactive oxygen species (ROS). Activating transcription factor-6 (ATF6) induces ER chaperones, improving protein folding in the ER. Activation of the pancreatic ER kinase-like ER kinase (PERK) arm results in Activating transcription factor-4 (ATF4) expression leading to the activation of antioxidant pathways. In addition, ATF4 also activates the proapoptotic protein DNA-damage-inducible transcript 3 (CHOP) when ER stress is unresolved. Inositol-requiring enzyme 1 (IRE1α) is primarily involved in ER-associated degradation (ERAD) of the damaged proteins. Nuclear factor-E2-related factor-2 (Nrf2), the key regulator of cellular antioxidant response, is strongly connected to the ER stress pathways, since PERK and sulfenylated IRE1α directly activates the Nrf2-mediated stress response. The different consequences of the crosstalk between ER stress and antioxidant response depend on the severity and duration of stress stimuli, as well as the target cell and organ.

Upon ER stress, PERK directly phosphorylates Nrf2, promoting dissociation from Keap-1 with the subsequent nuclear import of Nrf2 [52]. CO, a by-product of heme degradation by HOs, induces Nrf2 nuclear translocation via PERK activation [53]. In addition, Nrf2-ATF4 dimers bind to the StRE element in the HO-1 promoter, regulating its expression in a cell-specific manner [54,55]. On the other hand, Nrf2 also modulates the expression of certain ATF4-regulated genes such as CHOP, which is negatively regulated by Nrf2 and positively by ATF4 [56]. Another example of the crosstalk between ER residential proteins and antioxidant pathways is IRE1. Reactive oxygen species (ROS) generated by the ER or mitochondria facilitates the cysteine sulfenylation of IRE1 within the kinase loop, which inhibits IRE-1-induced UPR but initiates p38/Nrf2 antioxidant response [57].

Evidence suggests that ER stress pathways interact with Nrf2/HO-1 signaling. Upon ER stress, PERK directly phosphorylates Nrf2, promoting dissociation from Keap-1 with the subsequent nuclear import of Nrf2 [58]. CO, a by-product of heme degradation by HOs, induces Nrf2 nuclear translocation via PERK activation [53]. In addition, Nrf2-ATF4 dimers bind to the StRE element in the HO-1 promoter, regulating its expression in a cell-specific manner [54,55]. On the other hand, Nrf2 also modulates the expression of certain ATF4-regulated genes such as CHOP, which is negatively regulated by Nrf2 and positively by ATF4 [56]. Another example of the crosstalk between ER residential proteins and antioxidant pathways is IRE1. Reactive oxygen species (ROS) generated by the ER or mitochondria facilitates the cysteine sulfenylation of IRE1 within the kinase loop, which inhibits IRE-1-induced UPR but initiates p38/Nrdf2 antioxidant response [57].

4. Atherosclerosis

Atherosclerosis is a leading cause of death in developed countries. Labile free heme is one of the many known risk factors for atherosclerosis and contributes to the pathophysiology of this complex disease.

4.1. Heme Stress in Atherosclerosis

Hemorrhaged atherosclerotic plaques represent a highly oxidative milieu where invading RBCs are rapidly lysed with subsequent Hb and heme release. Hemorrhage is a frequent complication of plaque development after the rupture of the fibrous cap or intraplaque hemorrhage from the neovasculature budded from vasa plaquorum. Li and co-workers have demonstrated that hemorrhaged atherosclerotic plaques represent a "death zone", characterized by lipid peroxidation products, which are extremely toxic to the invading cells, including RBCs [59]. In this oxidative scenario, RBCs easily lyse, followed by Hb release and oxidation to form metHb and ferrylHb [60]. MetHb facilitates the oxidant-mediated killing of endothelial cells [61] as well as LDL oxidation via heme release [62,63]. In addition, ferrylHb is a strong proinflammatory agonist that increases endothelial cell permeability and monocyte adhesion [30,64]. These findings corroborate the hypothesis that the oxidation products of free Hb are involved in the progression of atherosclerosis.

Importantly, the lysis of RBCs is not the only fate of erythrocytes during atherosclerosis. ROS can also trigger senescence signals on RBCs, which has been reported in carotid atherosclerosis patients [65]. These senescent state of RBCs possesses remarkable immunomodulatory effects by influencing T cell integrity and function and by affecting the dendritic cell maturation contributing to plaque progression [66].

Recent evidence also shows that heme directly targets the ER of aortic smooth muscle cells, which might be implicated in the pathogenesis of atherosclerosis [29]. This study has demonstrated that heme induces the expression of Grp78 and activates the canonical ER stress pathways, namely PERK, IRE1α/XBP1, and ATF6. In accordance with these findings, higher ER stress marker (Grp78 and CHOP) expression was detected in hemorrhaged atherosclerotic lesions, compared to either atheromas or healthy arteries. Importantly, heme induced ER (HIER) stress is effectively attenuated by the heme scavenger proteins hemopexin and alpha-1-microglobulin. This report highlights that heme directly targets the ER, which might be involved in heme-driven cell damage.

Another mechanism that mediates heme toxicity in atherosclerosis is the oxidative modification of low-density lipoprotein (LDL). Heme, owing to its hydrophobic nature, prefers to associate not only with biological membranes, but also with LDL, representing a physiological mediator of LDL oxidation catalyzed by lipid hydroperoxides that is implicated in the pathogenesis of atherosclerosis [67–69]. LDL particles entering the subendothelial area of the arteries are exposed to rapid oxidation, which recruits macrophages and generates foam cells after binding to scavenger receptors on macrophages. Oxidized LDL (oxLDL) is directly cytotoxic to the residential cells of atherosclerotic plaques, such as endothelial cells [70], vascular smooth muscle cells (VSMCs) [71], and macrophages [72]. This oxidative interaction facilitates heme degradation with the subsequent iron release, which further accelerates heme degradation (Figure 2).

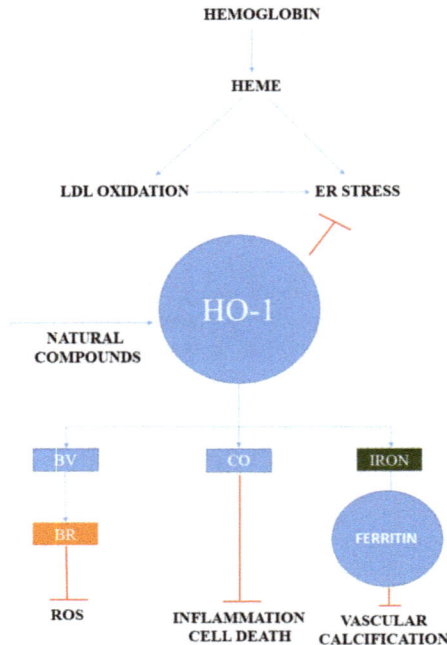

Figure 2. The central role of heme in vascular pathologies. In atherosclerosis, free heme is liberated during hemoglobin oxidation in the vessel wall inducing ER stress and low density lipoprotein (LDL) oxidation. Heme and oxidized LDL induced ER stress is an important factor in the pathogenesis of heme stress, while heme degrading heme oxygenase-1 (HO-1) initiates endogenous protective responses. Biliverdin (BV) is converted into is a natural antioxidant, bilirubin (BR) by biliverdin reductase, carbon monoxide (CO) possesses anti-inflammatory and anti-apoptotic capabilities. Redox-active iron is detoxified and stored by ferritin having ferroxidase activity. In addition, ferritin also mitigates vascular calcification. Natural compounds that induce HO-1 represent a potential therapeutic approach in vascular diseases.

4.2. HO-1 and Ferritin in Atherosclerosis

LDL oxidized either with heme+hydrogen peroxide or with Cu2+ strongly induces HO-1 in endothelial cells and macrophages [73,74]. In addition, HO-1 is upregulated in the endothelium and foam cells/macrophages of the intima in humans and in apolipoprotein E-deficient (ApoE−/−) mice [74]. The importance of HO-1 in vascular diseases is underlined by the discovery of the first human case of HO-1 deficiency [75]. The lymphoblastoid cell line derived from the HO-1-deficient patient has increased sensitivity to LDL oxidized by heme, compared to control lymphoblast cells with intact HO-1 [76], suggesting that HO-1 is an important factor to mitigate oxLDL-induced oxidative damage.

The pivotal role of HO-1 is also suggested by Cheng et al., who have reported the HO-1 induction reverses plaque progression from a vulnerable to a more stable phenotype [77], which might be mediated by heme catabolism by-products. In addition, plasma HO-1 levels are higher in patients with carotid plaques compared to healthy subjects, which probably indicates a possible protective response against carotid atherosclerosis [78].

It is postulated that pharmacologic HO-1 inducers might be potential candidates for improving atherosclerosis and reducing death in cardiovascular diseases. Mounting evidence suggests that the induction of HO-1 can ameliorate cell injury involved in atherogenesis. Zedoarondiol, an active compound extracted from *Curcuma zedaria*, has been reported to attenuate oxLDL-induced endothelial cell injury and inflammation via upregulation of HO-1 by Nrf2 [79]. Atractylenolide I (AO-I), isolated from *Atractylodes macrocephala*, inhibits ox-LDL-induced VSMCs migration and inflammatory responses partly in an HO-1-dependent manner [80].

4.3. BR and CO in Atherosclerosis

BR inhibits monocyte migration across the activated endothelial cells, by scavenging ROS and affecting adhesion molecule signaling, and prevents plaque formation in LDL receptor-deficient mice [81]. In addition, decreased serum BR is associated with the severity of atherosclerosis [82] and low serum levels of indirect BR is an independent predictor of subclinical atherosclerosis [83]. These are in good agreement with a study presenting that chronic moderate hyperbilirubinemia in Gilbert syndrome prevents the development of ischemic heart disease [84]. These findings show that a low serum BR level might be associated with increased risk for atherosclerosis.

CO, due its anti-inflammatory and anti-apoptotic properties, might be a promising candidate to mitigate inflammation and apoptotic cell death in vascular diseases. This hypothesis is supported by Soni et al., who have shown that exogenous CO mitigates doxorubicin-induced cardiotoxicity [85]. CORM-3 improves structural and functional cardiac recovery after myocardial injury [86]. In addition, CORM-2 reduces endothelial cell apoptosis triggered by ox-LDL [87].

Whereas CO might be beneficial in doxorubicin-induced damage, a recent report has shown that exogenous CO may disturb lipid metabolism in macrophages promoting foam cell formation, which is a hallmark of atherosclerosis [88]. To provide a more comprehensive understanding of the potential therapeutic application of CO releasing agent in vascular diseases, the safe therapeutic range of CORMs should be determined because of the toxicity of CO.

Ferritin, which is abundantly induced during heme catabolism by HOs, plays a pivotal role in the detoxification of redox-active iron releases by HOs. Induction of ferritin is also dependent upon the microenvironment within the vessels, because ferritin expression is also regulated by inflammation [89]. Interestingly, ferritin is more than an iron storage protein. Ferritin has been reported to inhibit the calcification of both aortic smooth muscle cells [90] and valvular interstitial cells [91]. Moreover, the pharmacological induction of ferritin also prevents smooth muscle cell calcification [92]. This raises an interesting hypothesis that cardiovascular complications might be mitigated not only by HO but also by ferritin, the other important component of the HO-ferritin system.

Protection against heme and ROS-mediated toxicity by HO-1 is important to maintain vascular homeostasis. Therefore, using pharmacologic approaches, the induction of HO-1 might be beneficial in atherosclerosis.

4.4. ER Stress in Atherosclerosis

Numerous reports have shown that ER stress is involved in the pathogenesis of vascular diseases. Elevated intracellular ox-LDL and 7-ketocholesterol provoke ER stress, characterized by a marked CHOP expression with subsequent apoptosis in macrophages and endothelial cells [93–95]. Moreover, CHOP expression is also high in unstable atherosclerotic plaques, which is likely to contribute to plaque vulnerability [94]. Additionally, ER stress also promotes foam cell formation in response to ox-LDL [96]. Thus, based on these findings, it is postulated that by mitigating ER stress, atherosclerosis might be resolved or, at least, its progression can be slowed down. In fact, recent studies corroborate this hypothesis. Lipid-mediated toxicity of macrophages is reduced by the chemical chaperone 4-phenylbuturic-acid (4-PBA) [97]. A recent study revealed that another ER stress inhibitor, tauroursodeoxycholic acid (TUDCA), ameliorates atherosclerotic lesions and systemic lipid levels in ApoE−/− mice on a high-fat diet [98].

Even though ER stress and oxidative stress coexist in many pathologies [99], the link between them has not been clearly elucidated. However, it is postulated that Nrf2/HO-1 and ER stress pathways interact and oxidative stress-triggered ER stress might be implicated in atherosclerosis. This

Int. J. Mol. Sci. **2019**, *20*, 3675

theory is supported by Mozzini et al., who showed that GRP78/CHOP expression is increased while Nrf2/HO-1 is decreased in coronary artery disease patients [100]. The authors suggest that phospholipid 1-palmitoyl-2-arachidonyl-sn-glycero-3-phosphorylcholine (oxPAPC), which presumably derives from oxidized LDL, negatively regulates Nrf2/HO-1 expression with a parallel ER stress-inducing effect. In another study, equol, a specific metabolic product of daidzein in soybean, reduces atherosclerotic lesions in ApoE−/− mice fed with high-fat diet via attenuating ER stress in an Nrf2-dependent manner [101].

Isorhamnetin (Iso), a flavonoid compound extracted from *Hippophae rhamnoides L.*, inhibits ox-LDL-induced macrophage injuries by decreasing ROS levels and protecting against ox-LDL-induced apoptosis [102]. In addition, Iso also reduced atherosclerotic plaque size and macrophage apoptosis in ApoE−/− mice. These in vitro and in vivo effects of Iso are mediated by PI3K/AKT activation and HO-1 induction. Interestingly, a recent report shows that Iso also inhibits ER stress, but the possible interplay between this effect and the HO-1-inducing effect of Iso needs further investigation [103].

These results highlight that pharmacologic induction of Nrf2/HO-1 provides a potent antioxidant defense mechanism, which might mitigate the progression of atherosclerosis, at least partly, by attenuating ER stress. These results highlight that little is known about the crosstalk between HO-1 and ER stress in vascular diseases. Further studies may reveal the potential link between antioxidant and ER stress responses and identify molecular targets that govern the pathophysiological events at the cellular level in atherosclerosis.

5. Brain

Lesions of the central nervous systems are not only based on primer brain pathologies, but also on secondary events. Analyzing the nature of brain lesions, there are several ones where the presence of heme and ER stress is obvious and well documented. The intracerebral hemorrhage (ICH) is a classical type of brain vascular disease gaining importance presently, as the ageing of the population and the use of anti-coagulants rises. The initiation step of ICH is the accumulation of blood within the brain tissue with the subsequent lysis of RBCs resulting in massive Hb and heme release. The breakdown of heme by HOs yields a large amount of redox-active iron, (CO) and BV, which is converted to BR. Hard evidence show that Hb, heme, and heme catabolism by-products are all involved in the pathogenesis of ICH (Figure 3).

MetHb is an endogenous ligand of toll-like receptor 2 (TLR2) and toll-like receptor 4 (TLR4) that induces pro-inflammatory cytokine and tumor necrosis factor α (TNF-α) production after hemorrhage [104,105]. Heme is also pro-inflammatory for many cell types, such as macrophages and microglia and activates TLR4-mediated inflammatory injury in ICH, which is markedly reduced by knocking down TLR4 [106,107]. In addition, heme binds to TLR2 and severely triggers blood–brain barrier damage in astrocytes in a TLR2 dependent manner [108]. These findings suggest that the modulation of TLR signaling might be a promising therapeutic target in hemorrhagic brain injury. This hypothesis has been tested by Wang et al., who showed that the TLR4 antagonist Ethyl-(6 R)-6-(N-(2-chloro-4-fluorophenyl)sulfamoyl)cyclohex-1-ene-1-carboxylate (TAK242) reduces inflammatory injury and neurological deficits in mice after hemorrhage [109]. Although this is an interesting approach, further research is needed in this field to explore the effect of TLR4 antagonists in immunohistochemistry (IHC).

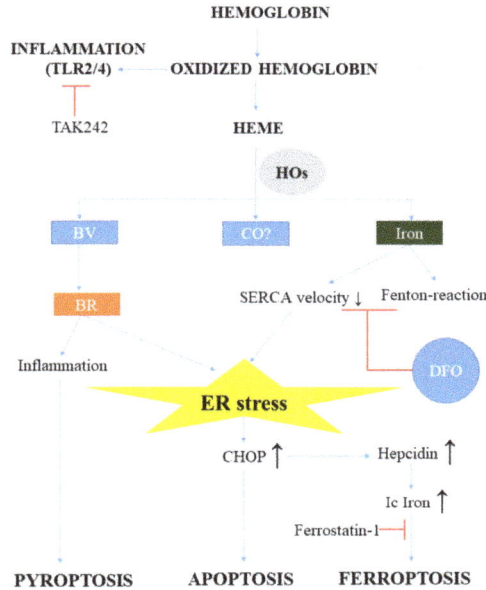

Figure 3. Free hemoglobin and heme participates in the pathophysiology of brain damage. During brain hemorrhage, massive amounts of free hemoglobin and heme are released, resulting in inflammation, heme-, and ER stress. Oxidized hemoglobin directly triggers inflammation via toll-like receptor 2 and 4 (TLR2, TLR4). Heme oxygenases (HOs) in the brain are Janus-faced and might induce both protective and adverse effects. Although bilirubin (BR) is a natural antioxidant, it can induce inflammation, pyroptosis, or even apoptosis through ER stress in the brain. Heme-derived iron also triggers ER stress. Elevated intracellular iron (Ic Iron) derives from HO activity and the disturbed cellular iron metabolism through hepcidin induction, leading to ferroptosis. (SERCA: sarco/endoplasmic reticulum Ca2+-ATPase; DFO: desferrioxamine; CHOP: C/EBP homologous protein; CO: carbon monoxide; BV: biliverdin; TAK: Ethyl-(6R)-6-(N-(2-chloro-4-fluorophenyl)sulfamoyl)cyclohex-1-ene-1-carboxylate; TLR: Toll-like receptor

5.1. ER Stress in Brain Injury After Hemorrhage

ER stress plays a significant role in brain injury after hemorrhage. Hemorrhage induced ER stress primarily targets neurons and the protein kinase R (PKR)-like endoplasmic reticulum kinase (PERK)/CHOP pathway might have predominant role in neuronal cell death. Several lines of evidence support this theory. In a rat model of cerebral ischemia/reperfusion injury, silencing of CHOP by the lentivirus-mediated transfer of short hairpin RNA mitigated inflammatory and apoptotic reactions in neuronal cells, suggesting that CHOP silencing is neuroprotective [110]. PERK signaling is activated following subarachnoid hemorrhage (SAH) and PERK inhibitor GSK2606414 reduces neuronal cell death via Akt activation [111]. This has also been corroborated by another study, which shows that the PERK pathway is involved in neuronal loss and apoptosis following ICH, which is attenuated by PERK inhibition [112]. Overall, these studies indicate that the PERK pathway plays an important role in brain injury after hemorrhage and PERK inhibitors might be potential candidates for mitigating ER stress-induced cell death and brain damage after SAH.

5.2. HO-1 and HO-2 in the Brain

The role of HO-1 in brain hemorrhage is highly controversial. In a murine model of SAH, Schallner and co-workers have demonstrated that inducible HO-1 in microglia is necessary to reduce neuronal cell death and to clear cerebral blood [113]. In contrast, the injury volume in HO-1 knockout (HO-1$^{-/-}$) mice is significantly smaller 24 and 72 h after the injury, compared to wildtype mice [114]. In addition, pharmacological inhibition of HO-1 with tin-mesoporphyrin also protects against neuronal loss in the rabbit model of ICH [115]. Others suggest that the effect of HO-1 is time dependent. In a collagenase-induced ICH murine model, Zhang and co-workers have suggested that HO-1 induction is Janus-faced. During the early-phase of ICH (day 1–3), HO-1 increases brain edema, white matter and neuronal damage, and elevates inflammation and iron deposition, but in the late-phase (day 28), HO-1 increases hematoma absorption and recovery of neurologic functions [116]. On the other hand, Wang et al. suggest that HO-1 is protective in the early-mid phase, but in the late phase, it might be toxic [117].

The role of HO-2 in brain hemorrhage is similarly controversial. In HO-2 knockout (−/−) mice, HO-2 deficiency worsens neurotoxicity mediated by stroma-free hemoglobin [118]. Others using HO-2 knockout mice support that HO-2 has a critical protective effect in IHC [12]. On the contrary, Chen-Roetling and co-workers suggest that, in the blood injection ICH model, neuronal survival is markedly increased in HO-2 knockout mice compared to wild-type mice [119].

Despite many years of research on the subject of how HO-1 and HO-2 may influence the pathomechanism and recovery after ICH, results are controversial, which might explain the diverse ICH models. Furthermore, not only heme but also its breakdown products, namely iron and BR, catalyze adverse reactions in the brain and contribute to brain damage after hemorrhage.

5.3. Iron Injury in the Brain

Free heme is rapidly taken up by microglia and invading macrophages, and within a few hours after hemorrhage it is catabolized into redox-active iron, CO, BV, and BR. One feasible mechanism of heme toxicity may be the release of free redox-active iron by HOs. Activation of HO-1 in response to heme is closely coupled with the induction of ferritin, an intracellular iron-storage protein with potent ferroxidase activity. It is likely that cells are not able to accumulate ferritin continually and this iron-scavenging homeostatic mechanism may fail after massive hemorrhage, resulting in a robust redox-active iron burden. Redox-active iron strongly promotes free radical formation by the Fenton-reaction, triggering oxidative damage in biological systems [120]. Importantly, free redox-active iron also promotes ER stress and this iron-induced ER stress is likely to contribute Hb/heme toxicity in the brain. In vitro and in vivo models support this tenable hypothesis, i.e., ER stress is a possible etiological factor in brain hemorrhage.

Ferrous iron induces both lipid-mediated and lipid-peroxidation independent changes in the ER membrane-associated proteins in brain [121]. Free radicals generated by redox-active iron decrease the velocity of sarco/endoplasmic reticulum Ca^{2+} ATPase (SERCA) [122]. This resembles the inhibitory mechanism of thapsigargin, a well-characterized ER stress inductor, which also inhibits SERCA. Therefore, one reasonable assumption might be that redox-active iron-induced cell damage is mediated via ER stress and factors regulating iron metabolism may influence brain damage. One of these factors might be hepcidin, which prevents iron export from cells by breaking down the iron transporter ferroportin. Remarkably, ER stress upregulates hepcidin expression [123]. This connection has been revealed in the brain by a recent study. Zhao and co-workers have shown the ER stress markers Grp78 and CHOP are induced in rats after subarachnoid hemorrhage and neuronal death is mediated by hepcidin [124]. Disturbed iron homeostasis might lead to an iron-dependent form of cell death, ferroptosis [125]. CHOP increases hepcidin levels and iron content in the brain, which localizes to the nuclei of neurons. Importantly, knocking-down CHOP improves neurological functions. These observations support the hypothesis that brain injury after hemorrhage is, at least partly, mediated by ER stress.

Previous studies have demonstrated that desferrioxamine (DFO) and other iron chelators are neuroprotective in brain hemorrhage [126,127]. Interestingly, a recent report makes this iron toxicity model more complex. LeBlanc and co-workers have shown in trans-well experiments with microglial and hippocampal neuronal cells that DFO markedly reduces RBC-mediated neuronal cell death, but this protective effect is highly dependent on microglial HO-1 [128].

5.4. BR Toxicity in Brain

BR is an important endogenous antioxidant [129,130]. However, BR neurotoxicity is well-founded, especially in newborns, and necessitates phototherapy [131,132] and, in extreme cases, exchange transfusion [133]. In addition, BR levels and toxicity should be more closely monitored in newborns with hemolytic disease [134]. Interestingly, inhibiting HO-1 catalytic activity by SnPP markedly mitigates hyperbilirubinemia in preterm newborns [135].

After hemorrhage, massive amounts of unconjugated bilirubin (UCB) might be produced by HOs during heme degradation. In hepatocytes, UCB is conjugated to glucuronides in the ER, which indicates that UCB has a high affinity towards the ER [136].

UCB has been reported to induce ER stress, inflammation, and even apoptosis in SH-SY5Y neuronal cells, while knocking down of CHOP or attenuating ER stress by 4-PBA markedly reduced cell death [137]. A current study supports this observation. Schiavon and co-workers showed that BR induces ER stress and inflammation, both in vitro and in vivo, and raised the auditory threshold together with behavioral impairment in the murine model [138]. Another form of UCB induced cell death, that is presumably pyroptosis, a highly inflammatory form of programmed cell death, has been revealed in rat cortical astrocytes [139].

These findings corroborate the fact that BR neurotoxicity and BR-induced ER stress might be involved in brain damage after hemorrhage.

5.5. CO in the Brain

Bioactivity of CO as the potential anti-inflammatory, anti-apoptotic, and antioxidant substance has been reviewed extensively in the past few years [4]. Based on these beneficial properties of CO, carbon monoxide-releasing molecules (CORMs) should be ideal candidates for reducing inflammation and cell death after brain hemorrhage. However, relevant data in this field are scarce.

CORM-3, a water-soluble CO-releasing molecule, promotes either neuroprotection or neuroinflammation, depending on the administration time in collagenase-induced ICH model [140]. It is neuroprotective when administered 5 min before the IHC and in the subacute phase, 3 days after the hemorrhage, but it aggravates brain injury and tumor necrosis factor-α production administered in the acute phase, 3 h after the IHC. Further research is needed to explore the possible beneficial effect of CORMs to mitigate brain injury after hemorrhage.

Overall, despite extensive research, the role of HOs in protecting brain homeostasis after hemorrhage remains elusive. However, it is likely that ER stress is deeply involved in heme toxicity in the brain and CHOP plays an important role in many aspects of this toxicity by regulating iron and BR induced cell damage.

6. Diabetes Mellitus

The long-term complications of diabetes mellitus, micro- and macro-vascular pathologies, largely influence quality of life and mortality. Diabetes poses a significant threat and economic burden, especially in developed countries. In this review, we will focus on the role of HO-1 in diabetic cardiovascular complications.

6.1. Diabetic Cardiovascular Complications

Whereas oxidative stress is significantly increased in diabetic patients, the total antioxidant status is decreased [141]. Hyperglycemia promotes oxidative stress by free radical generation and suppresses antioxidant defense [141,142]. Diabetes, together with hyperglycemia-induced oxidative stress, is a major risk factor for accelerated atherosclerosis [143]. Hyperglycemia lowers HO-1 activity and increases superoxide production in the vasculature, which is mitigated by CoPP [144,145]. In addition, both upregulation of HO-1 or production of CO with CORM-3 have beneficial effects on vascular relaxation in rats [144,146]. Interestingly, obesity also decreases HO-1 levels in both male and female rats, compared to lean animals, and the induction of HO-1 with CoPP reduces blood pressure and inflammatory cytokine levels [147].

Since the endogenous anti-oxidant response is impaired in diabetes, it is rational that resolving imbalances in the antioxidant system might restore redox homeostasis. This has been seen in the case of hydrogen sulfide (H_2S), which reduces aortic atherosclerotic plaque formation by detoxifying superoxide [148]. This protective effect of H_2S is mediated by HO-1 activation via the Keap-1/Nrf2 system. In addition, HO-1 induction by hemin restores the cardioprotective effect of ischemic preconditioning in diabetic rat heart [149]. Moreover, activation of HO-1 with hemin mitigates renal damage in streptozotocin (STZ)-induced diabetic nephropathy in rats, by reducing inflammation and apoptosis, and improves antioxidant response [150].

Another interesting therapeutic approach might be the epigenetic regulation of HO-1 expression. MicroRNAs (miRNAs) are small noncoding RNAs involved in the post-transcriptional regulation of protein expression. An interesting example by which HO-1 is regulated by miRNAs is microRNA-92a (miR-92a). The miR-92a expression is increased in diabetic endothelial cells. A recent report shows that miR92a negatively influences HO-1 expression [151] and the inhibition of miR-92a elevates HO-1 expression, mitigates oxidative stress, and improves endothelial function in diabetic mice. The importance of miRNAs in HO-1 regulation in diabetes is emphasized by the fact that other miRNAs, such as miR-218, induce cell death in podocytes by downregulation of HO-1 expression [152]. Together, an interesting epigenetic therapeutic approach would be the targeted induction of HO-1 in endothelial cells to prevent the diabetic dysfunction of these cells.

Data on the role of HO-2 in diabetes are scarce. It has been reported that HO-2 deficiency increases superoxide production and contributes to renal dysfunction in a diabetic rat model [153]. Very little is currently known about the protective role of HO-2 to cope with oxidative stress in diabetes, which necessitates further research in this field.

6.2. HO-1 Byproducts in Diabetes

BR provides a potent antioxidant defense mechanism in response to oxidative stress, suggesting that BR might mitigate ROS-induced cell- and tissue damage in diabetes. Elevated glucose concentrations trigger ROS generation and induce HO-1 expression both in vivo and in vitro. Inhibition of HO-1 activity exacerbates high glucose-induced oxidative cell injury, which is markedly attenuated by BR [154]. Diabetes and hyperglycemia induce oxidative stress, leading to endothelial dysfunction, which is a hallmark of diabetic vascular complications. It has been demonstrated that BR is implicated in HO-1-mediated restoration of impaired endothelial function in diabetic mice [155]. In addition, lower serum BR, together with the neutrophil-lymphocyte ratio, are independent predictors of subclinical atherosclerosis in prediabetes [156]. In murine model, BR increased insulin sensitivity by reducing ER stress and inflammation [157]. This suggests that BR is more than a potent antioxidant and it might be useful as an insulin sensitizer in type 2 diabetes by reducing hyperglycemia-induced ER stress and inflammation (Figure 4).

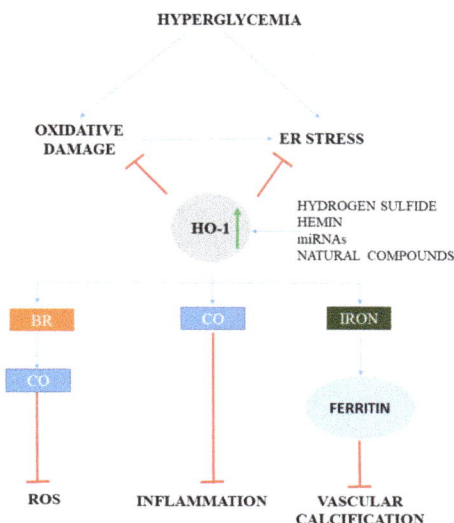

Figure 4. HO-1 mitigates oxidative and ER stress in diabetes. Hyperglycemia induces oxidative stress as well as ER stress, which is attenuated by HO-1. Bilirubin (BR), carbon monoxide (CO), and ferritin are all involved in decreasing diabetic vascular complications by reducing oxidative damage, inflammation, and vascular calcification. Pharmacologic and natural HO-1 inducers are potential candidates to reduce diabetic vascular complications. (HO: heme oxygenase; ROS: reactive oxygen species; miRNA: micro ribonucleic acid)

CO possesses remarkable anti-apoptotic and anti-inflammatory properties. Carbon monoxide-releasing molecules (CORMs) are valid and safe alternatives to the CO gas-based therapies and exert protective effects in diabetes-induced inflammation. In a mouse model of STZ-induced diabetes, CORM-3 markedly reduced hyperglycemia-induced inflammation by decreasing IL-1β production [158]. Others have demonstrated that hemin, as well as BR and the CO donor CORM-2, reduces hyperglycemia and improves the abnormality of endothelium-dependent vascular relaxation in STZ-induced diabetic rats [159]. These studies suggest that CORMs might have beneficial effects to reduce inflammation in diabetes as supportive therapy.

Coronary, carotid, and aortic valve calcification are frequent complications of diabetic patients [160,161], which might be mitigated by ferritin as discussed above.

Overall, a body of evidence suggests that HO-1, ferritin, BR, and CO are all capable of reducing oxidative stress and attenuating ROS-drive cell and tissue damage.

6.3. ER Stress in Diabetes

Mounting evidence suggests that ER stress is involved in the pathogenesis of diabetes and plays a pivotal role in diabetic complications, such as neuropathy [162], diabetic nephropathy [163], diabetic retinopathy [164], and cardiomyopathy [165,166]. HO-1 and ER stress pathways interact in diabetes (Figure 4). In type 2 diabetic patients, the ER stress marker expression with concomitant oxidative stress is significantly higher, while anti-inflammatory Inhibitor of κB-α (IκB-α) and Nrf2/HO-1 are significantly lower compared to healthy controls [167]. Parallel with that, prolonged hyperglycemia induces NFκB without a Nrf2 response. This demonstrates that hyperglycemia might trigger ER stress, inflammation, and oxidative stress without Nrf2 activation. Hyperglycemia also induces ER stress, inflammation, and apoptosis in endothelial cells, which is attenuated either by inducing HO-1 with Cobalt (III)-Protoporphyrin IX chloride (CoPPIX) or by reducing ER stress by 4-PBA [168]. In addition, both CoPPIX and 4-PBA reduce the angiogenic capacity of human umbilical vein endothelial cells (HUVECs) and increases vascular endothelial growth factor-A (VEGF-A) expression. This observation underlines that ER stress plays a role in diabetes-induced endothelial dysfunction and impaired angiogenesis and HO-1 induction might be a protective stratagem to mitigate the adverse effects of ER stress in diabetic cardiovascular complications.

Natural compounds that activate Nrf2/HO-1 pathways are promising therapeutics to reduce ER stress, as well as oxidative stress, in diabetes. Tangluoning, a traditional Chinese medicine, has been reported to activate the PERK/Nrf2 pathway upregulating antioxidant responsive element (ARE) elements including HO-1, thereby attenuating diabetic peripheral neuropathy and CHOP-mediated apoptosis [169]. Grape seed proanthocyanidins (GSP) mitigate early diabetic peripheral neuropathy by modulating endoplasmic reticulum stress and preventing calcium overload [170]. Interestingly, GSPs are effective activators of Nrf2 and HO-1, which suggests the protective effect of GSPs in diabetic neuropathy.

Chrysin, a naturally occurring flavonoid found in various herbs, reduces high glucose-induced ER stress in retinal pigment epithelial cells and might be beneficial to attenuate diabetes-associated visual cycle impairment implicated in diabetic retinopathy [171]. Another study revealed that chrysin is also a potent inductor of HO-1 [172].

Arctigenin (ATG), a lignan extract from *Fructus arctii*, has renoprotective effects on diabetes-related renal injury by inhibiting ER stress and apoptosis [173]. It is not surprising that ATG also induces HO-1 [174].

These studies possibly reveal that the crosstalk between ER stress and Nrf2/HO-1 mediates the beneficial effects of these natural compounds in hyperglycemia-induced complications. It still remains a remarkable question whether the effects of these compounds are attributed or not to HO-1 induction. Inhibition of HO-1 activity or by knocking down/out HO-1 might provide a mechanistic model by which the exact role of HO-1 can be revealed in ER stress in diabetes.

6.4. Brain and Diabetes

Another serious complication of diabetes is the diabetes-associated cognitive decline. In an STZ-induced diabetic mouse model, increased ER stress, JNK activation, and autophagy have been observed in hippocampal neurons [175]. Importantly, 4-PBA attenuates neuronal cell death in mice, while autophagy inhibitor bafilomycin A1 increases cell death in vitro.

A recent report has revealed that HO-1 expression in response to hyperglycemia might also have adverse effects [176]. Hyperglycemia induces HO-1 expression in rat astrocytes in vitro and conditioned medium from high glucose-treated astrocytes triggers cell death in neuronal cells. Interestingly, hemoglobin treatment, as a CO scavenger, prevents neuronal cell death provoked with high glucose conditioned medium, suggesting that HO-1/CO activation in astrocytes provokes neuronal cell death. These findings raise the hypothesis that HO-derived CO might play a negative role in diabetes-induced complications in the brain.

7. Discussion

Crosstalk between oxidative stress and ER stress markedly contributes to the pathogenesis of vascular diseases, as well as diverse pathologies associated with heme stress. Therefore, ER stress inhibitors might be ideal candidates to ameliorate the adverse pathophysiological consequences of these diseases when the endogenous homeostatic defense is overwhelmed. HO-1 is implicated in numerous cellular protective pathways and might be involved in managing ER stress due to its antioxidant, anti-inflammatory, and anti-apoptotic effects, mediated directly by HO-1 or by heme catabolism end-products, i.e., CO, bilirubin, and iron. However, HO-1 activity might pose a threat to the homeostasis of cells by releasing vast amounts of these agents when free heme is present. HO-1 induction by natural compounds is beneficial, in many aspects, by reducing ER stress and oxidative damage; however, the exact role of HO-1 in mitigating ER stress remains to be elucidated.

Author Contributions: Conceptualization: T.G., G.B., and J.B.; Writing—Original Draft Preparation: T.G., G.B.; Writing—Review & Editing: T.G., G.B., and J.B.; Supervision: J.B.; Funding Acquisition: G.B. and J.B.

Funding: This research was funded by by the Hungarian Academy of Sciences (11003), the European Union and the European Social Fund GINOP-2.3.2-15-2016-00043 (IRONHEARTH) and EFOP-3.6.2-16-2017-00006 (LIVE LONGER), Hungarian Government grants, OTKA-K112333 (J.B.), and the Thematic Excellence Programme of the Ministry for Innovation and Technology in Hungary, within the framework of the Space Sciences thematic programme of the University of Debrecen.

Acknowledgments: The research group of G.B. is supported by the Hungarian Academy of Sciences (11003). This research/project is supported by Hungarian Government grants, OTKA-K112333 (J.B.). The project is co-financed by the European Union and the European Social Fund GINOP-2.3.2-15-2016-00043 (IRONHEARTH) and EFOP-3.6.2-16-2017-00006 (LIVE LONGER). The research was financed by the Thematic Excellence Programme of the Ministry for Innovation and Technology in Hungary, within the framework of the Space Sciences thematic programme of the University of Debrecen.

Conflicts of Interest: The authors declare no conflict of interest.

References

1. Bunn, H.F.; Jandl, J.H. Exchange of heme among hemoglobins and between hemoglobin and albumin. *J. Biol. Chem.* **1968**, *243*, 465–475. [PubMed]
2. Balla, G.; Jacob, H.S.; Balla, J.; Rosenberg, M.; Nath, K.; Apple, F.; Eaton, J.W.; Vercellotti, G.M. Ferritin: A cytoprotective antioxidant strategem of endothelium. *J. Biol. Chem.* **1992**, *267*, 18148–18153. [PubMed]
3. Stocker, R. Antioxidant activities of bile pigments. *Antioxid. Redox Signal.* **2004**, *6*, 841–849. [PubMed]
4. Motterlini, R.; Otterbein, L.E. The therapeutic potential of carbon monoxide. *Nat. Rev. Drug Discov.* **2010**, *9*, 728–743. [CrossRef] [PubMed]
5. Maines, M.D. The heme oxygenase system: a regulator of second messenger gases. *Annu. Rev. Pharmacol. Toxicol.* **1997**, *37*, 517–554. [CrossRef] [PubMed]
6. Tenhunen, R.; Marver, H.S.; Schmid, R. The enzymatic conversion of heme to bilirubin by microsomal heme oxygenase. *Proc. Natl. Acad. Sci.* **1968**, *61*. [CrossRef]
7. Dunn, L.L.; Midwinter, R.G.; Ni, J.; Hamid, H.A.; Parish, C.R.; Stocker, R. New insights into intracellular locations and functions of heme oxygenase-1. *Antioxid. Redox Signal.* **2014**, *20*, 1723–1742. [CrossRef]
8. Ryter, S.W.; Alam, J.; Choi, A.M.K. Heme oxygenase-1/carbon monoxide: from basic science to therapeutic applications. *Physiol. Rev.* **2006**, *86*, 583–650. [CrossRef]
9. Lin, J.H.; Villalon, P.; Martasek, P.; Abraham, N.G. Regulation of heme oxygenase gene expression by cobalt in rat liver and kidney. *Eur. J. Biochem.* **1990**, *192*, 577–582. [CrossRef]
10. Sardana, M.K.; Kappas, A. Dual control mechanism for heme oxygenase: tin(IV)-protoporphyrin potently inhibits enzyme activity while markedly increasing content of enzyme protein in liver. *Proc. Natl. Acad. Sci. USA* **1987**, *84*, 2464–2468. [CrossRef]
11. Zhang, Y.; Furuyama, K.; Kaneko, K.; Ding, Y.; Ogawa, K.; Yoshizawa, M.; Kawamura, M.; Takeda, K.; Yoshida, T.; Shibahara, S. Hypoxia reduces the expression of heme oxygenase-2 in various types of human cell lines. A possible strategy for the maintenance of intracellular heme level. *FEBS J.* **2006**, *273*, 3136–3147. [CrossRef] [PubMed]
12. Wang, J.; Doré, S. Heme oxygenase 2 deficiency increases brain swelling and inflammation after intracerebral hemorrhage. *Neuroscience* **2008**, *155*, 1133–1141. [CrossRef] [PubMed]
13. Doré, S.; Sampei, K.; Goto, S.; Alkayed, N.J.; Guastella, D.; Blackshaw, S.; Gallagher, M.; Traystman, R.J.; Hurn, P.D.; Koehler, R.C.; et al. Heme oxygenase-2 is neuroprotective in cerebral ischemia. *Mol. Med.* **1999**, *5*, 656–663. [CrossRef] [PubMed]
14. Eizirik, D.L.; Cardozo, A.K.; Cnop, M. The role for endoplasmic reticulum stress in diabetes mellitus. *Endocr. Rev.* **2008**, *29*, 42–61. [CrossRef] [PubMed]
15. Xiang, C.; Wang, Y.; Zhang, H.; Han, F. The role of endoplasmic reticulum stress in neurodegenerative disease. *Apoptosis* **2017**, *22*, 1–26. [CrossRef] [PubMed]
16. Navid, F.; Colbert, R.A. Causes and consequences of endoplasmic reticulum stress in rheumatic disease. *Nat. Rev. Rheumatol.* **2017**, *13*, 25–40. [CrossRef] [PubMed]
17. Marciniak, S.J. Endoplasmic reticulum stress in lung disease. *Eur. Respir. Rev.* **2017**, *26*, 170018. [CrossRef]

18. Ivanova, E.A.; Orekhov, A.N. The Role of Endoplasmic Reticulum Stress and Unfolded Protein Response in Atherosclerosis. *Int. J. Mol. Sci.* **2016**, *17*, 193. [CrossRef]

19. Rosen, A.D.; Frumin, N. V Focal epileptogenesis after intracortical hemoglobin injection. *Exp. Neurol.* **1979**, *66*, 277–284. [CrossRef]

20. Sadrzadeh, S.M.; Anderson, D.K.; Panter, S.S.; Hallaway, P.E.; Eaton, J.W. Hemoglobin potentiates central nervous system damage. *J. Clin. Invest.* **1987**, *79*, 662–664. [CrossRef]

21. Orjih, A.U.; Banyal, H.S.; Chevli, R.; Fitch, C.D. Hemin lyses malaria parasites. *Science* **1981**, *214*, 667–669. [CrossRef] [PubMed]

22. Wijayanti, N.; Katz, N.; Immenschuh, S. Biology of heme in health and disease. *Curr. Med. Chem.* **2004**, *11*, 981–986. [CrossRef] [PubMed]

23. TAPPEL, A.L. Unsaturated lipide oxidation catalyzed by hematin compounds. *J. Biol. Chem.* **1955**, *217*, 721–733. [PubMed]

24. Gutteridge, J.M.; Smith, A. Antioxidant protection by haemopexin of haem-stimulated lipid peroxidation. *Biochem. J.* **1988**, *256*, 861–865. [CrossRef] [PubMed]

25. Schmitt, T.H.; Frezzatti, W.A.; Schreier, S. Hemin-induced lipid membrane disorder and increased permeability: a molecular model for the mechanism of cell lysis. *Arch. Biochem. Biophys.* **1993**, *307*, 96–103. [CrossRef] [PubMed]

26. Aft, R.L.; Mueller, G.C. Hemin-mediated DNA strand scission. *J. Biol. Chem.* **1983**, *258*, 12069–12072. [PubMed]

27. Suliman, H.B.; Carraway, M.S.; Velsor, L.W.; Day, B.J.; Ghio, A.J.; Piantadosi, C.A. Rapid mtDNA deletion by oxidants in rat liver mitochondria after hemin exposure. *Free Radic. Biol. Med.* **2002**, *32*, 246–256. [CrossRef]

28. Aft, R.L.; Mueller, G.C. Hemin-mediated oxidative degradation of proteins. *J. Biol. Chem.* **1984**, *259*, 301–305.

29. Gáll, T.; Pethő, D.; Nagy, A.; Hendrik, Z.; Méhes, G.; Potor, L.; Gram, M.; Åkerström, B.; Smith, A.; Nagy, P.; et al. Heme Induces Endoplasmic Reticulum Stress (HIER Stress) in Human Aortic Smooth Muscle Cells. *Front. Physiol.* **2018**, *9*, 1595. [CrossRef]

30. Silva, G.; Jeney, V.; Chora, A.; Larsen, R.; Balla, J.; Soares, M.P. Oxidized hemoglobin is an endogenous proinflammatory agonist that targets vascular endothelial cells. *J. Biol. Chem.* **2009**, *284*, 29582–29595. [CrossRef]

31. Wagener, F.A.; Eggert, A.; Boerman, O.C.; Oyen, W.J.; Verhofstad, A.; Abraham, N.G.; Adema, G.; van Kooyk, Y.; de Witte, T.; Figdor, C.G. Heme is a potent inducer of inflammation in mice and is counteracted by heme oxygenase. *Blood* **2001**, *98*, 1802–1811. [CrossRef] [PubMed]

32. Balla, G.; Vercellotti, G.M.; Muller-Eberhard, U.; Eaton, J.; Jacob, H.S. Exposure of endothelial cells to free heme potentiates damage mediated by granulocytes and toxic oxygen species. *Lab. Invest.* **1991**, *64*, 648–655. [PubMed]

33. Fortes, G.B.; Alves, L.S.; de Oliveira, R.; Dutra, F.F.; Rodrigues, D.; Fernandez, P.L.; Souto-Padron, T.; De Rosa, M.J.; Kelliher, M.; Golenbock, D.; et al. Heme induces programmed necrosis on macrophages through autocrine TNF and ROS production. *Blood* **2012**, *119*, 2368–2375. [CrossRef] [PubMed]

34. Heinecke, J.W. Mechanisms of oxidative damage of low density lipoprotein in human atherosclerosis. *Curr. Opin. Lipidol.* **1997**, *8*, 268–274. [CrossRef] [PubMed]

35. Ishii, T.; Itoh, K.; Takahashi, S.; Sato, H.; Yanagawa, T.; Katoh, Y.; Bannai, S.; Yamamoto, M. Transcription factor Nrf2 coordinately regulates a group of oxidative stress-inducible genes in macrophages. *J. Biol. Chem.* **2000**, *275*, 16023–16029. [CrossRef]

36. Kobayashi, A.; Ohta, T.; Yamamoto, M. Unique function of the Nrf2-Keap1 pathway in the inducible expression of antioxidant and detoxifying enzymes. *Methods Enzymol.* **2004**, *378*, 273–286.

37. Beck, K.; Wu, B.J.; Ni, J.; Santiago, F.S.; Malabanan, K.P.; Li, C.; Wang, Y.; Khachigian, L.M.; Stocker, R. Interplay between heme oxygenase-1 and the multifunctional transcription factor yin yang 1 in the inhibition of intimal hyperplasia. *Circ. Res.* **2010**, *107*, 1490–1497. [CrossRef]

38. Camhi, S.L.; Alam, J.; Otterbein, L.; Sylvester, S.L.; Choi, A.M. Induction of heme oxygenase-1 gene expression by lipopolysaccharide is mediated by AP-1 activation. *Am. J. Respir. Cell Mol. Biol.* **1995**, *13*, 387–398. [CrossRef]

39. Alam, J.; Igarashi, K.; Immenschuh, S.; Shibahara, S.; Tyrrell, R.M. Regulation of Heme Oxygenase-1 Gene Transcription: Recent Advances and Highlights from the International Conference (Uppsala, 2003) on Heme Oxygenase. *Antioxid. Redox Signal.* **2004**, *6*, 924–933.

40. Lee, P.J.; Jiang, B.H.; Chin, B.Y.; Iyer, N.V.; Alam, J.; Semenza, G.L.; Choi, A.M. Hypoxia-inducible factor-1 mediates transcriptional activation of the heme oxygenase-1 gene in response to hypoxia. *J. Biol. Chem.* **1997**, *272*, 5375–5381. [CrossRef]

41. Walter, P.; Ron, D. The Unfolded Protein Response: From Stress Pathway to Homeostatic Regulation. *Science (80-.).* **2011**, *334*, 1081–1086. [CrossRef] [PubMed]

42. Hamanaka, R.B.; Bennett, B.S.; Cullinan, S.B.; Diehl, J.A. PERK and GCN2 Contribute to eIF2 Phosphorylation and Cell Cycle Arrest after Activation of the Unfolded Protein Response Pathway. *Mol. Biol. Cell* **2005**, *16*, 5493–5501. [CrossRef] [PubMed]

43. Ameri, K.; Harris, A.L. Activating transcription factor 4. *Int. J. Biochem. Cell Biol.* **2008**, *40*, 14–21. [CrossRef] [PubMed]

44. Zinszner, H.; Kuroda, M.; Wang, X.; Batchvarova, N.; Lightfoot, R.T.; Remotti, H.; Stevens, J.L.; Ron, D. CHOP is implicated in programmed cell death in response to impaired function of the endoplasmic reticulum. *Genes Dev.* **1998**, *12*, 982–995. [CrossRef] [PubMed]

45. Oyadomari, S.; Mori, M. Roles of CHOP/GADD153 in endoplasmic reticulum stress. *Cell Death Differ.* **2004**, *11*, 381–389. [CrossRef]

46. Kim, I.; Xu, W.; Reed, J.C. Cell death and endoplasmic reticulum stress: disease relevance and therapeutic opportunities. *Nat. Rev. Drug Discov.* **2008**, *7*, 1013–1030. [CrossRef] [PubMed]

47. Endo, M.; Mori, M.; Akira, S.; Gotoh, T. C/EBP homologous protein (CHOP) is crucial for the induction of caspase-11 and the pathogenesis of lipopolysaccharide-induced inflammation. *J. Immunol.* **2006**, *176*, 6245–6253. [CrossRef]

48. Tabas, I.; Ron, D. Integrating the mechanisms of apoptosis induced by endoplasmic reticulum stress. *Nat. Cell Biol.* **2011**, *13*, 184–190. [CrossRef]

49. He, Y.; Sun, S.; Sha, H.; Liu, Z.; Yang, L.; Xue, Z.; Chen, H.; Qi, L. Emerging roles for XBP1, a sUPeR transcription factor. *Gene Expr.* **2010**, *15*, 13–25. [CrossRef]

50. Nishitoh, H.; Matsuzawa, A.; Tobiume, K.; Saegusa, K.; Takeda, K.; Inoue, K.; Hori, S.; Kakizuka, A.; Ichijo, H. ASK1 is essential for endoplasmic reticulum stress-induced neuronal cell death triggered by expanded polyglutamine repeats. *Genes Dev.* **2002**, *16*, 1345–1355. [CrossRef]

51. Todd, D.J.; Lee, A.-H.; Glimcher, L.H. The endoplasmic reticulum stress response in immunity and autoimmunity. *Nat. Rev. Immunol.* **2008**, *8*, 663–674. [CrossRef]

52. Cullinan, S.B.; Zhang, D.; Hannink, M.; Arvisais, E.; Kaufman, R.J.; Diehl, J.A. Nrf2 is a direct PERK substrate and effector of PERK-dependent cell survival. *Mol. Cell. Biol.* **2003**, *23*, 7198–7209. [CrossRef] [PubMed]

53. Kim, K.M.; Pae, H.-O.; Zheng, M.; Park, R.; Kim, Y.-M.; Chung, H.-T. Carbon Monoxide Induces Heme Oxygenase-1 via Activation of Protein Kinase R–Like Endoplasmic Reticulum Kinase and Inhibits Endothelial Cell Apoptosis Triggered by Endoplasmic Reticulum Stress. *Circ. Res.* **2007**, *101*, 919–927. [CrossRef] [PubMed]

54. He, C.H.; Gong, P.; Hu, B.; Stewart, D.; Choi, M.E.; Choi, A.M.K.; Alam, J. Identification of Activating Transcription Factor 4 (ATF4) as an Nrf2-interacting Protein. *J. Biol. Chem.* **2001**, *276*, 20858–20865. [CrossRef] [PubMed]

55. Dey, S.; Sayers, C.M.; Verginadis, I.I.; Lehman, S.L.; Cheng, Y.; Cerniglia, G.J.; Tuttle, S.W.; Feldman, M.D.; Zhang, P.J.L.; Fuchs, S.Y.; et al. ATF4-dependent induction of heme oxygenase 1 prevents anoikis and promotes metastasis. *J. Clin. Invest.* **2015**, *125*, 2592–2608. [CrossRef] [PubMed]

56. Zong, Z.-H.; Du, Z.-X.; Li, N.; Li, C.; Zhang, Q.; Liu, B.-Q.; Guan, Y.; Wang, H.-Q. Implication of Nrf2 and ATF4 in differential induction of CHOP by proteasome inhibition in thyroid cancer cells. *Biochim. Biophys. Acta - Mol. Cell Res.* **2012**, *1823*, 1395–1404. [CrossRef]

57. Hourihan, J.M.; Moronetti Mazzeo, L.E.; Fernández-Cárdenas, L.P.; Blackwell, T.K. Cysteine Sulfenylation Directs IRE-1 to Activate the SKN-1/Nrf2 Antioxidant Response. *Mol. Cell* **2016**, *63*, 553–566. [CrossRef] [PubMed]

58. Cullinan, S.B.; Diehl, J.A. PERK-dependent activation of Nrf2 contributes to redox homeostasis and cell survival following endoplasmic reticulum stress. *J. Biol. Chem.* **2004**, *279*, 20108–20117. [CrossRef]

59. Li, W.; Östblom, M.; Xu, L.-H.; Hellsten, A.; Leanderson, P.; Liedberg, B.; Brunk, U.T.; Eaton, J.W.; Yuan, X.-M. Cytocidal effects of atheromatous plaque components: the death zone revisited. *FASEB J.* **2006**, *20*, 2281–2290. [CrossRef]

60. Nagy, E.; Eaton, J.W.; Jeney, V.; Soares, M.P.; Varga, Z.; Galajda, Z.; Szentmiklosi, J.; Mehes, G.; Csonka, T.; Smith, A.; et al. Red Cells, Hemoglobin, Heme, Iron, and Atherogenesis. *Arterioscler. Thromb. Vasc. Biol.* **2010**, *30*, 1347–1353. [CrossRef]

61. Balla, J.; Jacob, H.S.; Balla, G.; Nath, K.; Eaton, J.W.; Vercellotti, G.M. Endothelial-cell heme uptake from heme proteins: induction of sensitization and desensitization to oxidant damage. *Proc. Natl. Acad. Sci. USA* **1993**, *90*, 9285–9289. [CrossRef] [PubMed]

62. Jeney, V.; Balla, J.; Yachie, A.; Varga, Z.; Vercellotti, G.M.; Eaton, J.W.; Balla, G. Pro-oxidant and cytotoxic effects of circulating heme. *Blood* **2002**, *100*, 879–887. [CrossRef] [PubMed]

63. Balla, G.; Jacob, H.S.; Eaton, J.W.; Belcher, J.D.; Vercellotti, G.M. Hemin: A possible physiological mediator of low density lipoprotein oxidation and endothelial injury. *Arterioscler. Thromb. J. Vasc. Biol.* **1991**, *11*, 1700–1711. [CrossRef]

64. Potor, L.; Bányai, E.; Becs, G.; Soares, M.P.; Balla, G.; Balla, J.; Jeney, V. Atherogenesis may involve the prooxidant and proinflammatory effects of ferryl hemoglobin. *Oxid. Med. Cell. Longev.* **2013**, *2013*, 676425. [CrossRef] [PubMed]

65. Buttari, B.; Profumo, E.; Cuccu, B.; Straface, E.; Gambardella, L.; Malorni, W.; Genuini, I.; Capoano, R.; Salvati, B.; Riganò, R. Erythrocytes from patients with carotid atherosclerosis fail to control dendritic cell maturation. *Int. J. Cardiol.* **2012**, *155*, 484–486. [CrossRef] [PubMed]

66. Profumo, E.; Buttari, B.; Petrone, L.; Straface, E.; Gambardella, L.; Pietraforte, D.; Genuini, I.; Capoano, R.; Salvati, B.; Malorni, W.; et al. Redox imbalance of red blood cells impacts T lymphocyte homeostasis: implication in carotid atherosclerosis. *Thromb. Haemost.* **2011**, *106*, 1117–1126.

67. Hennig, B.; Chow, C.K. Lipid peroxidation and endothelial cell injury: implications in atherosclerosis. *Free Radic. Biol. Med.* **1988**, *4*, 99–106. [CrossRef]

68. Steinberg, D.; Parthasarathy, S.; Carew, T.E.; Khoo, J.C.; Witztum, J.L.; Witztum, J.L. Beyond cholesterol. Modifications of low-density lipoprotein that increase its atherogenicity. *N. Engl. J. Med.* **1989**, *320*, 915–924.

69. Witztum, J.L.; Steinberg, D. Role of oxidized low density lipoprotein in atherogenesis. *J. Clin. Invest.* **1991**, *88*, 1785–1792. [CrossRef]

70. Thomas, J.P.; Geiger, P.G.; Girotti, A.W. Lethal damage to endothelial cells by oxidized low density lipoprotein: role of selenoperoxidases in cytoprotection against lipid hydroperoxide- and iron-mediated reactions. *J. Lipid Res.* **1993**, *34*, 479–490.

71. Guyton, J.R.; Lenz, M.L.; Mathews, B.; Hughes, H.; Karsan, D.; Selinger, E.; Smith, C. V Toxicity of oxidized low density lipoproteins for vascular smooth muscle cells and partial protection by antioxidants. *Atherosclerosis* **1995**, *118*, 237–249. [CrossRef]

72. Marchant, C.E.; Law, N.S.; van der Veen, C.; Hardwick, S.J.; Carpenter, K.L.; Mitchinson, M.J. Oxidized low-density lipoprotein is cytotoxic to human monocyte-macrophages: protection with lipophilic antioxidants. *FEBS Lett.* **1995**, *358*, 175–178. [CrossRef]

73. Agarwal, A.; Balla, J.; Balla, G.; Croatt, A.J.; Vercellotti, G.M.; Nath, K.A. Renal tubular epithelial cells mimic endothelial cells upon exposure to oxidized LDL. *Am. J. Physiol.* **1996**, *271*, F814–F823. [CrossRef] [PubMed]

74. Wang, L.J.; Lee, T.S.; Lee, F.Y.; Pai, R.C.; Chau, L.Y. Expression of heme oxygenase-1 in atherosclerotic lesions. *Am. J. Pathol.* **1998**, *152*, 711–720. [PubMed]

75. Yachie, A.; Niida, Y.; Wada, T.; Igarashi, N.; Kaneda, H.; Toma, T.; Ohta, K.; Kasahara, Y.; Koizumi, S. Oxidative stress causes enhanced endothelial cell injury in human heme oxygenase-1 deficiency. *J. Clin. Invest.* **1999**, *103*, 129–135. [CrossRef]

76. Nagy, E.; Jeney, V.; Yachie, A.; Szabó, R.P.; Wagner, O.; Vercellotti, G.M.; Eaton, J.W.; Balla, G.; Balla, J. Oxidation of hemoglobin by lipid hydroperoxide associated with low-density lipoprotein (LDL) and increased cytotoxic effect by LDL oxidation in heme oxygenase-1 (HO-1) deficiency. *Cell. Mol. Biol. (Noisy-le-grand)* **2005**, *51*, 377–385.

77. Cheng, C.; Noordeloos, A.M.; Jeney, V.; Soares, M.P.; Moll, F.; Pasterkamp, G.; Serruys, P.W.; Duckers, H.J. Heme oxygenase 1 determines atherosclerotic lesion progression into a vulnerable plaque. *Circulation* **2009**, *119*, 3017–3027. [CrossRef]

78. Kishimoto, Y.; Sasaki, K.; Saita, E.; Niki, H.; Ohmori, R.; Kondo, K.; Momiyama, Y. Plasma Heme Oxygenase-1 Levels and Carotid Atherosclerosis. *Stroke* **2018**, *49*, 2230–2232. [CrossRef]

79. Mao, H.; Tao, T.; Wang, X.; Liu, M.; Song, D.; Liu, X.; Shi, D. Zedoarondiol Attenuates Endothelial Cells Injury Induced by Oxidized Low-Density Lipoprotein via Nrf2 Activation. *Cell. Physiol. Biochem.* **2018**, *48*, 1468–1479. [CrossRef]

80. Li, W.; Zhi, W.; Liu, F.; He, Z.; Wang, X.; Niu, X. Atractylenolide I restores HO-1 expression and inhibits Ox-LDL-induced VSMCs proliferation, migration and inflammatory responses in vitro. *Exp. Cell Res.* **2017**, *353*, 26–34. [CrossRef]

81. Vogel, M.E.; Idelman, G.; Konaniah, E.S.; Zucker, S.D. Bilirubin Prevents Atherosclerotic Lesion Formation in Low-Density Lipoprotein Receptor-Deficient Mice by Inhibiting Endothelial VCAM-1 and ICAM-1 Signaling. *J. Am. Heart Assoc.* **2017**, *6*. [CrossRef] [PubMed]

82. Lapenna, D.; Ciofani, G.; Pierdomenico, S.D.; Giamberardino, M.A.; Ucchino, S.; Davì, G. Association of serum bilirubin with oxidant damage of human atherosclerotic plaques and the severity of atherosclerosis. *Clin. Exp. Med.* **2018**, *18*, 119–124. [CrossRef] [PubMed]

83. Tao, X.; Wu, J.; Wang, A.; Xu, C.; Wang, Z.; Zhao, X. Lower serum indirect bilirubin levels are inversely related to carotid intima-media thickness progression. *Curr. Neurovasc. Res.* **2019**, *16*. [CrossRef] [PubMed]

84. Vítek, L.; Jirsa, M.; Brodanová, M.; Kalab, M.; Marecek, Z.; Danzig, V.; Novotný, L.; Kotal, P. Gilbert syndrome and ischemic heart disease: a protective effect of elevated bilirubin levels. *Atherosclerosis* **2002**, *160*, 449–456. [CrossRef]

85. Soni, H.; Pandya, G.; Patel, P.; Acharya, A.; Jain, M.; Mehta, A.A. Beneficial effects of carbon monoxide-releasing molecule-2 (CORM-2) on acute doxorubicin cardiotoxicity in mice: role of oxidative stress and apoptosis. *Toxicol. Appl. Pharmacol.* **2011**, *253*, 70–80. [CrossRef] [PubMed]

86. Segersvärd, H.; Lakkisto, P.; Hänninen, M.; Forsten, H.; Siren, J.; Immonen, K.; Kosonen, R.; Sarparanta, M.; Laine, M.; Tikkanen, I. Carbon monoxide releasing molecule improves structural and functional cardiac recovery after myocardial injury. *Eur. J. Pharmacol.* **2018**, *818*, 57–66. [CrossRef] [PubMed]

87. Sun, H.-J.; Xu, D.-Y.; Sun, Y.-X.; Xue, T.; Zhang, C.-X.; Zhang, Z.-X.; Lin, W.; Li, K.-X. CO-releasing molecules-2 attenuates ox-LDL-induced injury in HUVECs by ameliorating mitochondrial function and inhibiting Wnt/β-catenin pathway. *Biochem. Biophys. Res. Commun.* **2017**, *490*, 629–635. [CrossRef] [PubMed]

88. Petrick, L.; Rosenblat, M.; Aviram, M. *In vitro* effects of exogenous carbon monoxide on oxidative stress and lipid metabolism in macrophages. *Toxicol. Ind. Health* **2016**, *32*, 1318–1323. [CrossRef]

89. Torti, S.V.; Kwak, E.L.; Miller, S.C.; Miller, L.L.; Ringold, G.M.; Myambo, K.B.; Young, A.P.; Torti, F.M. The molecular cloning and characterization of murine ferritin heavy chain, a tumor necrosis factor-inducible gene. *J. Biol. Chem.* **1988**, *263*, 12638–12644.

90. Zarjou, A.; Jeney, V.; Arosio, P.; Poli, M.; Antal-Szalmás, P.; Agarwal, A.; Balla, G.; Balla, J. Ferritin prevents calcification and osteoblastic differentiation of vascular smooth muscle cells. *J. Am. Soc. Nephrol.* **2009**, *20*, 1254–1263. [CrossRef]

91. Sikura, K.É.; Potor, L.; Szerafin, T.; Zarjou, A.; Agarwal, A.; Arosio, P.; Poli, M.; Hendrik, Z.; Méhes, G.; Oros, M.; et al. Potential Role of H-Ferritin in Mitigating Valvular Mineralization. *Arterioscler. Thromb. Vasc. Biol.* **2019**, *39*, 413–431. [CrossRef] [PubMed]

92. Becs, G.; Zarjou, A.; Agarwal, A.; Kovács, K.É.; Becs, Á.; Nyitrai, M.; Balogh, E.; Bányai, E.; Eaton, J.W.; Arosio, P.; et al. Pharmacological induction of ferritin prevents osteoblastic transformation of smooth muscle cells. *J. Cell. Mol. Med.* **2016**, *20*, 217–230. [CrossRef] [PubMed]

93. Hong, D.; Bai, Y.-P.; Gao, H.-C.; Wang, X.; Li, L.-F.; Zhang, G.-G.; Hu, C.-P. Ox-LDL induces endothelial cell apoptosis via the LOX-1-dependent endoplasmic reticulum stress pathway. *Atherosclerosis* **2014**, *235*, 310–317. [CrossRef] [PubMed]

94. Myoishi, M.; Hao, H.; Minamino, T.; Watanabe, K.; Nishihira, K.; Hatakeyama, K.; Asada, Y.; Okada, K.-I.; Ishibashi-Ueda, H.; Gabbiani, G.; et al. Increased Endoplasmic Reticulum Stress in Atherosclerotic Plaques Associated With Acute Coronary Syndrome. *Circulation* **2007**, *116*, 1226–1233. [CrossRef] [PubMed]

95. Tao, Y.K.; Yu, P.L.; Bai, Y.P.; Yan, S.T.; Zhao, S.P.; Zhang, G.Q. Role of PERK/eIF2α/CHOP Endoplasmic Reticulum Stress Pathway in Oxidized Low-density Lipoprotein Mediated Induction of Endothelial Apoptosis. *Biomed. Environ. Sci.* **2016**, *29*, 868–876.

96. Yao, S.; Miao, C.; Tian, H.; Sang, H.; Yang, N.; Jiao, P.; Han, J.; Zong, C.; Qin, S. Endoplasmic reticulum stress promotes macrophage-derived foam cell formation by up-regulating cluster of differentiation 36 (CD36) expression. *J. Biol. Chem.* **2014**, *289*, 4032–4042. [CrossRef]

97. Erbay, E.; Babaev, V.R.; Mayers, J.R.; Makowski, L.; Charles, K.N.; Snitow, M.E.; Fazio, S.; Wiest, M.M.; Watkins, S.M.; Linton, M.F.; et al. Reducing endoplasmic reticulum stress through a macrophage lipid chaperone alleviates atherosclerosis. *Nat. Med.* **2009**, *15*, 1383–1391. [CrossRef]

98. Sun, Y.; Zhang, D.; Liu, X.; Li, X.; Liu, F.; Yu, Y.; Jia, S.; Zhou, Y.; Zhao, Y. Endoplasmic Reticulum Stress Affects Lipid Metabolism in Atherosclerosis Via CHOP Activation and Over-Expression of miR-33. *Cell. Physiol. Biochem.* **2018**, *48*, 1995–2010. [CrossRef]

99. Cao, S.S.; Kaufman, R.J. Endoplasmic reticulum stress and oxidative stress in cell fate decision and human disease. *Antioxid. Redox Signal.* **2014**, *21*, 396–413. [CrossRef]

100. Mozzini, C.; Fratta Pasini, A.; Garbin, U.; Stranieri, C.; Pasini, A.; Vallerio, P.; Cominacini, L. Increased endoplasmic reticulum stress and Nrf2 repression in peripheral blood mononuclear cells of patients with stable coronary artery disease. *Free Radic. Biol. Med.* **2014**, *68*, 178–185. [CrossRef]

101. Zhang, T.; Hu, Q.; Shi, L.; Qin, L.; Zhang, Q.; Mi, M. Equol Attenuates Atherosclerosis in Apolipoprotein E-Deficient Mice by Inhibiting Endoplasmic Reticulum Stress via Activation of Nrf2 in Endothelial Cells. *PLoS ONE* **2016**, *11*, e0167020. [CrossRef] [PubMed]

102. Luo, Y.; Sun, G.; Dong, X.; Wang, M.; Qin, M.; Yu, Y.; Sun, X. Isorhamnetin attenuates atherosclerosis by inhibiting macrophage apoptosis via PI3K/AKT activation and HO-1 induction. *PLoS ONE* **2015**, *10*, e0120259. [CrossRef] [PubMed]

103. Zheng, Q.; Tong, M.; Ou, B.; Liu, C.; Hu, C.; Yang, Y. Isorhamnetin protects against bleomycin-induced pulmonary fibrosis by inhibiting endoplasmic reticulum stress and epithelial-mesenchymal transition. *Int. J. Mol. Med.* **2019**, *43*, 117–126. [CrossRef] [PubMed]

104. Wang, Y.-C.; Zhou, Y.; Fang, H.; Lin, S.; Wang, P.-F.; Xiong, R.-P.; Chen, J.; Xiong, X.-Y.; Lv, F.-L.; Liang, Q.-L.; et al. Toll-like receptor 2/4 heterodimer mediates inflammatory injury in intracerebral hemorrhage. *Ann. Neurol.* **2014**, *75*, 876–889. [CrossRef] [PubMed]

105. Kwon, M.S.; Woo, S.K.; Kurland, D.B.; Yoon, S.H.; Palmer, A.F.; Banerjee, U.; Iqbal, S.; Ivanova, S.; Gerzanich, V.; Simard, J.M. Methemoglobin is an endogenous toll-like receptor 4 ligand-relevance to subarachnoid hemorrhage. *Int. J. Mol. Sci.* **2015**, *16*, 5028–5046. [CrossRef] [PubMed]

106. Figueiredo, R.T.; Fernandez, P.L.; Mourao-Sa, D.S.; Porto, B.N.; Dutra, F.F.; Alves, L.S.; Oliveira, M.F.; Oliveira, P.L.; Graça-Souza, A.V.; Bozza, M.T. Characterization of heme as activator of Toll-like receptor 4. *J. Biol. Chem.* **2007**, *282*, 20221–20229. [CrossRef] [PubMed]

107. Lin, S.; Yin, Q.; Zhong, Q.; Lv, F.-L.; Zhou, Y.; Li, J.-Q.; Wang, J.-Z.; Su, B.; Yang, Q.-W. Heme activates TLR4-mediated inflammatory injury via MyD88/TRIF signaling pathway in intracerebral hemorrhage. *J. Neuroinflammation* **2012**, *9*, 46. [CrossRef]

108. Min, H.; Choi, B.; Jang, Y.H.; Cho, I.-H.; Lee, S.J. Heme molecule functions as an endogenous agonist of astrocyte TLR2 to contribute to secondary brain damage after intracerebral hemorrhage. *Mol. Brain* **2017**, *10*, 27. [CrossRef]

109. Wang, Y.-C.; Wang, P.-F.; Fang, H.; Chen, J.; Xiong, X.-Y.; Yang, Q.-W. Toll-like receptor 4 antagonist attenuates intracerebral hemorrhage-induced brain injury. *Stroke* **2013**, *44*, 2545–2552. [CrossRef]

110. Wang, F.; Zhang, Y.; He, C.; Wang, T.; Piao, Q.; Liu, Q. Silencing the gene encoding C/EBP homologous protein lessens acute brain injury following ischemia/reperfusion. *Neural Regen. Res.* **2012**, *7*, 2432–2438.

111. Yan, F.; Cao, S.; Li, J.; Dixon, B.; Yu, X.; Chen, J.; Gu, C.; Lin, W.; Chen, G. Pharmacological Inhibition of PERK Attenuates Early Brain Injury After Subarachnoid Hemorrhage in Rats Through the Activation of Akt. *Mol. Neurobiol.* **2017**, *54*, 1808–1817. [CrossRef] [PubMed]

112. Meng, C.; Zhang, J.; Dang, B.; Li, H.; Shen, H.; Li, X.; Wang, Z. PERK Pathway Activation Promotes Intracerebral Hemorrhage Induced Secondary Brain Injury by Inducing Neuronal Apoptosis Both in Vivo and in Vitro. *Front. Neurosci.* **2018**, *12*, 111. [CrossRef] [PubMed]

113. Schallner, N.; Pandit, R.; LeBlanc, R.; Thomas, A.J.; Ogilvy, C.S.; Zuckerbraun, B.S.; Gallo, D.; Otterbein, L.E.; Hanafy, K.A. Microglia regulate blood clearance in subarachnoid hemorrhage by heme oxygenase-1. *J. Clin. Invest.* **2015**, *125*, 2609–2625. [CrossRef] [PubMed]

114. Wang, J.; Doré, S. Heme oxygenase-1 exacerbates early brain injury after intracerebral haemorrhage. *Brain* **2007**, *130*, 1643–1652. [CrossRef] [PubMed]

115. Koeppen, A.H.; Dickson, A.C.; Smith, J. Heme oxygenase in experimental intracerebral hemorrhage: the benefit of tin-mesoporphyrin. *J. Neuropathol. Exp. Neurol.* **2004**, *63*, 587–597. [CrossRef] [PubMed]

116. Zhang, Z.; Song, Y.; Zhang, Z.; Li, D.; Zhu, H.; Liang, R.; Gu, Y.; Pang, Y.; Qi, J.; Wu, H.; et al. Distinct role of heme oxygenase-1 in early- and late-stage intracerebral hemorrhage in 12-month-old mice. *J. Cereb. Blood Flow Metab.* **2017**, *37*, 25–38. [CrossRef] [PubMed]

117. Wang, G.; Yang, Q.; Li, G.; Wang, L.; Hu, W.; Tang, Q.; Li, D.; Sun, Z. Time course of heme oxygenase-1 and oxidative stress after experimental intracerebral hemorrhage. *Acta Neurochir. (Wien)* **2011**, *153*, 319–325. [CrossRef] [PubMed]

118. Ma, B.; Day, J.P.; Phillips, H.; Slootsky, B.; Tolosano, E.; Doré, S. Deletion of the hemopexin or heme oxygenase-2 gene aggravates brain injury following stroma-free hemoglobin-induced intracerebral hemorrhage. *J. Neuroinflamm.* **2016**, *13*, 26. [CrossRef]

119. Chen-Roetling, J.; Cai, Y.; Regan, R.F. Neuroprotective effect of heme oxygenase-2 knockout in the blood injection model of intracerebral hemorrhage. *BMC Res. Notes* **2014**, *7*, 561. [CrossRef]

120. Graf, E.; Mahoney, J.R.; Bryant, R.G.; Eaton, J.W. Iron-catalyzed hydroxyl radical formation. Stringent requirement for free iron coordination site. *J. Biol. Chem.* **1984**, *259*, 3620–3624.

121. Kaplán, P.; Doval, M.; Majerová, Z.; Lehotský, J.; Racay, P. Iron-induced lipid peroxidation and protein modification in endoplasmic reticulum membranes. Protection by stobadine. *Int. J. Biochem. Cell Biol.* **2000**, *32*, 539–547. [CrossRef]

122. Kaplan, P.; Babusikova, E.; Lehotsky, J.; Dobrota, D. Free radical-induced protein modification and inhibition of Ca2+-ATPase of cardiac sarcoplasmic reticulum. *Mol. Cell. Biochem.* **2003**, *248*, 41–47. [CrossRef] [PubMed]

123. Vecchi, C.; Montosi, G.; Zhang, K.; Lamberti, I.; Duncan, S.A.; Kaufman, R.J.; Pietrangelo, A. ER stress controls iron metabolism through induction of hepcidin. *Science* **2009**, *325*, 877–880. [CrossRef] [PubMed]

124. Zhao, J.; Xiang, X.; Zhang, H.; Jiang, D.; Liang, Y.; Qing, W.; Liu, L.; Zhao, Q.; He, Z. CHOP induces apoptosis by affecting brain iron metabolism in rats with subarachnoid hemorrhage. *Exp. Neurol.* **2018**, *302*, 22–33. [CrossRef] [PubMed]

125. Li, Q.; Han, X.; Lan, X.; Gao, Y.; Wan, J.; Durham, F.; Cheng, T.; Yang, J.; Wang, Z.; Jiang, C.; et al. Inhibition of neuronal ferroptosis protects hemorrhagic brain. *JCI insight* **2017**, *2*, e90777. [CrossRef] [PubMed]

126. Selim, M. Deferoxamine Mesylate: A New Hope for Intracerebral Hemorrhage: From Bench to Clinical Trials. *Stroke* **2009**, *40*, S90–S91. [CrossRef] [PubMed]

127. Li, Q.; Wan, J.; Lan, X.; Han, X.; Wang, Z.; Wang, J. Neuroprotection of brain-permeable iron chelator VK-28 against intracerebral hemorrhage in mice. *J. Cereb. Blood Flow Metab.* **2017**, *37*, 3110–3123. [CrossRef] [PubMed]

128. LeBlanc, R.H.; Chen, R.; Selim, M.H.; Hanafy, K.A. Heme oxygenase-1-mediated neuroprotection in subarachnoid hemorrhage via intracerebroventricular deferoxamine. *J. Neuroinflammation* **2016**, *13*, 244. [CrossRef] [PubMed]

129. Stocker, R.; Yamamoto, Y.; McDonagh, A.F.; Glazer, A.N.; Ames, B.N. Bilirubin is an antioxidant of possible physiological importance. *Science* **1987**, *235*, 1043–1046. [CrossRef]

130. Stocker, R.; Glazer, A.N.; Ames, B.N. Antioxidant activity of albumin-bound bilirubin. *Proc. Natl. Acad. Sci. USA* **1987**, *84*, 5918–5922. [CrossRef]

131. Cremer, R.J.; Perryman, P.W.; Richards, D.H. Influence of light on the hyperbilirubinÆmia of infants. *Lancet* **1958**, *271*, 1094–1097. [CrossRef]

132. McDonagh, A.F.; Palma, L.A.; Trull, F.R.; Lightner, D.A. Phototherapy for neonatal jaundice. Configurational isomers of bilirubin. *J. Am. Chem. Soc.* **1982**, *104*, 6865–6867. [CrossRef]

133. Diamond, L.K.; Allen, F.H.; Thomas, W.O. Erythroblastosis Fetalis. *N. Engl. J. Med.* **1951**, *244*, 39–49. [CrossRef] [PubMed]

134. Newman, T.B.; Maisels, M.J. Evaluation and treatment of jaundice in the term newborn: a kinder, gentler approach. *Pediatrics* **1992**, *89*, 809–818. [PubMed]

135. Valaes, T.; Petmezaki, S.; Henschke, C.; Drummond, G.S.; Kappas, A. Control of jaundice in preterm newborns by an inhibitor of bilirubin production: studies with tin-mesoporphyrin. *Pediatrics* **1994**, *93*, 1–11. [PubMed]

136. Kamisako, T.; Kobayashi, Y.; Takeuchi, K.; Ishihara, T.; Higuchi, K.; Tanaka, Y.; Gabazza, E.C.; Adachi, Y. Recent advances in bilirubin metabolism research: the molecular mechanism of hepatocyte bilirubin transport and its clinical relevance. *J. Gastroenterol.* **2000**, *35*, 659–664. [CrossRef]

137. Qaisiya, M.; Brischetto, C.; Jašprová, J.; Vitek, L.; Tiribelli, C.; Bellarosa, C. Bilirubin-induced ER stress contributes to the inflammatory response and apoptosis in neuronal cells. *Arch. Toxicol.* **2017**, *91*, 1847–1858. [CrossRef]

138. Schiavon, E.; Smalley, J.L.; Newton, S.; Greig, N.H.; Forsythe, I.D. Neuroinflammation and ER-stress are key mechanisms of acute bilirubin toxicity and hearing loss in a mouse model. *PLoS ONE* **2018**, *13*, e0201022. [CrossRef]

139. Feng, J.; Li, M.; Wei, Q.; Li, S.; Song, S.; Hua, Z. Unconjugated bilirubin induces pyroptosis in cultured rat cortical astrocytes. *J. Neuroinflammation* **2018**, *15*, 23. [CrossRef]

140. Yabluchanskiy, A.; Sawle, P.; Homer-Vanniasinkam, S.; Green, C.J.; Foresti, R.; Motterlini, R. CORM-3, a carbon monoxide-releasing molecule, alters the inflammatory response and reduces brain damage in a rat model of hemorrhagic stroke*. *Crit. Care Med.* **2012**, *40*, 544–552. [CrossRef]

141. Maritim, A.C.; Sanders, R.A.; Watkins, J.B. Diabetes, oxidative stress, and antioxidants: a review. *J. Biochem. Mol. Toxicol.* **2003**, *17*, 24–38. [CrossRef] [PubMed]

142. Giacco, F.; Brownlee, M. Oxidative stress and diabetic complications. *Circ. Res.* **2010**, *107*, 1058–1070. [CrossRef] [PubMed]

143. Regidor, D.L.; Kopple, J.D.; Kovesdy, C.P.; Kilpatrick, R.D.; McAllister, C.J.; Aronovitz, J.; Greenland, S.; Kalantar-Zadeh, K. Associations between changes in hemoglobin and administered erythropoiesis-stimulating agent and survival in hemodialysis patients. *J. Am. Soc. Nephrol.* **2006**, *17*, 1181–1191. [CrossRef] [PubMed]

144. Quan, S.; Kaminski, P.M.; Yang, L.; Morita, T.; Inaba, M.; Ikehara, S.; Goodman, A.I.; Wolin, M.S.; Abraham, N.G. Heme oxygenase-1 prevents superoxide anion-associated endothelial cell sloughing in diabetic rats. *Biochem. Biophys. Res. Commun.* **2004**, *315*, 509–516. [CrossRef] [PubMed]

145. Ahmad, M.; Turkseven, S.; Mingone, C.J.; Gupte, S.A.; Wolin, M.S.; Abraham, N.G. Heme oxygenase-1 gene expression increases vascular relaxation and decreases inducible nitric oxide synthase in diabetic rats. *Cell. Mol. Biol. (Noisy-le-grand)* 2005, *51*, 371–376.

146. Di Pascoli, M.; Rodella, L.; Sacerdoti, D.; Bolognesi, M.; Turkseven, S.; Abraham, N.G. Chronic CO levels have [corrected] a beneficial effect on vascular relaxation in diabetes. *Biochem. Biophys. Res. Commun.* **2006**, *340*, 935–943. [CrossRef] [PubMed]

147. Burgess, A.; Li, M.; Vanella, L.; Kim, D.H.; Rezzani, R.; Rodella, L.; Sodhi, K.; Canestraro, M.; Martasek, P.; Peterson, S.J.; et al. Adipocyte heme oxygenase-1 induction attenuates metabolic syndrome in both male and female obese mice. *Hypertens. (Dallas, Tex. 1979)* **2010**, *56*, 1124–1130. [CrossRef]

148. Xie, L.; Gu, Y.; Wen, M.; Zhao, S.; Wang, W.; Ma, Y.; Meng, G.; Han, Y.; Wang, Y.; Liu, G.; et al. Hydrogen Sulfide Induces Keap1 S-sulfhydration and Suppresses Diabetes-Accelerated Atherosclerosis via Nrf2 Activation. *Diabetes* **2016**, *65*, 3171–3184. [CrossRef]

149. Gupta, I.; Goyal, A.; Singh, N.K.; Yadav, H.N.; Sharma, P.L. Hemin, a heme oxygenase-1 inducer, restores the attenuated cardioprotective effect of ischemic preconditioning in isolated diabetic rat heart. *Hum. Exp. Toxicol.* **2017**, *36*, 867–875. [CrossRef]

150. Ali, M.A.M.; Heeba, G.H.; El-Sheikh, A.A.K. Modulation of heme oxygenase-1 expression and activity affects streptozotocin-induced diabetic nephropathy in rats. *Fundam. Clin. Pharmacol.* **2017**, *31*, 546–557. [CrossRef]

151. Gou, L.; Zhao, L.; Song, W.; Wang, L.; Liu, J.; Zhang, H.; Huang, Y.; Lau, C.W.; Yao, X.; Tian, X.Y.; et al. Inhibition of miR-92a Suppresses Oxidative Stress and Improves Endothelial Function by Upregulating Heme Oxygenase-1 in db/db Mice. *Antioxid. Redox Signal.* **2018**, *28*, 358–370. [CrossRef] [PubMed]

152. Yang, H.; Wang, Q.; Li, S. MicroRNA-218 promotes high glucose-induced apoptosis in podocytes by targeting heme oxygenase-1. *Biochem. Biophys. Res. Commun.* **2016**, *471*, 582–588. [CrossRef] [PubMed]

153. Goodman, A.I.; Chander, P.N.; Rezzani, R.; Schwartzman, M.L.; Regan, R.F.; Rodella, L.; Turkseven, S.; Lianos, E.A.; Dennery, P.A.; Abraham, N.G. Heme oxygenase-2 deficiency contributes to diabetes-mediated increase in superoxide anion and renal dysfunction. *J. Am. Soc. Nephrol.* **2006**, *17*, 1073–1081. [CrossRef] [PubMed]

154. He, M.; Nitti, M.; Piras, S.; Furfaro, A.L.; Traverso, N.; Pronzato, M.A.; Mann, G.E. Heme oxygenase-1-derived bilirubin protects endothelial cells against high glucose-induced damage. *Free Radic. Biol. Med.* **2015**, *89*, 91–98. [CrossRef] [PubMed]

155. Liu, J.; Wang, L.; Tian, X.Y.; Liu, L.; Wong, W.T.; Zhang, Y.; Han, Q.-B.; Ho, H.-M.; Wang, N.; Wong, S.L.; et al. Unconjugated Bilirubin Mediates Heme Oxygenase-1–Induced Vascular Benefits in Diabetic Mice. *Diabetes* **2015**, *64*, 1564–1575. [CrossRef] [PubMed]

156. Hamur, H.; Duman, H.; Demirtas, L.; Bakirci, E.M.; Durakoglugil, M.E.; Degirmenci, H.; Kalkan, K.; Yildirim, E.; Vuruskan, E. Total Bilirubin Levels Predict Subclinical Atherosclerosis in Patients With Prediabetes. *Angiology* **2016**, *67*, 909–915. [CrossRef]

157. Dong, H.; Huang, H.; Yun, X.; Kim, D.; Yue, Y.; Wu, H.; Sutter, A.; Chavin, K.D.; Otterbein, L.E.; Adams, D.B.; et al. Bilirubin Increases Insulin Sensitivity in Leptin-Receptor Deficient and Diet-Induced Obese Mice Through Suppression of ER Stress and Chronic Inflammation. *Endocrinology* **2014**, *155*, 818–828. [CrossRef]

158. Lee, D.W.; Shin, H.Y.; Jeong, J.H.; Han, J.; Ryu, S.; Nakahira, K.; Moon, J.-S. Carbon monoxide regulates glycolysis-dependent NLRP3 inflammasome activation in macrophages. *Biochem. Biophys. Res. Commun.* **2017**, *493*, 957–963. [CrossRef]

159. Wang, Y.; Ying, L.; Chen, Y.; Shen, Y.; Guo, R.; Jin, K.; Wang, L. Induction of heme oxygenase-1 ameliorates vascular dysfunction in streptozotocin-induced type 2 diabetic rats. *Vascul. Pharmacol.* **2014**, *61*, 16–24. [CrossRef]

160. Yahagi, K.; Kolodgie, F.D.; Lutter, C.; Mori, H.; Romero, M.E.; Finn, A.V.; Virmani, R. Pathology of Human Coronary and Carotid Artery Atherosclerosis and Vascular Calcification in Diabetes Mellitus. *Arterioscler. Thromb. Vasc. Biol.* **2017**, *37*, 191–204. [CrossRef]

161. Boon, A.; Cheriex, E.; Lodder, J.; Kessels, F. Cardiac valve calcification: characteristics of patients with calcification of the mitral annulus or aortic valve. *Heart* **1997**, *78*, 472–474. [CrossRef] [PubMed]

162. El-Horany, H.E.-S.; Watany, M.M.; Hagag, R.Y.; El-Attar, S.H.; Basiouny, M.A. Expression of LRP1 and CHOP genes associated with peripheral neuropathy in type 2 diabetes mellitus: Correlations with nerve conduction studies. *Gene* **2019**, *702*, 114–122. [CrossRef] [PubMed]

163. Brosius, F.C.; Kaufman, R.J. Is the ER stressed out in diabetic kidney disease? *J. Am. Soc. Nephrol.* **2008**, *19*, 2040–2042. [CrossRef] [PubMed]

164. Sánchez-Chávez, G.; Hernández-Ramírez, E.; Osorio-Paz, I.; Hernández-Espinosa, C.; Salceda, R. Potential Role of Endoplasmic Reticulum Stress in Pathogenesis of Diabetic Retinopathy. *Neurochem. Res.* **2016**, *41*, 1098–1106. [CrossRef] [PubMed]

165. Liu, Z.-W.; Zhu, H.-T.; Chen, K.-L.; Dong, X.; Wei, J.; Qiu, C.; Xue, J.-H. Protein kinase RNA-like endoplasmic reticulum kinase (PERK) signaling pathway plays a major role in reactive oxygen species (ROS)-mediated endoplasmic reticulum stress-induced apoptosis in diabetic cardiomyopathy. *Cardiovasc. Diabetol.* **2013**, *12*, 158. [CrossRef] [PubMed]

166. Sun, S.; Yang, S.; An, N.; Wang, G.; Xu, Q.; Liu, J.; Mao, Y. Astragalus polysaccharides inhibits cardiomyocyte apoptosis during diabetic cardiomyopathy via the endoplasmic reticulum stress pathway. *J. Ethnopharmacol.* **2019**, *238*, 111857. [CrossRef] [PubMed]

167. Mozzini, C.; Garbin, U.; Stranieri, C.; Pasini, A.; Solani, E.; Tinelli, I.A.; Cominacini, L.; Fratta Pasini, A.M. Endoplasmic reticulum stress and Nrf2 repression in circulating cells of type 2 diabetic patients without the recommended glycemic goals. *Free Radic. Res.* **2015**, *49*, 244–252. [CrossRef] [PubMed]

168. Maamoun, H.; Zachariah, M.; McVey, J.H.; Green, F.R.; Agouni, A. Heme oxygenase (HO)-1 induction prevents Endoplasmic Reticulum stress-mediated endothelial cell death and impaired angiogenic capacity. *Biochem. Pharmacol.* **2017**, *127*, 46–59. [CrossRef]

169. Yang, X.; Yao, W.; Liu, H.; Gao, Y.; Liu, R.; Xu, L. Tangluoning, a traditional Chinese medicine, attenuates in vivo and in vitro diabetic peripheral neuropathy through modulation of PERK/Nrf2 pathway. *Sci. Rep.* **2017**, *7*, 1014. [CrossRef]

170. Ding, Y.; Dai, X.; Zhang, Z.; Jiang, Y.; Ma, X.; Cai, X.; Li, Y. Proanthocyanidins protect against early diabetic peripheral neuropathy by modulating endoplasmic reticulum stress. *J. Nutr. Biochem.* **2014**, *25*, 765–772. [CrossRef]

171. Kang, M.-K.; Lee, E.-J.; Kim, Y.-H.; Kim, D.Y.; Oh, H.; Kim, S.-I.; Kang, Y.-H. Chrysin Ameliorates Malfunction of Retinoid Visual Cycle through Blocking Activation of AGE-RAGE-ER Stress in Glucose-Stimulated Retinal Pigment Epithelial Cells and Diabetic Eyes. *Nutrients* **2018**, *10*, 1046. [CrossRef] [PubMed]

172. Huang, C.-S.; Lii, C.-K.; Lin, A.-H.; Yeh, Y.-W.; Yao, H.-T.; Li, C.-C.; Wang, T.-S.; Chen, H.-W. Protection by chrysin, apigenin, and luteolin against oxidative stress is mediated by the Nrf2-dependent up-regulation of heme oxygenase 1 and glutamate cysteine ligase in rat primary hepatocytes. *Arch. Toxicol.* **2013**, *87*, 167–178. [CrossRef] [PubMed]

173. Zhang, J.; Cao, P.; Gui, J.; Wang, X.; Han, J.; Wang, Y.; Wang, G. Arctigenin ameliorates renal impairment and inhibits endoplasmic reticulum stress in diabetic db/db mice. *Life Sci.* **2019**, *223*, 194–201. [CrossRef] [PubMed]

174. Jeong, Y.-H.; Park, J.-S.; Kim, D.-H.; Kim, H.-S. Arctigenin Increases Hemeoxygenase-1 Gene Expression by Modulating PI3K/AKT Signaling Pathway in Rat Primary Astrocytes. *Biomol. Ther. (Seoul)* **2014**, *22*, 497–502. [CrossRef] [PubMed]

175. Kong, F.-J.; Ma, L.-L.; Guo, J.-J.; Xu, L.-H.; Li, Y.; Qu, S. Endoplasmic reticulum stress/autophagy pathway is involved in diabetes-induced neuronal apoptosis and cognitive decline in mice. *Clin. Sci. (Lond.)* **2018**, *132*, 111–125. [CrossRef] [PubMed]

176. Yang, C.-M.; Lin, C.-C.; Hsieh, H.-L. High-Glucose-Derived Oxidative Stress-Dependent Heme Oxygenase-1 Expression from Astrocytes Contributes to the Neuronal Apoptosis. *Mol. Neurobiol.* **2017**, *54*, 470–483. [CrossRef] [PubMed]

MDPI

St. Alban-Anlage 66

4052 Basel

Switzerland

Tel. +41 61 683 77 34

Fax +41 61 302 89 18

www.mdpi.com

International Journal of Molecular Sciences Editorial Office

E-mail: ijms@mdpi.com

www.mdpi.com/journal/ijms

www.ingramcontent.com/pod-product-compliance
Lightning Source LLC
Chambersburg PA
CBHW051843210326
41597CB00033B/5757